Nanotechnology

Understanding Small Systems

SECOND EDITION

Mechanical Engineering Series
Frank Kreith, Series Editor

Nanotechnology

Understanding Small Systems

SECOND EDITION

Ben Rogers
Sumita Pennathur
Jesse Adams

CRC Press
Taylor & Francis Group
Boca Raton   London   New York

CRC Press is an imprint of the
Taylor & Francis Group, an **informa** business

CRC Press
Taylor & Francis Group
6000 Broken Sound Parkway NW, Suite 300
Boca Raton, FL 33487-2742

© 2011 by Taylor & Francis Group, LLC
CRC Press is an imprint of Taylor & Francis Group, an Informa business

No claim to original U.S. Government works

Printed and bound in India by Replika Press Pvt. Ltd.
Version Date: 20110518

International Standard Book Number: 978-1-4398-4920-0 (Hardback)

Visit the Taylor & Francis Web site at
http://www.taylorandfrancis.com

and the CRC Press Web site at
http://www.crcpress.com

Contents

Preface

We did not want this to be a book that glosses over the nitty-gritty stuff, assuming you already know everything, nor a book that uses "hand-waving" to magically skirt around real explanations of the complex stuff. The tone of the book is intended to make it more readable—which is to say that it is not too "textbook-y." Having used hundreds of textbooks ourselves, we knew how we did not want this one to be, and that was stodgy.

This book is about nanotechnology, a gigantic topic about small things. It is a book that is intended to excite, inspire, and challenge you. We want to uncover the most important things about nanotechnology and give you the tools you need to dig deeper on your own. We want you to enjoy learning (maybe even laugh) and for you to find out a lot in a short time. There will be plenty of rigorous scientific support, but concepts will be conveyed in clear, simple language that you can digest and apply immediately. We do "back-of-the-envelope" calculations together throughout the process so that you get a good feeling for the numbers of nanotechnology. Creative problem sets (Homework Exercises) follow each chapter to test your understanding of new concepts.

Nanotechnology represents a convergence of many sciences and technologies at the nanometer scale. In fact, it is becoming its own discipline altogether. It requires the ability to apply various scientific principles to system-level design and analysis. The multidisciplinary nature of nanotechnology—which draws from physics, chemistry, biology, and engineering—has the inherent challenge of teaching students with backgrounds in different knowledge domains.

And because the synthesis of disciplines is at the core of nanotechnology, we focus on *systems* in this book. A system is a set of interacting, interrelated, or interdependent elements that are put together to form a complex whole. We discuss nanotechnology on a system-by-system basis to foster both an appreciation and an understanding of this multifaceted topic.

We start with an overview treatment of nanotechnology, with special emphasis on the history, key personalities, and early milestones. Then on to the issues, promises, and fundamentals of nanotechnology. In fact, Chapter 1, "Big Picture and Principles of the Small World," stands alone as a comprehensive introduction, intended to answer your first questions as to what nanotechnology really is and could be. This chapter is self-contained and comprehensive; there is enough information for a freshman or general public course. It includes a discussion of the effects this new industry could have on human life, careers, education, and the environment.

Chapter 2 discusses scaling laws, giving us intuition about the physical ramifications of miniaturization. (While we think this is a useful chapter, be warned: it could bore you. If so, feel free to skip or skim it and use it as a reference.)

Then we dive headlong into nanotechnology. We begin with an "Introduction to Nanoscale Physics" (Chapter 3). Then we tackle the seven main disciplines: nanomaterials (Chapter 4), nanomechanics (Chapter 5), nanoelectronics (Chapter 6), nanoscale heat transfer (Chapter 7), nanophotonics (Chapter 8), nanoscale fluid mechanics (Chapter 9), and nanobiotechnology (Chapter 10). In these "nano" chapters, we provide the specific, fundamental differences between macroscale and nanoscale phenomena and devices, using applications to teach key concepts.

Welcome!

Acknowledgments

For their help in bringing this book to life, the authors thank the National Science Foundation, Melodi Rodrigue, David Bennum, Jonathan Weinstein, Joe Cline, Jeff LaCombe, Michael Hagerman, Seyfollah Maleki, Palma Catravas, Roop Mahajan, Frank Kreith, Daniel Fletcher, Katherine Chen, Todd Sulchek, Nevada Nanotech Systems, Inc., Nevada Ventures, Robb Smith, Stuart Feigin, Chris Howard, Ian Rogoff, Ralph Whitten, David Burns, and the University of Nevada, Reno—especially the Library Department, the College of Engineering, and the Department of Mechanical Engineering.

We would also like to make specific acknowledgments.

Ben Rogers

For their guidance and support, I thank the love of my life, Jill; my vivacious little ladies, Sydney and Quinn; the patient pug, Grace; the nurturing parents, Jim and Sandra; the best of brothers, Judd and Tyler; the big-hearted Hamilton clan; the fresh-thinking Eric Wang; the inimitable Brad Snyder; the late, great Travis Linn; the demanding and kind Jake Highton; the familial Reynolds School of Journalism; and my wise, wise-cracking friends.

Sumita Pennathur

I thank Anthony T. Chobot III and Anthony T. Chobot IV for their undying love and support, and for allowing me the opportunity to be part of this work. Additionally, I thank all my educators throughout the years for giving me the guidance and motivation to create such a textbook.

Jesse Adams

Thanks to all my family, friends, and mentors. You are all the best and this is dedicated to your hard work.

AN INVITATION

One more thing. We put a lot of work into making this book useful for you. So we invite every reader to comment on this book and tell us how we can make it even better. We want your suggestions for future editions and corrections to any errors you may discover. Please e-mail suggestions, questions, comments, and corrections to: michael.slaughter@

taylorandfrancis.com. We plan to list the names of helpful readers in the Acknowledgments section of future editions of this book.

Here are those who have already contributed: Ongi Englander, Ed Hodkin, Morteza Mahmoudi, Aaron S. Belsh, Eva Wu, Brett Pearson, John C. Bean, Darryl Wu, Lia Hankla, Alec Hendricks, and the Davidson Academy.

That said, let us get started.

Authors

Ben Rogers is a writer and an engineer (BS 2001; MS 2002, University of Nevada, Reno). He has done research at Nanogen, the Oak Ridge National Laboratory, and NASA's Jet Propulsion Laboratory, and published many technical papers, as well as fictional works and essays. He is currently the Principal Engineer at NevadaNano and lives in Reno with his wife and two daughters.

Sumita Pennathur is currently an assistant professor of mechanical engineering at the University of California, Santa Barbara (BS 2000, MS 2001, Massachusetts Institute of Technology; PhD 2005, Stanford University). She has been actively contributing to the fields of nanofluidics and nanoelectromechanical systems (NEMS), and has spent some time at both Sandia National Laboratories in Livermore, California, and the University of Twente MESA+ research facility in the Netherlands. When not enveloped in her research work, she can be found either spending time with her husband and son or at a local club wailing on her saxophone.

Jesse Adams (BS 1996, University of Nevada; MS 1997 and PhD 2001, Stanford University) is the Vice President and CTO of NevadaNano. He is working to bring multifunctional microsensor technology to the chemical sensing market space.

CHAPTER 1

Big Picture and Principles of the Small World

Nanotechnology means putting to use the unique physical properties of atoms, molecules, and other things measuring roughly 0.1–1000 nm. We are talking about engineering the smallest-ever structures, devices, and systems.

Nanotechnology is also a promise.

A big one. Nobel laureates, novelists, and news anchors alike tell us on a daily basis that nanotechnology will completely change the way we live. They have promised us microscopic, cancer-eating robots swimming through our veins! Self-cleaning glass! Digital threads! Electronic paper! Palm-sized satellites! The cure for deafness! Molecular electronics! Smart dust! What the heck *is* smart dust—and when can we get our hands on some? A promise is a promise ...

Such things are actually down the road. Nanotechnology has been hyped by techies who cannot wait to order a wristwatch with the entire Library of Congress stored inside; while others bespeak the hysteria of rapidly self-replicating gray goo. Much that is *nano* is burdened with over-expectations and misunderstanding. As usual, the reality lives somewhere between such extremes. Nanotechnology is like all technological development: inevitable. It is not so much a matter of what remains to be seen; the fun question is, who will see it? Will we? Will our children? Their children? Turns out, we will all get to see some. Nanotechnology is already changing the way we live, and it is just getting started.

The "nano" from which this relatively new field derives its name is a prefix denoting 10^{-9}. "Nano" comes from *nanos*, a Greek word meaning dwarf. In the case of nanotechnology, it refers to things in the ballpark that are one-billionth of a meter in size. When Albert Einstein was in graduate school in 1905, he took experimental data on the diffusion of sugar in water and showed that a single sugar molecule is about one nanometer in diameter. Prefixes can be applied to any unit of the International System of Units (SI) to give multiples of that unit. Some of the most common prefixes for the various powers of 10 are listed in Table 1.1.

TABLE 1.1   Some Prefixes for SI Units

yotta (Y)	10^{24}	1 septillion
zetta (Z)	10^{21}	1 sextillion
exa (E)	10^{18}	1 quintillion
peta (P)	10^{15}	1 quadrillion
tera (T)	10^{12}	1 trillion
giga (G)	10^{9}	1 billion
mega (M)	10^{6}	1 million
kilo (k)	10^{3}	1 thousand
hecto (h)	10^{2}	1 hundred
deka (da)	10	1 ten
deci (d)	10^{-1}	1 tenth
centi (c)	10^{-2}	1 hundredth
milli (m)	10^{-3}	1 thousandth
micro (μ)	10^{-6}	1 millionth
nano (n)	10^{-9}	1 billionth
pico (p)	10^{-12}	1 trillionth
femto (f)	10^{-15}	1 quadrillionth
atto (a)	10^{-18}	1 quintillionth
zepto (z)	10^{-21}	1 sextillionth
yocto (y)	10^{-24}	1 septillionth

The word "nanotechnology" was first used in 1974 by Norio Taniguchi in a paper titled "On the Basic Concept of Nano-Technology" (with a hyphen) (Taniguchi, 1974). He wrote:

> In the processing of materials, the smallest bit size of stock removal, accretion or flow of materials is probably of one atom or one molecule, namely 0.1–0.2 nm in length. Therefore, the expected limit size of fineness would be of the order of 1 nm. …"Nano-technology" mainly consists of the processing … separation, consolidation and deformation of materials by one atom or one molecule.

By the 1980s, people were regularly using and spreading the word "nanotechnology."

The late Richard Smalley (Figure 1.1), who shared the 1996 Nobel Prize in Chemistry with Harry Kroto and Robert Curl, was a champion of the nanotech cause. In 1999 he told Congress that "the impact of nanotechnology on the health, wealth, and lives of people will be at least the equivalent of the combined influences of microelectronics, medical imaging, computer-aided engineering and manmade polymers" (full written statement available at http://www.er.doe.gov/bes/House/smalley.htm). Many scientists share Smalley's bullish assessment.

Nanotechnology has ambitiously been called the next industrial revolution, a wholly different approach to the way human beings rearrange matter. People have always tinkered with what the earth has to offer—there is nothing else with which to work. Technology is in many respects just the rearrangement of chunks of the Earth to suit our needs and our

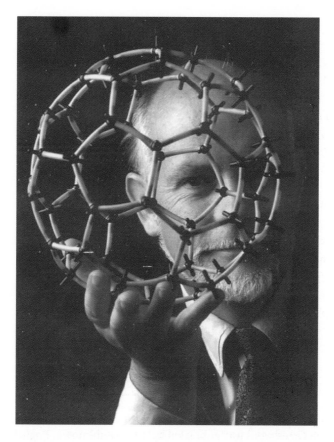

FIGURE 1.1 Richard Smalley. Until 1985, graphite and diamond were believed the only naturally occurring forms of carbon. Then Dr. Smalley, Harold Kroto, James Heath, Sean O'Brien, and Robert Curl discovered another one. It was a soccer-ball-type arrangement they called buckminsterfullerenes ("buckyballs," for short) after Richard Buckminster Fuller, the renowned architect credited with popularizing the geodesic dome. Similar molecules were soon discovered, including nanotubes. These new forms of carbon are called fullerenes. (Photo used with permission of Dr. Richard E. Smalley and Rice University.)

wants. And the Earth is nothing more than atoms. Ever since we dwelled in caves, we have put atoms over fires to heat them, bashed them against rocks to regroup them, and swallowed them for lunch. We just did not know about them, and we certainly could not see them, nor control them one at a time.

Those days are over.

1.1 UNDERSTANDING THE ATOM: *EX NIHILO NIHIL FIT*

Take a block of gold. If we slice the block in two, it is still gold. Half it again, and again, and again—still gold. But how many times can we divide the chunk and still have gold? And is it made up of *only* gold, or is there also empty space in the block?

In the fifth century B.C., Greek philosophers Democritus (Figure 1.2) and his teacher, Leucippus, were asking questions like these. They posited that all matter was composed of undividable particles called *atomos*, which in Greek means "unbreakable" or "not sliceable."

Rubens delin.   *Blake sculp.*

FIGURE 1.2   Greek philosopher Democritus. In the fifth century, he proposed the general idea of atoms, or *atomos*, the Greek word for "unbreakable." (Artist unknown, English eighteenth century Democritus, from *Essays on Physiognomy*, Vol. I by Johann Kaspar Lavater [London: John Murray et al., 1792] engraving and letterpress text, 19.5 × 20.3 cm [plate], 33.6 × 27.0 cm [page]; National Gallery of Victoria Library, Gift of John Cotterell, 1952, National Gallery of Victoria, Melbourne, Australia.) (With permission from National Gallery of Victoria, Melbourne, Australia.)

These particles were completely solid, homogeneous, and varied in size, shape, and weight. Between the atoms was void, Democritus said. The famous expression, *ex nihilo nihil fit* (nothing comes from nothing) was his. Although he is credited with writing about 60 books on his theories, none survived.

Plato (ca. 427–347 B.C.) and Aristotle (384–322 B.C.) disagreed with the atom idea and stuck with the prevailing belief that all matter was composed of the four basic elements: earth, water, air, and fire. Epicurus (341–270 B.C.), however, adopted "atomism" as the foundation of his teachings and wrote hundreds of books on the topic. These, like Democritus' works, were lost. But the idea was not, and a Roman named Titus Lucretius Carus (96–55 B.C.) wrote poetry extolling atomism. These writings were unpopular with the Romans and later considered atheistic by many Christians. Carus' books of poetry, unlike the writings of his predecessors, were saved and passed on. French philosopher Pierre Gassendi (1592–1655) read these books and spread the word, penning persuasive treatises on atomism. But the writings were just that—words. His were convincing arguments, but there was no proof.

Robert Boyle (1627–1691), a British scientist, read Gassendi's work and was interested in it. He later provided the arguments that would be the first in a string of physical proofs: Boyle's law. This law states that there is an inverse relationship between the pressure of a gas and its volume (so long as the temperature and the quantity of the gas do not change). The existence of atoms explains this behavior. The pressure in a container of gas is caused by tiny particles (atoms) colliding over and over again with the container walls, exerting a force. If you make the volume of a container larger, the particles collide with the walls less frequently, and the pressure decreases.

Next to come was the French chemist Louis Proust (1754–1826). Proust noticed that copper carbonate—be it native or prepared in a lab—always broke down into the same proportions of copper, carbon, and oxygen by mass. Like Boyle's law, Proust's Law of Definite Proportions jived well with Democritus' concept of indivisible pieces of matter.

Building on the ideas of those who came before him, it would be John Dalton (1766–1844) who at last set the record straightest. Dalton was not formally schooled past the age of 12. (In fact, he began teaching at that age!) It would be Dalton who, in experiments with carbon monoxide and carbon dioxide, realized that one of the gases had one oxygen atom while the other had two oxygen atoms. He eventually expanded Proust's Law of Definite Proportions into a Law of Multiple Proportions. Molecules, Dalton found, were made from fixed numbers of different atoms. Elements combine in ratios of small whole numbers. While carbon and oxygen can react to form CO or CO_2, they cannot form $CO_{2.6}$. As for the elements everyone had been trying to get to the bottom of—well, those were atoms. Different elements are just different atoms with different masses. Modern atomic theory had been established.

Our understanding of the atom has since been refined by the likes of Ernest Rutherford, Niels Bohr, Albert Einstein, and countless others. We are definitely not finished understanding yet. *Ex nihilo nihil fit*—nothing comes from nothing, and the opposite is also true: everything comes from something. The modern model of the atom did not manifest out of thin air. It is the culmination of centuries of work done by creative and diligent thinkers, and those soon to follow.

The significance of atoms cannot be overstated: their form, behavior, and relationships with one another can be used to explain much of the universe. *You* are atoms. So is everything else. Think about a carbon atom in a neuron in your brain, an atom you are using this very second to help you read this sentence and understand it. That same atom was likely once part of an asteroid, then a tree, then a piece of fruit, then maybe a dinosaur, then dirt—on down the line until eventually you came to borrow it for a while.

BACK-OF-THE-ENVELOPE 1.1

Democritus said matter was composed of undividable particles called *atomos*, Greek for "unbreakable." Was he right?

No, atoms are indeed divisible; however, to split one is a messy endeavor and they do not stay divided for long. A process of splitting atoms into smaller pieces is nuclear fission. This process releases tremendous amounts of energy—used in nuclear weapons and nuclear power generation.

Okay, so if atoms can be broken down into smaller parts, will picotechnology and femto-technology be next?

During fission (and fusion also), certain subatomic particles are released but stable atoms are reformed immediately. Subatomic things—neutrons, protons, electrons—are of great consequence; and as our level of understanding about them deepens, entirely new technological possibilities will emerge. Still, picotechnology and femtotechnology do not make sense in the way that nanotechnology does. The nanoscale is the realm of the atom. With today's technology, you cannot build anything that lasts with smaller stuff. Atoms represent a fundamental frontier. Nanotechnology is not just another step in an ongoing technological trend toward miniaturization. We have reached a boundary.

If, in some cataclysm, all of scientific knowledge were to be destroyed, and only one sentence passed on to the next generation of creatures, what statement would contain the most information in the fewest words? I believe it is the *atomic hypothesis* (or the *atomic fact* or whatever you wish to call it) that *all things are made of atoms— little particles that move around in perpetual motion, attracting each other when they are a little distance apart, but repelling upon being squeezed into one another.* In that one sentence, you will see, there is an *enormous* amount of information about the world, if just a little imagination and thinking are applied.

(RICHARD FEYNMAN, 1963)

The composition and behavior of this most important unit of matter have been completely rethought and refined many times over. Two examples of early atom models are

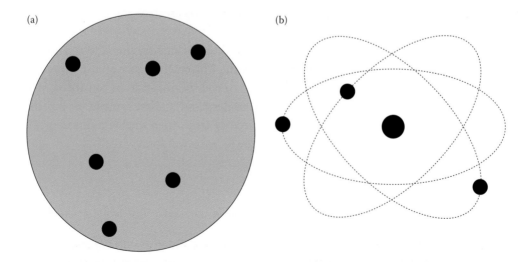

FIGURE 1.3 Early attempts to model the atom. J. J. Thompson's model was that of a positively charged volume embedded with electrons like watermelon seeds (a). Later, Ernest Rutherford hit upon the idea of a nucleus and electrons that orbited it like planets around the Sun (b).

shown in Figure 1.3. J. J. Thompson suggested that the atom was a volume of positive charge embedded with negatively charged electrons similar to seeds in a watermelon. Experiments by Ernest Rutherford in 1911 disproved this model and suggested that there must be a concentration of positive charges in the center of the atom, which he called the nucleus, around which the electrons moved in stable orbits much like planets orbiting the Sun. This model was later refined by Niels Bohr, Werner Heisenberg, and others into an atom looking more like the one depicted in Figure 1.4.

Here is what we understand about the atom:

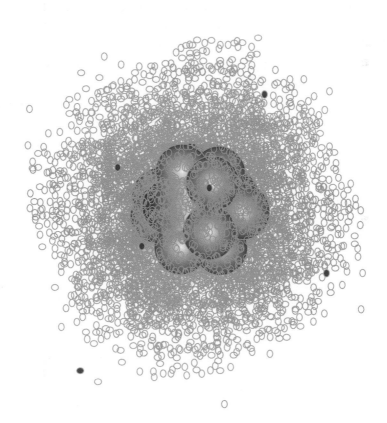

FIGURE 1.4   The atom. The building block of the universe, an atom has a positively charged centralized mass called the nucleus, which is a cluster of neutrons (having no electrical charge) and protons (having positive charge). Negatively charged electrons in perpetual motion enclose the nucleus. In this depiction, the empty dots represent locations where electrons are likely to be found, mostly near the nucleus, and the black dots are locations of the electrons at a particular instant in time.

The nucleus occupies the center. It is made of a dense packing of protons and neutrons. As their names indicate, protons have a positive charge and neutrons have a neutral charge. The two particles weigh about the same. Neutrons hold the nucleus together because without them the positively charged protons would repulse each other and the nucleus would break up. The nucleus contains almost all the atom's mass. Negatively charged electrons in perpetual motion enclose the nucleus. Electrons are most likely to be found within regions of space called orbitals. These orbitals have differing shapes and sizes, depending on how many electrons there are and how much energy they have. In the case of the simplest atom, hydrogen, which has only one electron, the shape of this orbital is a sphere.

Now consider this: we are the first generation in the history of the world to look right at atoms, to pick them up one at a time, and to put them back down where we like.

Stop! Go back, go back. Reread that last sentence. It may have been too easily swallowed whole. Without tasting anything. Without sliding the i-n-d-i-v-i-d-u-a-l atoms around on your tongue like caviar and bursting them one by one between your molars.

Atoms. Mmmmmmmm.

To get a feel for the significance of this achievement, think of it this way: if you were to take an apple and make it as big as the Earth, then the original apple would be the size of an atom in the earth-sized apple.

We can see and move atoms using the scanning tunneling microscope (STM). This is "bottom-up" engineering, creating something by arranging atoms one by one exactly where we want them, as opposed to "top-down" engineering, where a piece of raw material is drilled, milled, chipped away—reformed—until what is left is what is needed. Nanotechnology is the epitome of bottom-up engineering.

The STM belongs to a family of versatile, small systems tools called scanning probe microscopes (SPMs). Figure 1.5 shows a ring of atoms arranged and imaged using an STM. This tool is a great example of a small system that bridges the gap between size scales—it has microscale parts used to do nanoscale work. We learn more about the STM later.

BACK-OF-THE-ENVELOPE 1.2

How much does an atom weigh?

This prompts another question: which kind of atom? The Periodic Table has 116 different elements in it. Each atom has a different number of protons and neutrons, and so their atomic masses are different. Also, since atoms are so minute, we often perform atomic computations using large groups of them. One particularly useful group is called a mole (mol), which consists of Avogadro's number, N_A, of whatever one is counting—atoms, molecules, apples. Avogadro's number is 6.02×10^{23}. It is defined as the number of carbon-12 atoms in exactly 12 g. Therefore, 12 g/mol is the atomic mass of carbon-12. The atomic mass of lead is 207 g/mol, and that of aluminum is 27 g/mol.

To determine the mass of a single atom, divide the atomic mass by Avogadro's number:

$$\text{Mass of 1 aluminum atom} = \frac{27\,\text{g/mol}}{N_A} = 4.5 \times 10^{-23}\ \text{g/atom}$$

So, an aluminum atom weighs 45 yoctograms.

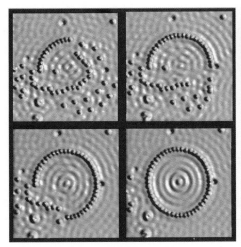

FIGURE 1.5   IBM's Quantum Corral. In 1993, a ring of 48 iron atoms was arranged one at a time (four steps are shown) on a copper surface using the tip of a low-temperature STM. (We will learn more about the STM in Chapters 5 and 6.) The STM was then used to capture an image of the ring, which measures about 14.3 nm across. The iron atoms confine some of the copper's surface electrons, and this barrier forces the electrons into quantum states, visible as concentric standing waves inside the corral. A three-dimensional rendering of this astounding achievement appeared on the cover of *Science*. (Corral image originally created by IBM Corporation. Journal cover from *Science* Vol. 262, No. 5131, October 8, 1993. Reprinted with permission from AAAS.)

BACK-OF-THE-ENVELOPE 1.3

Proust's work with copper carbonate showed it always broke down in a specific ratio, by mass. This led him to draw his historic conclusion, the Law of Definite Proportions, and furthered our understanding of atoms. What is the ratio Proust measured for this particular substance?

The chemical formula of copper(II) carbonate is $CuCO_3$. Its molecular weight is

$$63.55 \text{g/mol} + 12.01 \text{g/mol} + 3(16.00 \text{g/mol}) = 123.56 \text{g/mol}$$

The fractional composition by mass of each element is therefore

$$Cu: \frac{63.55 \text{g/mol}}{123.56 \text{g/mol}} = 0.5143$$

$$O: \frac{3(16.00 \text{ g/mol})}{123.56 \text{ g / mol}} = 0.3885$$

$$C: \frac{12.01 \text{g/mol}}{123.56 \text{ g/mol}} = 0.0972$$

Dividing through by the smallest fraction (that of C) gives the relative elemental composition by mass:

Cu: 5.2912

O: 3.9967

C: 1.0000

These are the very same ratios Proust measured over two centuries ago: 5.3 parts copper, 4 parts oxygen, and 1 part carbon.

First, however, let us introduce the first in nanotechnology's cast. If nanotechnology can be said to have a history, then it probably starts with Richard Feynman.

1.2 NANOTECHNOLOGY STARTS WITH A DARE: FEYNMAN'S BIG LITTLE CHALLENGES

Perhaps it was Richard Feynman's enigmatic reputation. Not quite the scientific celebrity he would soon become—he had not won his Nobel Prize for quantum electrodynamics yet—the physicist's ideas and motives were perhaps still enshrouded in enough mystery to keep his colleagues on their toes. When the American Physical Society asked him to serve as the featured speaker during its annual banquet, he accepted. The Society's 1959 meeting was in Pasadena, California—Feynman's backyard. It was hosted by Caltech on a warm December 29th, a few days before the New Year, a fitting time of the year for predicting the future.

The banquet was downtown at Pasadena's Huntington-Sheraton. For $4.50 a plate, roughly 300 members of the scientific community ate and drank together. Among the topics of discussion was the title of their guest speaker's talk: "There's Plenty of Room at the Bottom." Some were embarrassed to admit they did not know what the title meant. Had Feynman selected a topic over their heads? Others guessed that Feynman was going to tell them about how there were plenty of lousy jobs left for the taking in the physics industry. When Society President George Uhlenbek welcomed Dr. Feynman to the stage amid the clinking of busy silverware, attendees readied themselves for some kind of elaborate put-on, the kind only Dr. Feynman could conjure.

However, Feynman was utterly serious. And he was talking about a field for which a name had yet to be coined. He was talking about nanotechnology.

Feynman had spent a lot of his own time mulling over the possibilities of small things. Small as in atoms. His tone suggested something akin to disappointment and at the same time hope. His questions: Why has someone not already done this? Why have we yet to think big about the very small? I will tell you what needs to be done, and the best part is: it is doable!

"Now the name of the talk is 'There's *Plenty* of Room at the Bottom'—not just 'There's Room at the Bottom,'" Feynman said. "I will not discuss how we are going to do it, but only that it is possible in principle—in other words, what is possible according to the laws of physics. I am not inventing antigravity, which is possible someday only if the laws are not what we think. I am telling you what could be done if the laws *are* what we think; we are not doing it simply because we haven't gotten around to it" (Feynman, 1960).

In his speech, he would pose challenge after challenge—such as writing the entire 24 volumes of the *Encyclopedia Britannica* on a pinhead, making an electron microscope that could see individual atoms, or building a microscopic computer—and then he would outline the parameters of that challenge. Feynman made many references to examples in nature such as DNA and the human brain where miniaturization was already wildly successful.

Much of what he said has since come to fruition and is commonplace, although in a year when computers took up entire rooms it all sounded quite like fantasy. He explained many of the physical issues and challenges inherent in moving to the small scale, including quantum behavior, van der Waals forces, heat transport, and, of course, fabrication. Still he was intrepid: "I am not afraid to consider the final question as to whether, ultimately—in the great future—we can arrange atoms the way we want; the very *atoms*, all the way down!"

In parting, Feynman offered a pair of $1000 prizes. One was for the first person to create an operating electrical motor no larger than 1/64 in.[3]. To Feynman's dismay, William McLellan tediously did just that, using tweezers and a microscope, within four months of the speech. The motor had 13 parts, weighed 250 μg, and rotated at 2000 rpm. Feynman had been home from his honeymoon less than a week when he had to explain to his wife

---

BACK-OF-THE-ENVELOPE 1.4

What does fitting the *Encyclopedia Britannica* on a pinhead entail?
The head of a pin measures about 1/16th of an inch across; therefore,

$$\text{Area} = \pi\left(\frac{0.0625}{2}\right)^2 = 0.00307\,\text{in.}^2$$

The *Encyclopedia Britannica* has approximately 30 volumes and each volume has about 1000 pages, each measuring 9 in. × 11 in. Thus,

$$\text{Area} = (30)(1000)(9)(11) = 2{,}970{,}000\ \text{in.}^2$$

A pinhead large enough to fit the regular-sized *Encyclopedia Britannica* would need to have a diameter X times bigger than a real pinhead:

$$\text{Area} = \pi\left(\frac{0.0625X}{2}\right)^2 = 2{,}970{,}000\ \text{in.}^2$$

Solving for X shows that the encyclopedia would need to be about 32,000 times smaller in order to fit on the head of a pin. Is that feasible? The diameter of the dot on an "i" in a fine printing of a book is about 1/120 of an inch wide (which, by the way, is about the smallest feature the human eye can resolve). If that dot is reduced 32,000 times, it would be about 7 nm across. In an ordinary metal, an atom is about 0.25 nm in diameter, so the dot would be about 28 atoms wide. The whole dot would contain about 1000 atoms. So there are definitely enough atoms on the head of a pin to accommodate all the letters in the encyclopedia.

THE PRESCIENT PHYSICIST: RICHARD FEYNMAN (1918–1988)

Richard Feynman (Figure 1.6) was a zany, independent thinker who restlessly challenged anything he could not prove to himself. As a boy, he often disregarded his textbooks so as to derive the formulas in his own way—sometimes sloppily, and yet sometimes more elegantly than the established derivations. By ignoring so much of what had been posited by his scientific predecessors, he discovered yet untried ways to unravel the thorny conundrums of quantum physics and was for his contributions to the field awarded the 1965 Nobel Prize in Physics.

Feynman was a member of the exclusive Manhattan Project team developing the first atomic bomb, and resisted his scientific quarantine by regularly picking Los Alamos safes full of classified secrets. He communicated with his wife through letters written in code, all for his deciphering amusement and the annoyance of the government censors. He chose to view the first test of the atomic bomb through the windshield of a truck instead of through the standard-issue dark glasses donned by fellow researchers. His own quick calculations had convinced him the blast's ultraviolet light probably would not permanently blind him—making him the only one, he figured, to directly witness the explosion.

When the space shuttle *Challenger* exploded on January 28, 1986, killing all seven astronauts aboard, Ronald Reagan added the iconoclastic physicist to the commission charged with ferreting out the cause of the accident. After many interviews with NASA officials, in which he favored engineers over managers, Feynman concluded that the mechanical fault lay in a tiny O-ring—a rubbery gasket used to seal the solid-fuel rockets. These O-rings lost their malleability at low temperatures and failed to adequately form a seal. This he demonstrated before a throng of reporters during public testimony at the commission's hearings. He dunked a rubbery O-ring in a small cup of frigid water and demonstrated its acquired rigidity. Despite typically warm weather at the Kennedy Space Center in Florida, the day of the launch had been characterized by near-freezing temperatures. Feynman's "minority report" conclusions about the accident were only included in the appendix of the commission's findings.

When he left his Caltech office for good in 1988, he left behind an epigram on the blackboard: "What I cannot create I do not understand."

the promise he had made. They were not exactly financially prepared to divvy out prize money, but he did. (The motor, on display at Caltech, no longer spins.)

The second prize was won by Tom Newman, a Stanford grad student who met Feynman's challenge to "take the information on the page of a book and put it on an area 1/25,000 smaller in linear scale in such a manner that it can be read by an electron microscope" (roughly the scale at which the entire *Encyclopedia Britannica* could squeeze onto a pinhead). Newman and colleague Ken Polasko were at the forefront of electron-beam lithography, approaching the level of quantum effects. They sent a letter to Feynman asking if the prize was still unclaimed, and he phoned them personally in their lab to encourage them to pursue it. They did, using a specialized electron beam writing program to transcribe the full first page of *A Tale of Two Cities* by Charles Dickens.

Feynman's visions were being realized, and at a pace quicker than perhaps even he had expected. The unfathomable was getting more mundane everyday.

FIGURE 1.6   Richard Feynman. His prophetic talk, "There's Plenty of Room at the Bottom," is credited with inspiring the development of nanotechnology. Feynman was a charismatic teacher (and student) of physics. (Photograph by Floyd Clark, courtesy of Caltech Archives, used with permission of Melanie Jackson Agency, LLC.)

As with the gradual refinement of atomic theory, the breakthroughs that continue to enable development in nanotechnology are the result of thousands of years of scientific inquiry. Like children, scientists have always been curious, optimistic, and fearless. They continue to tackle nature's toughest puzzles; yet when it comes to nanotechnology, they do it under new auspices. The prospects for prestige, government funding, and the excitement of working at the forefront of a scientific uprising have convinced many scientists and engineers to relabel their nanoscale work—be it thin films, fine fibers, submicron lithography, colloidal particles, and so forth—as nanotech. Condensed-matter physicists and organic chemists are nowadays likely to go by the common job title of nanoscientist.

Nanotechnology is engineering. It is about the practical application of science. Nanotechnology borrows liberally from physics, chemistry, materials science, and biology—ranking it among the most all-encompassing engineering fields. Figure 1.7a shows the relative occurrences of a few nanotechnology topics on the Internet—a quick and intriguing way to estimate the size of these subjects. Figure 1.7b shows international interest in nanotechnology based on Google© searches.

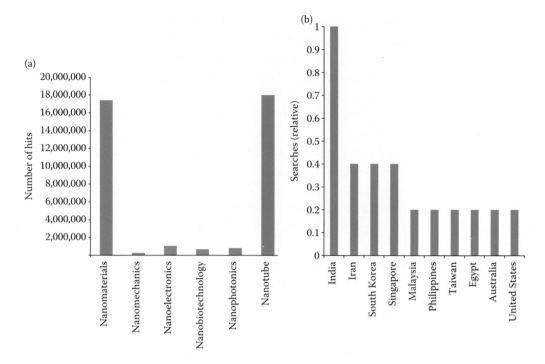

FIGURE 1.7   Googling nano. An interesting way to gauge the relative "size" of a topic is to count the number of hits using the Internet search engine Google©—which on August 25, 2010, gave the results shown in (a). We can also get a sense of the international interest in nanotechnology with data available from Google Trends about where search queries originated. Graph (b) shows the relative number of searches for the term "nanotechnology" from the top 10 regions for the years 2004–2010.

In this book, the cluster of nanotech disciplines is broken into seven main areas:

1. Nanomaterials
2. Nanomechanics
3. Nanoelectronics
4. Nanoscale heat transfer
5. Nanophotonics
6. Nanoscale fluid mechanics
7. Nanobiotechnology

1.3 WHY ONE-BILLIONTH OF A METER IS A BIG DEAL

At one-billionth of a meter, the world does not behave in quite the ways we are used to. We are familiar with the physical behavior of things we can hold in our hands, like a baseball, or, more accurately, the trillions of trillions (not trillions *and* trillions—trillions *of* trillions) of atoms of which the baseball is comprised. Toss it up, and it comes down. Put it near another ball, and that is probably where it will stay—near the other ball.

However, an atom does not behave like a baseball. It is beholden to laws of physics that really only kick in below 100 nm, where matter becomes a foreign substance we do not understand quite as well. Gravity, for example, has little effect. However, intermolecular forces, such as van der Waals forces, play a large role when the atom nears its neighbors. Quantum effects manifest also. In the quantum regime, electrons no longer flow through conductors like water through a hose, but behave instead more like waves; electrons can hop, or "tunnel," across insulating layers that would have barred passage if conventional, macro-scale physics were in charge. Strange phenomena such as size confinement effects and Coulomb blockage are relevant below 30 nm.

When you move through a swimming pool, your body feels like it is gliding through the water. The wetness is continuous. However, if you were a chlorine atom in the same pool, you would find a discreteness to your motion as you went bumping around between molecules.

As we gain unprecedented control over matter, entirely new devices are possible. Think of atoms as bricks and electrons as mortar, and begin tailoring things from the bottom up, using nature's building blocks. The human body is nothing but a bunch of oxygen, hydrogen, nitrogen, carbon, and other atoms; but when you combine these ingredients properly, the resulting object is exponentially more significant and miraculous than the sum of its parts. The primary challenge here is the integration of objects over diverse size scales, the overlapping of bottom-up engineering with top-down approaches to realize real devices with nanoscale functionality. Right now, it is difficult to work at the nanoscale. It is expensive and tedious and hit-or-miss. It is difficult. Imagine building something as mechanically straightforward as the gear assembly shown in Figure 1.8—a simple machine on the macroscale but quite a challenge when the teeth of the gears are molecules, or even individual atoms.

However, it is not impossible. Nanosized particles have for a century been used to reinforce tire rubber, long before nanotechnology went by that name. Chemists design nanoscale catalysts to accelerate thousands of chemical transformations every day, including the conversion of crude oil into gasoline for cars, organic molecules into pills for fighting disease, graphite into diamond for cutting tools. Engineered vaccines contain proteins with nanoscale dimensions. Computer disk drives have nanometer layers for memory storage.

These applications are a glimpse into how one might use nanotechnology in the future. The tenet of this gradual conversion to bottom-up thinking is: Do More with Less. Molecular-scale manufacturing ensures that very little raw material is wasted and that we make only what we intend to make, no more. Factories begin to look more like clean rooms.

All manufactured products already consist of atoms. Theoretically, this fact leaves no product unaffected by nanotechnology. Whereas machines have always been a means of outperforming human strength, nanotechnology will enable us to overcome our limitations in the opposite direction—below our physical size limits.

Nanotechnology could make for healthier, wealthier nations. Governmental preparedness and foresight are crucial and should coincide with scientific and engineering progress. James Canton, President of the Institute for Global Futures, has said of nanotechnology's disruptive economic potential that "those nations, governments, organizations, and citizens

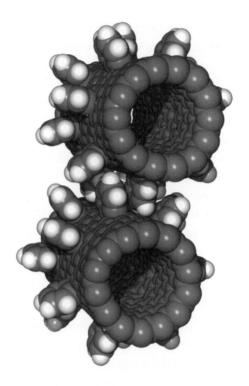

FIGURE 1.8 Molecular gears. This schematic representation shows how molecules might be employed in a gear mechanism. (Image courtesy of NASA.)

who are unaware of this impending power shift must be informed and enabled so that they may adequately adapt" (Canton, 1999). Materials, manufacturing, electronics, medicine, health care, energy, information technology, national security—the list of industries nanotechnology will influence is long.

The sooner we understand what nanotechnology is, the quicker we can temper any expectations or fears we may have. It is going to take teamwork to realize its many brilliant promises.

1.4 THINKING IT THROUGH: THE BROAD IMPLICATIONS OF NANOTECHNOLOGY

Like all technological trends throughout history, nanotechnology has been subjected to vigorous overstatement, speculation, and opposition. The chain of events has been typical. A handful of paradigm-shifting scientific breakthroughs in otherwise disparate disciplines (such as the invention of the STM and the discovery of carbon buckminsterfullerenes— both of which we will discuss in more detail later in this book) leave scientists at once awed by new possibilities and at the same time suspicious. It has been said that extraordinary claims require extraordinary proof. Bolstered by a breakthrough, but scientifically skeptical, scientists worldwide work quickly to verify or refute a new idea. The ideas that take hold are those that emerge intact from the spotlight of scrutiny.

This is not to say that skeptics cannot be wrong, or that they are completely objective in their evaluation of new ideas; the history of technological development is littered with people, including scientists, eventually proven wrong by persistent optimists with great ideas. The Wright brothers and the inventors of the STM were both hounded by criticism from their peers before and after their breakthroughs. Copernicus' belief about the Sun being the center of the solar system, and not the Earth, met with great opposition from religious leaders.

Technological and scientific breakthroughs represent giant leaps in our understanding of the world. At the same time, they reveal giant gaps. Discovery is not only about finding answers, but new questions.

Some of the new ideas regarding nanotechnology have weathered decades of skepticism. Some, such as nanotube semiconductance, have been demonstrated enough times to warrant widespread acceptance and now application development. Others, such as the possibility of a molecular assembler that could piece together specific molecules over and over again, are still hotly and publicly debated. (See "Gray Goo" below.)

However, it is not only the scientific merit of a new idea that must face opposition. The ethical implications are considered also, or at least they should be. As is often the case, the excitement of discovery and our innate desire to alter the world to better suit ourselves (sometimes at the expense of the planet) quickens the pace of development. Before anyone has had time to think through the ramifications of a new idea, it is not new anymore. The world has already absorbed it, for better or for worse.

During World War II, scientists secretly building and testing the first nuclear weapon were among the only ones who knew about the technology. The billions of people affected by the emergence of such a weapon had no say in how it was developed or deployed until after it was dropped on Hiroshima toward the end of the war. So began the nuclear arms race, with nations competing to build better and better weapons in the hope that no other nation would ever use them: the only foreseeable outcome of nuclear war was mutually assured destruction, or MAD. Was this the best course of development for nuclear technology?

Vehement backlashes against biotechnologies such as stem cell research, cloning, and genetically modified foods suggest that people find such things important enough to debate, or outright oppose them.

That which is new can be frightening—oftentimes because it is unknown. The more we learn about new technologies, the better equipped we are to decide how to use (or not use) them. The potential uses of nanotechnology, both good and bad, remain topics of debate. Ethical issues such as these do not necessarily have right or wrong answers. Since the topic is the future, the debate is fueled by speculation. For solutions that benefit the most people in the end, the discussion needs to be steered by people who understand the technology. Education across the board—from scientists to journalists to kindergartners—seems the best approach to ensure that nanotechnology is harnessed and used for the greatest good. Museums, libraries, laboratories, schools, and members of industry each play a role in helping everyone understand what exactly nanotechnology means on both a global and a neighborhood scale. Governments must also intervene where necessary and appropriate in order to mitigate potentially harmful outcomes of nanotechnology, and facilitate positive ones.

The 21st Century Nanotechnology Research and Development Act, signed by President George W. Bush in December 2003, authorized billions of dollars in funding for nanotechnology research and development. The act appropriated money not only for research and development, but also to ensure that "ethical, legal, environmental, and other appropriate societal concerns ... are considered during the development of nanotechnology" and are called for "public input and outreach to be integrated into the program by the convening of regular and ongoing public discussions, through mechanisms such as citizens' panels, consensus conferences, and educational events." Groups such as the Center for Responsible Nanotechnology and the Foresight Institute often weigh in on political and policy-related issues.

There is a pressing need to answer, or at least discuss, the many ethical, legal, and environmental questions surrounding nanotechnology. Such questions tend to arise before, during, and after the creation of new technologies and play a role in how they are developed. It is important not to pick a side, but rather to be able to understand both sides of the debate. By exploring and acknowledging the potential risks of nanotechnology, a track record of honesty and openness can be established.

Here are some questions about nanotechnology still being answered:

1. How will the military use nanotechnology?

2. How might pervasive, undetectable surveillance affect our privacy?

3. Might nanotechnology be used in acts of mass terror?

4. How do we safeguard workers from potentially dangerous fabrication processes?

5. How will our attempts to better our bodies with nanotechnology affect later generations and society as a whole?

6. Who will define what is ethical about nanotechnology?

7. Who will regulate nanotechnology?

8. Should we limit research in areas that could be dangerous, even if this prevents beneficial technologies from being developed as well?

9. How will the benefits (financial, health, military, etc.) of nanotechnology be distributed among the world's nations?

10. To what extent will the public be involved in decision making? Legislators? Scientists? Businesspeople?

11. Will nanotechnology reduce the need for human workers and cause unemployment?

12. How will intellectual property (patents) be handled?

13. Who will profit from nanotech innovations? Universities? Businesses? Individuals?

1.4.1 Gray Goo

It has been suggested that self-replicating "nanobots" could become a new parasitic lifeform that reproduces uncontrollably, remaking everything on Earth into copies of itself.

The result would be undifferentiated "gray goo." The gray goo idea is one that has received much press and has sparked vehement arguments within the scientific community. K. Eric Drexler (Figure 1.9), one of nanotechnology's earliest advocates and thinkers, once quipped that gray goo's catchy, alliterative name fuels the interest (as with the oft-mentioned "digital divide") and that there is really nothing gray or gooey about the scenario.

The majority of scientists who have weighed in on gray goo discount its possibility as no possibility at all. Most importantly, no self-replicating machine has ever been built, on any scale, let alone the nanoscale. Richard Smalley penned an editorial in *Scientific American* in 2001 that fueled an ongoing debate with Drexler about the feasibility of a nanoscale assembler that could self-replicate. When would we see such a thing, Dr. Smalley asked rhetorically. "The simple answer is never" (Smalley, 2001).

1.4.2 Environmental Impact

Dichlorodiphenyltrichloroethane (DDT) was first synthesized in 1873. In 1939, the Swiss scientist Paul Hermann Müller discovered that DDT was an effective insecticide, earning him the 1948 Nobel Prize in Physiology and Medicine. During the 1940s and 1950s, DDT was widely used to kill mosquitoes and thereby prevent the spread of malaria, typhus, and

FIGURE 1.9   K. Eric Drexler. He is founder and chairman of the Foresight Institute, a think tank and public interest institute on nanotechnology founded in 1986. (Photo courtesy of K. E. Drexler.)

other insect-borne human diseases. It seemed a perfect synthetic pesticide: toxic to a broad range of insects, seemingly safe for mammals, and inexpensive.

In 1962, a book by Rachel Carson, *Silent Spring*, described the ill effects of pesticides on the environment, including birds, whose reproduction was harmed by thinning egg shells. Studies revealed that DDT was toxic to fish and mammals, including humans, and that it accumulated through the food chain. It is now banned in the United States and most other countries. The publication of Carson's book is considered one of the first events in the environmental movement.

The plight of DDT provides a valuable lesson about how a synthetic product—bestowed with accolades, proven effective, and seemingly safe—can wreak havoc in the environment. It also illustrates how scientists and businesspeople tend to focus on breakthroughs and bottom lines, sometimes at nature's expense.

Nanotechnology has been touted as an eco-friendly approach to making things. Little raw material would be needed and very little would be wasted by nanotech processes. However, there are valid concerns as to how nanosized products such as nanoparticles would accumulate in nature. For example, could large amounts be ingested by fish? And if so, would it be harmful? Would the particles be passed along the food chain like DDT? Could genetically modified microorganisms reproduce uncontrollably into a *green* goo that clogs the environment?

The effects of nanomaterials in the environment remain mostly unknown. Manufactured nanoparticles—which will become increasingly prevalent in consumer products—are one example of a nanotechnology for which environmental and health impact studies are lacking. The fate, transport, transformation, and recyclability of nanoparticles in the environment must be studied; exposure thresholds must be determined. It is important to find out how to remove or simply detect nanomaterials if they become problematic. Research to assess the toxicological effects of inhaled nanotubes is currently underway using test animals. Preliminary results with rats suggest that nanotubes can be toxic when inhaled. Another early study reported brain damage in a large-mouth bass whose aquarium water contained the common carbon molecules known as buckyballs. While not yet conclusive, such findings raise red flags, and more studies are needed to ensure that nanomaterials are employed safely.

1.4.3 The Written Word

Nanotechnology's reputation has been authored not only by scientists and technologists, but also by novelists. By surveying our present capabilities and then extrapolating to the extreme of possible outcomes, fantastic fictions emerge. Such stories create public fascination and help focus attention on the subject of nanotechnology—both good and bad. Publicity of this sort can stimulate ideological debate, foster industry growth, or generate fear.

Engines of Creation by K. Eric Drexler was published in 1986. Although not fiction, the book was imaginatively forward-looking and discussed the potential of nanotechnology to give us unprecedented control of matter. Self-replicating nanomachines, Drexler said, might be used to produce any material good, pause global warming, cure diseases, and dramatically prolong human life spans. These visions were the stuff of science fiction, but also fodder for ridicule. Still, the book stirred the imagination of many of today's leading nanotechnologists and is considered a seminal work.

Novelists also were intrigued by Drexler's book. *Prey*, a novel by Michael Crichton published in 2002 (Figure 1.10), quotes Drexler in the opening of the book. *Prey* does for nanotechnology what Crichton's *Jurassic Park* did to heighten concerns over cloning. The story is about a company developing nanoscale surveillance technology. The spying nanoparticles eventually become a predatory swarm, evolve rapidly, run amok, and feed on people in the Nevada desert—not exactly good public relations for nanotechnology but an awareness booster nonetheless. Other science fiction books involving nanotechnology include *The Diamond Age* by Neal Stephenson, *Slant* by Greg Bear, and *Idoru* by William Gibson.

FIGURE 1.10 The novel *Prey*. Author Michael Crichton: "The creators of technology often do not seem to be as concerned about the effects of their work as outsiders think they ought to be. But this attitude is changing. Just as war is too important to be left to the generals, science is too important to be left to the scientists." (Copyright © 2002 by Michael Crichton. Reprinted by permission of HarperCollins Publishers.)

Novelists are not the only writers weighing in. In a widely quoted article from *Wired Magazine* (April 2000), "Why the Future Doesn't Need Us," Bill Joy, cofounder of Sun Microsystems, wrote that nanotech, genetic engineering, and robotics together make for a bleak outlook for the human race, and possibly our extinction. The arguments have met strong criticism. *Fantastic Voyage* by Ray Kurzweil and Terry Grossman (who met at a Foresight Institute conference in 1999) discusses the use of nanobots to improve and repair the human body.

1.5 THE BUSINESS OF NANOTECH: PLENTY OF ROOM AT THE BOTTOM LINE TOO

As technology drivers, curiosity and the thrill of innovation are often no match for the lure of big profits. Money is the primary force pushing nanotechnology development world-wide. Faster, smaller, cheaper—like a mantra, these long-standing engineering goals aimed at displacing existing products are the same reasons nanotechnology will figure into the balance sheets of so many companies in the coming decades. If a better product can be built in a small clean room instead of a gigantic factory, using less raw material, and shipped in envelopes instead of on pallets, it is going to earn the company money. In terms of products, "nano" also spells "new"—new ideas, new markets, and new ways to turn a profit.

The investment in tech development, paradoxically, costs a fortune. Governments around the world are among the only investors that can afford to take the long-term financial plunge. Private industry usually expects investments to pay off in 5–10 years, sometimes even faster. As such, federal nanotech funding in the United States, Europe, and Asia tends to support fundamental research, intended to build the scientific knowledge base upon which businesses can later capitalize. Governmental investments in science do not necessarily require *immediate* economic benefits.

Still, large multinational research and development giants such as IBM, General Electric, and Hewlett-Packard have made significant internal investments in research and development programs. While the outcomes are historically difficult to predict, basic research is vital to technological growth. It is where the lucrative scientific surprises come from. It is not surprising, then, that more than half of the 30 companies in the Dow Jones Industrial Index have launched some type of nano initiative.

Estimates pegged the number of people working in nanotech research in 2005 at around 20,000. Within 15 years, this number is expected to reach 2 million. Your next job may be nanorelated.

In 2004, manufactured goods incorporating nanotechnology accounted for just over $10 billion in sales—the equivalent of a rounding error on the global economy. But it is a ripple that over the coming years is expected to grow into a more disruptive wave, especially as the prices of nanoengineered materials drop. Figure 1.11 shows how the price of nanotubes has come down since their introduction as a commercial product. There are now numerous companies selling all varieties of nanotubes. Sales of products using nanotechnology could surpass $500 billion by 2010.

Whereas the Internet has proved a singular phenomenon with definite uses and boundaries, the marketplace for nanotechnology does not appear so easily constrained; it

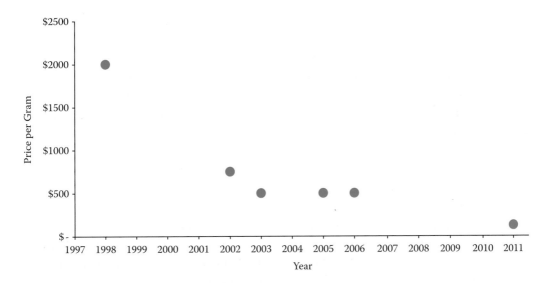

FIGURE 1.11 Carbon nanotube prices over time. Nanotube production processes originally developed at Rice University were later used for production and sale of nanotubes by Tubes@Rice (1998) and subsequently by Carbon Nanotechnologies Incorporated (CNI). The price shown is for a single gram of pure, single-walled nanotubes (SWNTs). All data for 1998–2006 are from Tubes@Rice and CNI. The 2011 data point ($97) is the average price of a single gram of >90% pure SWNTs from three suppliers—Nano-C, mknano, and CheapTubes.

is difficult to draw a boundary around it because it affects no single industry, but rather the scale of engineering in general. It offers thousands of new possibilities for materials and devices that already exist. This makes market speculation even more speculative. Burned by dot com failures, investors have been weary of overhype and simultaneously allured by the promise of new markets, new customers. Venture capitalists have thus far invested about $1 billion in nano companies and more than 1000 start-ups have formed around the world, about half of them in the United States.

However, investors are scrupulous. They have found themselves asking: What does nano even mean? In April 2004, Merrill Lynch introduced a Nano Index, featuring 25 companies in the nanotech business. Critics pounced on the index, debating the nanoness of, for example, the pharmaceutical companies on the roster. These companies make molecules for drugs, a common industry practice, but is that truly *nano*? Merrill revised the index to include only companies with publicly disclosed nanotech initiatives representing a significant component of their future business strategy. Reshuffling, they dropped six companies and added three new ones.

Meanwhile, using the same old periodic table of raw materials, nanoscientists are inventing new things. A deluge of patent applications has resulted in what some are calling a "land grab" for intellectual property. This gold-rush mentality has led to thousands of U.S. patents issued, with thousands awaiting judgment. Most of these fall into five nanomaterials categories: (1) dendrimers, (2) quantum dots, (3) carbon nanotubes, (4) fullerenes, and (5) nanowires. Reviews of these patents have revealed much overlap and fragmentation, meaning that to avoid infringement, many entrepreneurs will first need to strike agreements with numerous patent holders before cashing in on the technology.

There are also other challenges facing nanotech companies. Things taken for granted with traditional size scales—quality control, inventory—are difficult to manage when the products are practically invisible. Industry standards and enforcement practices are needed, as are assurances that the products sold are safe in the body and in the environment.

1.5.1 Products

Chances are that you have already bought something that uses nanotechnology. Companies of all sizes are hurrying small science from labs to the marketplace. The primary markets are for nanoparticles that bolster products such as the rubber in tires and the silver used in photography. Specialized coatings on glass reduce glare and make it easier to clean. Other products incorporating nanotechnology include burn and wound dressings, water filtration systems, dental bonding agents, car parts including bumpers and catalytic converters, sunscreens and cosmetics, tennis balls, golf clubs and tennis rackets, stain-free clothing and mattresses, and ink.

Further along the nanotech timeline, expect to see more ubiquitous energy generation. Ideas in the works include solar cells in roofing tiles and siding. Computers also will become more pervasive as nanotech breakthroughs make circuits small and cheap enough to build into fabrics and other materials. Look for better drug delivery systems, including implantable devices that release drugs and measure drug levels while inside the body.

HOMEWORK EXERCISES

1.1 What is the definition of nanotechnology?

1.2 The number 1,234,567 written out is one million, two hundred thirty-four thousand, five hundred sixty seven. Write out this number: 1,200,300,400,500,600,700,800,901.

1.3 Rank the following things from largest to smallest: polio virus, drop of water, mercury atom, *E. coli* bacterium, helium atom, human red blood cell.

1.4 Approximately how long ago was the concept of the atom introduced?

1.5 What are the three main components of an atom?

1.6 What are the main components of an atom's nucleus?

1.7 What is the Law of Definite Proportions?

1.8 What is the Law of Multiple Proportions?

1.9 Boyle's law states that $PV = C$, where P is the pressure of a gas, V is the volume, and C is a constant (assuming constant temperature). Consider a gas held in a 4 m^3 container at 1 kPa. The volume is then slowly doubled.
   a. What is the new pressure?
   b. Use atoms to explain how a larger container leads to a lower pressure.

1.10 a. What is the mass of a square piece of aluminum foil 100 μm thick and 10 cm wide (aluminum = 2.7 g/cm^3)?
   b. How many atoms are in the piece of foil (aluminum = 27 g/mol)?

1.11 Calculate the mass of an atom of
   a. Hydrogen (1.0 g/mol)
   b. Silver (107.87 g/mol)
   c. Silicon (28.09 g/mol)

1.12 What are the mass ratios of the elements in these chemical compounds? (*Note:* nitrogen = 14 g/mol; hydrogen = 1 g/mol; carbon = 12 g/mol; oxygen = 16 g/mol.)
   a. Ammonia, NH_3
   b. Ethanol, C_2H_6O
   c. Toluene, C_7H_8

1.13 What now-famous talk did Feynman give to stimulate development in nanotechnology? What year did he give it?

1.14 What less optimistic topic did some of those in the audience suspect was meant by the title of Feynman's talk?

1.15 Many computers use one byte (8 bits) of data for each letter of the alphabet. There are 44 million words in the *Encyclopedia Britannica*.
   a. What is the bit density (bits/in^2) of the head of a pin if the entire encyclopedia is printed on it? Assume the average word is five letters long.
   b. What is the byte density?
   c. What is the area of a single bit in nm^2?
   d. A CD-ROM has a storage density of 46 megabytes/in.^{2} and a DVD has a storage density of 329 megabytes/in.^{2}. Is the pinhead better or worse than these two storage media? How much better or worse?

1.16 What is meant by "gray goo"?

1.17 For a gray goo scenario to play out, what entirely new type of machine would be necessary?

1.18 What year was the word "nanotechnology" first used?

1.19 A baseball is made of trillions of trillions of atoms.
   a. Write out the number for one trillion trillion.
   b. NASA estimates that there are about 10^{21} stars in the universe. Is this number higher or lower than the number of atoms in a baseball?

1.20 The distance between the nuclei of two iron atoms is about 4 Å ($1 Å = 10^{-10}$ m).
   a. How many nanometers is that?
   b. How many iron atoms at this spacing would it take to reach 2 μm ($1 μm = 10^{-6}$ m)?

1.21 What five categories are the most popular areas for nanotechnology patents in the United States?

Short Answers

1.22 Name three things you are familiar with (easy for you, personally, to identify with) that are roughly 1 mm (millimeter) in size. Name something that is approximately 1 μm (micrometer) in size. Name something that is 1 nm (nanometer) in size.

1.23 Based on your education and interests, describe the role you might be best suited to play in the multidisciplinary arena of nanotechnology.

1.24 Perform your own topic search using an Internet search engine. Use the same search terms as Figure 1.7 and reconstruct the chart. How have the results changed, and what does this suggest?

1.25 Make a list of at least five name-brand products that incorporate nanotechnology.

1.26 Search *Science* magazine's online table of contents. Find the percentage of issues from the previous year with at least one article whose title contains the prefix "nano."

1.27 The concept of the atom was ridiculed by Romans; the idea that the Earth revolved around the Sun was initially shunned also. What scientific ideas are at the heart of controversy these days? What are the implications of these ideas? Which groups are at odds? How much is proven about the idea and how much is conjecture?

Writing Assignments

1.28 Nanotechnology is multidisciplinary; it draws from, and requires expertise in, numerous scientific and engineering fields. So the question becomes: Is there such a thing as nanotechnology? Are there any applications, research fronts, concepts, or overarching goals that are unique to nanotechnology and not just an advancement in another field (chemistry, physics, medicine, biology, etc.)? Or is nanotechnology really just the name for where all these other fields overlap? Citing and quoting evidence from credible sources (including at least two that are nontechnical in nature such as a newspaper article, a book review, or a governmental document) and those more geared toward scientists and engineers (e.g., an editorial or an article from a scientific journal or a speech from a convention) take one side of this issue and argue it in 500 words. It would also certainly be worth interviewing an expert on the topic (a professor or government official perhaps).

1.29 Technological progress in nuclear power and biotechnology has been thwarted to a degree by public distrust, misinformation, and resistance to change. There are very real dangers and ethical issues involved in such technological progress, and at the same time, very real advantages. How is nanotechnology similar? How is it different? What lessons can be taken from the manner in which nuclear power and biotechnology are understood by the general public to make for a safer, more productive transition period in the case of nanotechnology?

1.30 Richard Feynman thought that the atomic hypothesis was the best single sentence to summarize all of scientific knowledge. Write your own sentence at the top of a page and use the rest of the page to convince the reader that your choice makes sense.

REFERENCES

J. Canton. 1999. *Nanotechnology: Shaping the World Atom by Atom*, National Science and Technology Council, Washington, D.C., p. 8.

R. Feynman. 1960. *Engineering and Science*, XXIII(5), and now widely available online.

R. Feynman. 1963. *Six Easy Pieces*, chap. 1, Perseus Books, Cambridge, MA.

R. E. Smalley. 2001. Of chemistry, love and nanobots, *Scientific American*, September, 68–69.

N. Taniguchi. 1974. On the basic concept of nano-technology, *Proc. Intl. Conf. Prod. Eng.*, Tokyo, Part II, pp. 18–23.

RECOMMENDATIONS FOR FURTHER READING

1. I. Asimov. 1991. *Atom: Journey across the Subatomic Cosmos.* Truman Talley Books.
2. Foresight Institute: http://www.foresight.org.
3. E. Regis. 1995. *Nano: The Emerging Science of Nanotechnology.* Diane Publishing Company.
4. Nanotech: The science of the small gets down to business. *Scientific American* Special Issue, September 2001.
5. Lynn E. Foster. 2009. *Nanotechnology: Science, Innovation, and Opportunity.* Prentice Hall.
6. The Interagency Working Group on Nanoscience, Engineering and Technology. 1991. *Nanotechnology: Shaping the World Atom by Atom.* National Science and Technology Council, Committee on Technology, Washington, D.C.
7. National Nanotech Initiative: http://www.nano.gov.
8. Small Tech 101: An introduction to micro and nanotechnology. *Small Times Magazine*, 2003.
9. *Small Times Magazine*: http://www.smalltimes.com.
10. Mihail C. Roco and William S. Bainbridge. 2001 and 2006. *Societal Implications of Nanoscience and Nanotechnology.* Springer.

CHAPTER 2

Introduction to Miniaturization

2.1 BACKGROUND: THE SMALLER, THE BETTER

Why is it that engineers are so gung ho about making things smaller? What spurs the relentless push toward compactness in so many modern things? Of course there are exceptions, like monster trucks, skyscrapers, and passenger planes that we are always trying to make bigger—but even large things like these are possible because so many other things have been made smaller. The computers required to navigate a 747 used to take up the same space as a 747.

Cell phones are becoming tinier, laptops are getting lighter, and a thin strand of optical fiber is replacing thick bundles of copper telephone wire. Meanwhile, products do more with less. This is a central tenet and the primary motivation for miniaturization; for with miniaturization comes multifunctionality. Consider what happens when two different products can each be made half their size. A cell phone becomes part digital camera, part music player. A wristwatch keeps track of your schedule and your heart rate (let alone the time). When we marry devices this way, new, multifunctional products are born. The functional density—or the number of functions a product has per volume—improves exponentially with smaller parts. Consider the ramifications for functionality offered by molecular-scale electronics, which would be about 100 times smaller across than today's computer transistors. This 100-fold improvement on the linear size scale translates into 100^3 by volume, meaning that one million times more circuitry could be stuffed into the same volume.

The engineering benefits of smaller systems can justify the cost inherent in designing and building them. First, shrinking something means that less material is required to build it. Material is like excess baggage. It costs money, it adds weight, and it takes up space. These considerations weigh heavily on all engineering decisions and are of utmost importance for certain applications—for example, satellite and spacecraft systems, which must be as small and light as possible. After all, it is expensive and inefficient to take heavy things into space: the cost of launching the space shuttle in 2004 was about $10,000 per pound. Smaller devices are imperative in the medical field and have enabled unprecedented surgical and imaging techniques. Bulky tools and big cameras simply do not fit inside the delicate pathways of the human body.

Smaller systems perform quicker because they have less mass and therefore less inertia (the tendency of mass to resist acceleration). This improved speed leads to products that perform tasks faster, just as a fly can flap its wings much faster than a bird. Or consider an assembly robot in a factory. It might perform 10 welds in a second, while an enzyme in your body performs as many as a million chemical operations in the same amount of time.

Thermal distortions and vibrations do not perturb smaller devices as much as large ones because the resonant vibration of a system is inversely proportional to its mass. Generally, the smaller the system, the higher its resonance frequency; and the low-frequency vibrational disturbances that affect large systems are less of an issue. Higher motional exactness and dimensional stability are other important advantages of smaller devices; highly precise measurements or movements are possible on a small scale. Finally, smaller things tend to need less energy in order to function. Power consumption concerns can make or break a new product design, and miniaturization is one way to minimize the "fuel" factor. Power density, or the amount of power that can be generated per unit volume, also favors miniaturization. This is discussed in the next section.

Figure 2.1 shows the dimensions in meters and some representative things on the scale from macroscale, to microscale, to nanoscale.

2.2 SCALING LAWS

While the physical characteristics of miniaturized systems tend to vary a great deal from macroscale systems, engineers have found ways to effectively estimate how the characteristics of something will change as its dimensions change. These generalizations are known as the scaling laws, and they are quite useful. Scaling affects the volume of an object and the way it deals with forces, conducts electricity, and handles liquid flow.

At the nanoscale, modeling such physical behaviors means elaborate calculations; the more accurate the model, the more computationally demanding. Scaling laws are hand calculations that can help orient us to unfamiliar scales and highlight the ways a smaller design might improve or inhibit performance.

2.2.1 The Elephant and the Flea

While it is now possible to build things at the smallest imaginable level, the atomic scale, we will first explore the implications of scaling on larger objects—sizes between an elephant and a flea.

Hopefully you fall into that category.

Let us take a look at you. You are a human being and so it would be appropriate to measure you in meters, as opposed to millimeters or kilometers. The meter, therefore, can be used as the unit by which we define your characteristic dimension, a metric used to simplify your size to a representative measurement for comparison purposes. The characteristic dimension, D, takes on new significance when scaling something down or up.

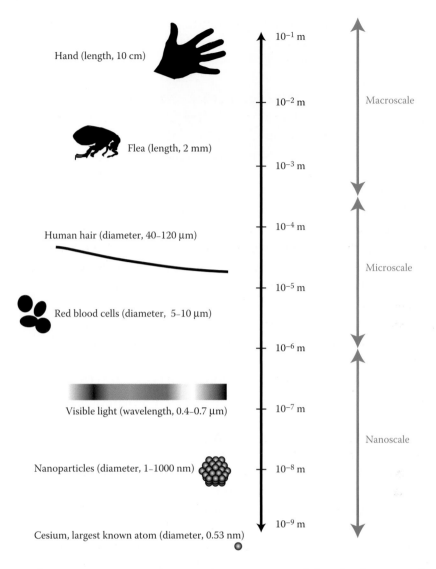

FIGURE 2.1 The macro-, micro-, and nanoscales. Keep in mind that the dimensional range of these scales varies, depending on whom you ask. Some say the nanoscale truly begins below 100 nm. The blue arrows therefore serve as general guides and should be thought to overlap.

Let us consider a simple rectangular solid, with length l, height h, and thickness t, as shown in Figure 2.2. Each of these dimensions can be generalized as a characteristic dimension, D. (Note that D does not have to be equal on all sides. It merely represents dimensions of the same scale.) The volume of the solid, V, and the surface area, S, are then

$$V = htl \propto D \cdot D \cdot D \propto D^3 \qquad (2.1)$$

$$S = 2(ht + tl + hl) = 2(D^2 + D^2 + D^2) \propto D^2 \qquad (2.2)$$

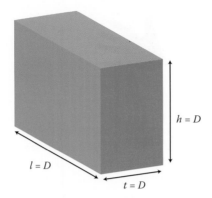

FIGURE 2.2   A rectangular solid with length *l*, height *h*, and thickness *t*. For order-of-magnitude scaling law calculations, we can express the size of this solid instead using a characteristic dimension *D*.

The surface-to-volume ratio, *S/V*, then, scales as follows:

$$S/V = \frac{D^2}{D^3} \propto \frac{1}{D} \qquad (2.3)$$

This relationship is one way to characterize things and to predict, for example, the energy and power they need. An elephant and a car each have a characteristic dimension, *D*, of about 1 m and so *S/V* is on the order of 1 m^{-1}; on the other hand, *D* of a flea or a bullet is about 1 mm and so *S/V* is on the order of 1000 m^{-1}.

BACK-OF-THE-ENVELOPE 2.1

What can scaling laws and the characteristic dimension tell us about the differences between the elephant and the flea shown in Figure 2.3? Among the differentiators between the elephant and the flea is their strength-to-weight ratio. Both elephants and fleas have muscles that allow them to apply forces, so let us look at the difference between these two animals and the differences between their strength-to-weight ratios.

Since the strength of a muscle (the force it can produce) is roughly proportional to the cross-sectional area of that muscle, we can relate strength to the characteristic dimension:

$$\text{Strength} \propto D^2 \qquad (2.4)$$

Meanwhile, a muscle's weight varies proportionately with volume:

$$\text{Weight} \propto D^3 \qquad (2.5)$$

The strength-to-weight ratio is therefore given by

$$\frac{\text{Strength}}{\text{Weight}} \propto \frac{D^2}{D^3} \propto \frac{1}{D} \qquad (2.6)$$

Thus, for an elephant, the strength-to-weight ratio is 1 m^{-1}, versus about 1000 m^{-1} for a flea—looks like a flea has a lot more strength comparatively. The strength-to-weight ratio is a crude indicator of how high these two organisms can jump. An elephant does not have sufficient strength to overcome its weight and cannot jump at all. A flea, meanwhile, can jump hundreds of times the length of its body. Of course, this is a simplified explanation that does not take into account the shape, orientation, and structure of these animals' muscles and bodies—note that the flea's body would collapse under its own weight if it were the size of an elephant. Nonetheless, the scaling law explains quite a bit.

If a flea falls from a second-story window, it lands undisturbed but lost—not the elephant! Again, the explanation lies in characteristic dimension. The elephant's characteristic dimension (m) is 1000 times that of the flea (mm). The elephant is about 1000^3 (1 billion) times heavier than his companion. Based on the strength-to-weight ratio determined above, if a flea can support something 100 times its body weight, then an elephant can only support something 1/10 its body weight.

The example we just worked illustrates how small-scale devices can be relatively stronger than their macroscopic counterparts. Just imagine the strength-to-weight ratio of a nanomachine that is about a million times smaller than the flea.

The effect of scaling on the strength-to-weight ratio is also apparent in human beings, which comes in all shapes and sizes. Figure 2.4 shows men's gold medal winning lifts from the 2004 Olympic Games in Athens as a percentage of the competitors' weights. The smaller the competitor, the greater the percentage of his own body weight he is able to lift. One of the greatest weightlifters in history is Naim Süleymanoğlu of Turkey, a weightlifter whose 4-foot-11-inch-tall body, short arms, and short legs were a major advantage in his three Olympic championships, seven World Championships, and six European Championships, earning him the nickname "The Pocket Hercules."

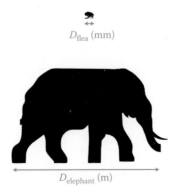

D_{flea} (mm)

D_{elephant} (m)

FIGURE 2.3 The flea and the elephant. The flea's characteristic dimension is best measured in millimeters, the elephant's in meters. This difference gives rise to very different strength-to-weight ratios for these very different creatures.

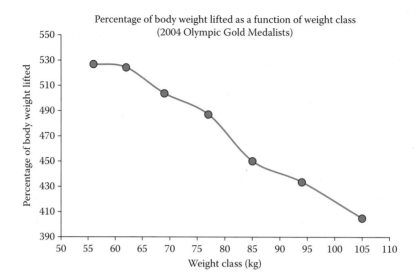

Percentage of body weight lifted as a function of weight class
(2004 Olympic Gold Medalists)

FIGURE 2.4   The advantage of scaling is evident in humans. Based on men's gold medal lifts from the 2004 Olympics in Athens, we see that the bigger athletes competing in the heavier weight classes cannot lift as great a percentage of their own body weight as the competitors in smaller weight classes. For example, the gold medalist in the 55-kg weight class lifted nearly 530% of his own body weight while the larger 105-kg medalist lifted 405% of his own body weight.

2.2.2 Scaling in Mechanics

Here we address the scaling laws pertaining to classical mechanical systems. In all cases, we use the characteristic dimension, D. We assume that all material properties are held constant (density, modulus, strength) and that all the linear dimensions vary proportionally to D.

We begin with mass, m. Mass is equal to density times volume, $m = \rho V$. In this approximation, let us assume density is constant, so that mass is directly proportional to volume, which in turn is proportional to D^3. Therefore,

$$m \propto D^3 \tag{2.7}$$

Among the most common forces is gravity, which at Earth's surface is a constant: $g = 9.8 \text{ m/s}^2$. The force of gravity, F_g, on an object is therefore

$$F_g = mg \propto D^3 \tag{2.8}$$

A dust particle measuring 10 μm across is 10,000 times smaller than a rock 10 cm across. The force of gravity is therefore $10{,}000^3$ or 1 trillion times less for the dust, which helps explain why it can remain airborne.

At the macroscopic level, two surfaces sliding across one another are subject to friction. This frictional force, $F_f = \mu_f F_g = \mu_f mg$, where μ_f is the coefficient of friction. Therefore, at this level, $F_f \propto D^3$ and is independent of the contact area. However, at small scale, this is not the case. Surface roughness and adhesive forces between atoms and molecules are considerable (especially van der Waals forces, discussed in Chapter 4, "Nanomaterials"). We will discuss nanoscale friction in greater detail in Chapter 5, "Nanomechanics." The combination of adhesion and friction is called striction, and the striction force, F_{str}, scales with the contact area:

$$F_{str} \propto D^2 \tag{2.9}$$

As for wear, the rate of erosion is independent of scale but the amount of material available to erode away decreases as the system gets smaller. This means that the wear life of a system, defined as the thickness of a system divided by the erosion rate, is proportional to the characteristic dimension:

$$\text{Wear life} \propto D \tag{2.10}$$

Consider a part measuring a few centimeters and subject to conventional wear mechanisms. If this part has a 10-year wear life, it would be expected to have a lifetime of mere seconds if it were scaled to nanometer dimensions. However, as we scale down systems, we may also be building devices with atomically smooth surfaces. The covalent chemical bonds binding atoms together are strong and would not be broken by the amount of heat energy generated by the rubbing surfaces. Such devices have the potential to operate with near-zero friction or wear.

BACK-OF-THE-ENVELOPE 2.2

Assuming that it is possible to substantially reduce the size of an engine, would 1000 engines occupying the same volume as one engine be more or less powerful, according to the scaling laws?

This is a matter of power density. Power, P, at one moment in time is force times velocity. If we assume constant velocity (i.e., both engines operating at the same speed), power is proportional only to force (we can use Equation 2.4, assuming that force is proportional to strength):

$$P \propto D^2 \tag{2.11}$$

An engine twice the size of another will have 2^2, or four times the power. The volume of the engine, as shown in Equation 2.1, scales as D^3. The power density is therefore

$$\text{Power density} \propto \frac{D^2}{D^3} \propto \frac{1}{D} \tag{2.12}$$

Consider an engine 1 m on a side that produces 100,000 W of power. An engine one-tenth the size (i.e., 10 cm on a side) would then produce one-hundredth the power (1000 W). But in the volume occupied by the large engine, there is room for 1000 of the smaller engines, which all together would produce 1,000,000 W according to the scaling laws. While this is not feasible with an internal combustion engine, other types of motors using electrostatic forces can be miniaturized to the nanometer scale for ultrahigh power densities: an electrostatic motor at that scale could produce as much as a million watts per cubic millimeter.

One structure that remains applicable at all scales is the spring. The force a spring exerts, $F_{\text{spring}} = -kx$, where k is the stiffness or spring constant (assumed here to remain constant) and x is the elongation of the spring. This elongation is proportional to the characteristic dimension, D, and so we can say

$$F_{\text{spring}} \propto D \tag{2.13}$$

Note that this relationship is also valid for the bending stiffness of a rod or beam. The oscillating frequency, f_{spring}, is

$$f_{\text{spring}} = \frac{1}{2\pi}\sqrt{\frac{k}{m}} \propto D^{-3/2} \tag{2.14}$$

BACK-OF-THE-ENVELOPE 2.3

A microcantilever beam 100 μm long and a diving board 5 m long are both deflected and then released simultaneously. Estimate how many oscillations the microcantilever beam completes in the time the diving board completes a single oscillation.

The period of oscillation of the spring, T_{spring}, is the inverse of f_{spring}:

$$T_{\text{spring}} \propto D^{3/2} \tag{2.15}$$

The diving board's characteristic dimension is 50,000 times that of the microcantilever. Its period of oscillation is $50,000^{3/2}$ times larger. So in the time the diving board takes to oscillate once, the microcantilever beam oscillates about 11 million times.

Kinetic energy is $KE = mv^2/2$, so the following relation applies:

$$KE \propto D^3 \tag{2.16}$$

Gravitational potential energy is $PE_{\text{gravity}} = mgh$, so we have

$$PE_{\text{gravity}} \propto D^3 \tag{2.17}$$

In the case of a spring, the potential energy is $PE_{\text{spring}} = kx^2$. Because the elongation, x, is proportional to D:

$$PE_{\text{spring}} \propto D^2 \tag{2.18}$$

Note that this relationship is also valid for the deflection of a rod or beam.

2.2.3 Scaling in Electricity and Electromagnetism

Electricity is the primary source of power and control for small systems. Applications include electrostatic and piezoelectric actuation, thermal resistance heating, electro-dynamic pumping, and electromechanical transduction. Some of the scaling laws for electrical systems are provided here in terms of their characteristic dimension, D.

Here we assume that voltage, V, is a constant, while electrostatic field strength, E_S, varies with the characteristic length over which the electric field is applied, as follows:

$$E_S = \frac{V}{D} \propto \frac{1}{D} \tag{2.19}$$

The resistance of a conductor, R, is a function of the material's resistivity, ρ, its characteristic length, D, and its cross-sectional area, $A = D^2$. If the resistivity is constant, then we can write the following:

$$R = \frac{\rho D}{A} \propto \frac{1}{D} \tag{2.20}$$

Thus, a 10-nm block of gold ($\rho = 2.44 \times 10^{-8}$ Ω m) would have a resistance of about 2 Ω, while a 1-cm block would have practically no resistance: about 2 $\mu\Omega$.

Using Ohm's law, the current, I, is

$$I = \frac{V}{R} \propto D \tag{2.21}$$

Note that in very large electric fields, Ohm's law loses validity. For example, a voltage of 10 V across an element 2 μm wide creates a field, $E_S = 5 \times 10^6$ V/m. In electric fields this large, the material's resistivity can vary with E_S.

The power dissipated by an electrical element, W, is defined by Joule's law and scales as

$$W = RI^2 \propto D \tag{2.22}$$

BACK-OF-THE-ENVELOPE 2.4

In the microelectronics industry, it is well known that the smaller and more densely packed the electrical elements on a chip, the more power the chip will dissipate. How can the scaling laws explain this phenomenon?

Consider a microprocessor chip. The number of elements per unit area is proportional to the characteristic dimension:

$$\frac{\text{elements}}{\text{area}} = \frac{\text{elements}}{D^2} \propto \frac{1}{D^2} \tag{2.23}$$

Equation 2.22 provides us with a relationship between the characteristic dimension of a single element and the power it dissipates. To determine the power dissipated per unit area, we multiply this equation by Equation 2.23. So the power dissipated per unit area varies as

$$W\left(\frac{\text{elements}}{D^2}\right) \propto \frac{D}{D^2} \propto \frac{1}{D} \qquad (2.24)$$

If the area of a transistor can be made half the size of the previous design, the number of transistors in a given area would double, but so would the power dissipated by the processor. This can make heat management challenging. Engineers can try to offset this effect by reducing the operating voltage.

To determine how scaling affects capacitors, let us consider two parallel plates of area, A, separated by a distance, d. The capacitance, C, scales as

$$C = \frac{\varepsilon_o \varepsilon_r A}{d} \propto \frac{\varepsilon_o D^2}{D} \propto D \qquad (2.25)$$

ε_o is the permittivity of free space ($8.85418782 \times 10^{-12}$ m^{-3} kg^{-1} s^4 A^2) and ε_r is the relative permittivity of the dielectric material between the plates.

Electrostatic forces are often used for actuation at small scales. The electrostatic force between capacitive plates is a special case where the breakdown voltage can vary with the gap distance due to what is known as the Paschen effect. For our purposes, this simply means that the largest voltage we are able to safely apply, V, is proportional to D. The expression for electrostatic force, F_e, between parallel plates scales as

$$F_e = -\frac{1}{2}\frac{\varepsilon_o \varepsilon_r A V^2}{d^2} \propto \frac{(D^2)(D^2)}{D^2} \propto D^2 \qquad (2.26)$$

Electromagnetic forces are typically not attractive for small systems. The magnetic force, F_{mag}, is given by

$$F_{mag} = \frac{1}{2}I^2\frac{\partial L}{\partial x} \qquad (2.27)$$

where ($\partial L / \partial x$) is a dimensionless term relating to the inductance of the conductor. We know from Equation 2.21 that $I \propto D$; therefore,

$$F_{mag} \propto I^2 \propto D^2 \qquad (2.28)$$

From this relationship we can see that a 10 times reduction in size translates to a 100 times reduction in the electromagnetic force.

BACK-OF-THE-ENVELOPE 2.5

Electromagnetic forces are not large enough for meaningful application at the nanoscale. To illustrate this, compare the strength of a typical covalent bond—about 1 nN—to the electromagnetic force between a pair of parallel nanowires each carrying 10 nA of current. Both wires are 10 nm long and they are separated by a single nanometer.

The electromagnetic force per length, l, between parallel wires is given by

$$\frac{F_{mag}}{l} = \frac{\mu_0 I_1 I_2}{2\pi d} \tag{2.29}$$

The term μ_0 in the equation is a constant, $4\pi \times 10^{-7}$ T · m/A, and d is the distance separating the two wires. Substituting gives

$$F_{mag} = \frac{\left(10 \times 10^{-9}\text{m}\right)\left(4\pi \times 10^{-7}\text{ T} \cdot \text{m/A}\right)\left(10 \times 10^{-9}\text{A}\right)^2}{2\pi\left(1 \times 10^{-9}\text{ m}\right)} = 2 \times 10^{-22}\text{N}$$

This force is 2×10^{-13} nN, about 12 orders of magnitude smaller than the force of a single covalent bond.

2.2.4 Scaling in Optics

Optical devices are useful not only as components of small systems, but also for investigating small systems themselves, such as using microscopes. In addition, small devices can be fabricated using optics, or photolithographic techniques, whereby photosensitive materials are selectively etched using light.

If a plane wave (light) of wavelength, λ, shines upon an object with a characteristic dimension, D, the reflected wave diverges at an angle, $\theta \approx \lambda/D$. Such is also the case if light passes through an opening with characteristic dimension, D, as depicted in Figure 2.5. This means that

$$\theta \propto \frac{1}{D} \tag{2.30}$$

If the opening is much bigger than the wavelength, the waves hardly diverge at all. When shining light on a microscopic object, the light diffracts and spreads out. The more the amount of this diffracted light that can be captured in the lens of the microscope, the more information it will have for creating an image.

In photolithography, light is focused through a lens onto a surface to "write" patterns. In one type of photolithography, a thin polymer film is spread over a surface and wherever the light hits, the polymer hardens. When the unexposed polymer is washed away in a solvent, the hardened features remain. The minimum diameter of the spot size, or irradiated zone,

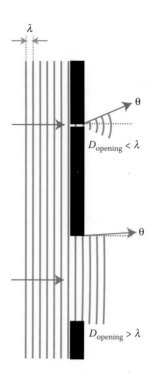

FIGURE 2.5 Waves of light with wavelength, λ, pass through a small and a large opening in a wall. The width of each opening is represented by the characteristic dimension $D_{opening}$. As the waves pass through the opening, they diverge at an angle, $\theta \approx \lambda/D$. If $D_{opening}$ is much larger than the wavelength, the waves hardly diverge at all.

is also the smallest feature we can write. The characteristic dimension, D, of this feature is given by

$$D = \frac{2\lambda}{\pi(NA)} \tag{2.31}$$

Here, λ is the wavelength of light used and NA is the numerical aperture of the system, a metric that describes how sharp an angle a particular lens can focus light in a particular medium. See Figure 2.6. Therefore,

$$D \propto \lambda \tag{2.32}$$

So the smallest features we can make with optical lithography are limited by the wavelength of light used. It is no wonder that engineers have found ways to use ultraviolet light (which has smaller wavelengths than visible and infrared light) for photolithography. This helps them build much smaller electronic components on chips.

The resolution, Δ, of optical elements (including your eyes, microscopes, and telescopes) is the smallest distinguishable distance between two objects. (This is discussed in greater

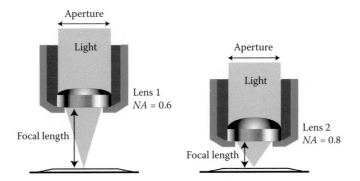

FIGURE 2.6   Numerical apertures of two lenses. The cone of light captured by lens 1 is not as wide as lens 2. In the case of a microscope, this means that lens 2 will gather more light diffracting off the sample for a brighter, higher resolution image. In the case of photolithography, the smallest spot we can illuminate with a given lens depends on the wavelength of light used.

detail in Chapter 8, "Nanophotonics.") It is determined in part by the diameter of the circular aperture (shown in Figure 2.6) because the size of this aperture dictates how much light can be gathered. This diameter can be thought of as D. Therefore,

$$\Delta \propto \frac{1}{D} \tag{2.33}$$

The focal length of a lens, FL (as shown in Figure 2.6) is given by the lens-maker's formula:

$$\frac{1}{FL} = (n-1)\left(\frac{1}{R_f} - \frac{1}{R_b}\right)$$

Here, n is the refractive index of the lens (in this case a constant), and R_f and R_b are the radii of curvature of the lens' front and back surfaces, respectively. If a lens retains its shape while scaling up or down, the radii (considered the lens' characteristic dimension, D) must vary proportionally. This means that the lens' focal length is proportional to D:

$$FL \propto D \tag{2.34}$$

Note that these optical scaling laws hold true as long as the dimensions of the optical components under consideration remain much larger than the wavelength ($D \gg \lambda$).

2.2.5 Scaling in Heat Transfer

Heat transfer plays a significant role in small systems. One common application involves transferring heat away from densely packed electrical components in a computer processor to keep it from overheating.

The energy, E_{th}, needed to heat a system of characteristic dimension, D, to a given temperature is directly related to the system's mass and therefore its volume:

$$E_{th} \propto D^3 \tag{2.35}$$

This relationship holds true for the heat capacity, Q_{cap}, of a system—also proportional to mass (i.e., volume):

$$Q_{cap} \propto D^3 \tag{2.36}$$

Heat can be transferred by conduction, convection, and radiation. Conduction and convection are the more common modes of heat transmission, especially in microsystems, and are therefore discussed in greater detail here. As you learn about the scaling laws for conduction and convection, pay attention to the fact that the scaling laws are often valid in two distinct size regimes: (1) one being the submicrometer scale, which includes systems with $D \leq 1\ \mu m$, and (2) the other being systems larger than that, with $D > 1\ \mu m$.

The rate of heat conduction, $Q_{conduct}$, in a solid is given by

$$Q_{conduct} = -k_c A \frac{\Delta T}{\Delta x}$$

In this equation, heat moves in one dimension, x, through the solid, while k_c is the thermal conductivity of the solid, A is the cross-sectional area of the heat flow, and ΔT is the temperature difference across the solid. At most scales, we can assume that k_c is constant throughout the solid and therefore can draw a relationship between the heat conduction and the characteristic dimension:

$$Q_{conduct} = -k_c A \frac{\Delta T}{\Delta x} \propto D^2 \frac{1}{D} \propto D \quad \left[D > 1\ \mu m \right] \tag{2.37}$$

In submicrometer-sized systems, however, thermal conductivity varies with the characteristic dimension, $k_c \propto D$. (This is due to molecular heat transfer mechanisms that vary with specific heat, molecular velocity and average mean free path. These concepts are beyond the scope of this chapter and are discussed in Chapter 7, "Nanoscale Heat Transfer.") Therefore, for submicrometer systems

$$Q_{conduct} = -k_c A \frac{\Delta T}{\Delta x} \propto (D)(D^2)\frac{1}{D} \propto D^2 \quad \left[D \leq 1\ \mu m \right] \tag{2.38}$$

To determine the amount of time, τ_{th}, needed to achieve thermal equilibrium in a system (an equal temperature throughout), we can use a dimensionless parameter called the Fourier number, $F_o = \alpha t/D^2$, where α is the thermal diffusivity of the material and t is the time it takes heat to flow across the system's characteristic length, D. By rearranging this equation and defining t as τ_{th}, we see that the characteristic time for thermal equilibrium is proportional to the square of the characteristic dimension:

$$\tau_{th} = \frac{F_o}{\alpha} D^2 \propto D^2 \tag{2.39}$$

Heat transferred by the motion of fluid is convection, and is governed by Newton's cooling law. This law gives the rate of heat convection, $Q_{convect}$, between two points in the fluid and can be stated in a generic form as

$$Q_{convect} = h_t A \Delta T$$

Here, h_t is the heat transfer coefficient, A is the cross-sectional area of the heat flow, and ΔT is the difference in temperature between the two points in the fluid. The heat transfer coefficient, h_t, is a function of how fast the fluid moves, which does not significantly impact the scaling of the heat flow. This means that the area, A, is the primary factor in the scaling of $Q_{convect}$.

$$Q_{convect} = h_t A \Delta T \propto D^2 \quad \left[D > 1\,\mu m \right] \tag{2.40}$$

Note that this scaling relationship does not apply to submicrometer-sized systems. (The convective heat transfer in such systems is not governed by continuum fluid theories; in fact, convective heat transfer becomes conductive among gas molecules in tight confinement. Again, the mean free path of the molecules is taken into account. More on this topic is available in Chapter 7.)

2.2.6 Scaling in Fluids

All bodies move in fluid environments such as air and water, the only exception being fluidless vacuum environments. So it is worthwhile to examine some of the ways fluid parameters vary with D.

A simple way to approach scaling in fluid mechanics can be illustrated by examining fluid flow in a section of small circular pipe. As shown in Figure 2.7, the pipe's radius, r, serves as the characteristic dimension, D, of the pipe ($D = r$). The volumetric flow, Q_{vol}, through the pipe is given by

$$Q_{vol} = \frac{\pi D^4 \Delta P}{8\,\mu l}$$

Here, μ is the dynamic viscosity of the fluid, l is the pipe length (which could be centimeters or kilometers and is not part of this scaling discussion), and ΔP is the pressure drop

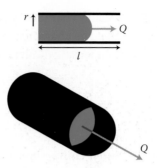

FIGURE 2.7 Fluid flowing through a small circular pipe. Changes in pipe radius affect the volumetric flow significantly more than changes in any of the other variables.

over the length of the pipe. As you can see, changes in pipe radius affect the volumetric flow significantly more than changes to any of the other variables. As such

$$Q_{vol} \propto D^4 \tag{2.41}$$

The volumetric flow, Q_{vol}, can be also expressed as

$$Q_{vol} = v\pi D^2 \tag{2.42}$$

Here, v is the average velocity of the fluid and πD^2 is the cross-sectional area of the pipe (D serving as the pipe radius). We can substitute to get the pressure gradient (pressure drop per unit length), such that

$$\frac{\Delta P}{l} = \frac{Q_{vol}8\mu}{\pi D^4} = \frac{(v\pi D^2)8\mu}{\pi D^4} = \frac{8\mu v}{D^2} \tag{2.43}$$

The scaling law for pressure gradient in the pipe is therefore

$$\frac{\Delta P}{l} \propto \frac{1}{D^2} \tag{2.44}$$

At the microscale, reducing the tube radius by a factor of 10 leads to a 1000-fold increase in the pressure drop per unit length—an important design problem and the reason why engineers have devised alternative methods of pumping at that scale. These techniques are often based on surface pumping forces. Like the flea and the elephant described earlier, a microscale flow has a much larger surface-to-volume ratio compared to a flow through a pipe in your bathroom, for example. Electrohydrodynamic pumping techniques, including electroosmotic and electrophoretic pumping, focus on the surface of the fluid; they are widely used in the biotechnology and pharmaceutical industries.

A separate scaling phenomenon involving fluids is the motion of falling bodies. Let us consider a sphere (like a particle or a ball) falling through the air. Viscous friction effects on the sphere cause it to eventually fall with a constant velocity, v_{lim}, sometimes called terminal velocity. If the flow past the sphere is laminar (not turbulent), the limiting velocity is

$$v_{lim} = \frac{2gr^2(\rho_{sphere} - \rho_{fluid})}{9\eta}$$

Here, ρ_{fluid} is the density of the fluid, ρ_{sphere} is the density of the sphere, η is the fluid's viscosity, g is gravity, and r is the radius of the sphere. In this case, $r = D$, and so

$$v_{lim} \propto D^2 \qquad (2.45)$$

The time, t_{lim}, that the sphere takes to reach this velocity is equal to v_{lim}/g, which means that

$$t_{lim} \propto D^2 \qquad (2.46)$$

BACK-OF-THE-ENVELOPE 2.6

Raindrops evolve from clouds composed of tiny droplets, about 20 μm in diameter. These droplets agglomerate within the cloud to form larger drops that fall to the ground. Let us compare the terminal velocity of a droplet to that of a drop formed from 1000 agglomerated droplets. The viscosity and density of air are 1.8×10^{-5} N s/m^2 and 1.29 kg/m^3, respectively, and the density of water is $\rho_{water} = 1000$ kg/m^3.

The 20-μm droplet reaches a terminal velocity of

$$v_{lim} = \frac{2gr^2(\rho_{sphere} - \rho_{fluid})}{9\eta} = \frac{2(9.8 \text{ m/s}^2)(10 \times 10^{-6} \text{ m})^2(1000 \text{ kg/m}^3 - 1.29 \text{ kg/m}^3)}{9(1.8 \times 10^{-5} \text{ Ns/m}^2)} = 0.01 \text{ m/s}$$

This is only 1 cm/s, which shows how droplets of water can be kept aloft inside a cloud by a slight updraft of air. The volume, V, of this droplet is

$$V = \frac{4}{3}\pi r^3 = \frac{4}{3}\pi(10 \times 10^{-6} \text{ m})^3 = 4.2 \times 10^{-15} \text{ m}^3$$

Therefore, 1000 agglomerated droplets would have a volume of $1000(4.2 \times 10^{-15} \text{ m}^3) = 4.2 \times 10^{-12}$ m^3. The radius of a drop with that volume is

$$r = \left(\frac{3V}{4\pi}\right)^{1/3} = \left(\frac{3(4.2 \times 10^{-12} \text{ m}^3)}{4\pi}\right)^{1/3} = 1 \times 10^{-4} \text{ m}$$

That is, 100 μm. A drop with that radius reaches a terminal velocity of

$$v_{\text{lim}} = \frac{2gr^2\left(\rho_{\text{sphere}} - \rho_{\text{fluid}}\right)}{9\eta} = \frac{2\left(9.8 \text{ m/s}^2\right)\left(100 \times 10^{-6} \text{ m}\right)^2\left(1000 \text{ kg/m}^3 - 1.29 \text{ kg/m}^3\right)}{9\left(1.8 \times 10^{-5} \text{ N s/m}^2\right)} = 1.2 \text{ m/s}$$

The agglomerated drop falls over 100 times faster than the tiny droplet.

2.2.7 Scaling in Biology

Recently, scientists have begun applying scaling laws to living things in interesting ways. We have already discussed the strength-to-weight ratio and how it affects the capabilities of elephants, fleas, and humans. This is not the only scaling phenomenon in the animal kingdom. In general, smaller animals have quick pulse rates and short lives, while larger animals have slow pulse rates and long lives. And, as it turns out, the number of heartbeats during an animal's time on the Earth tends to always be the same—about one billion beats. Be you shrew, you, or ewe, the number's the same.

Pulse rate and life span are just two of the biological characteristics that vary with body size. The mathematical principle governing these phenomena is sometimes referred to as quarter-power scaling, and it plays a role in life span, heartbeat, and metabolic rate.

The life span, LS, of an animal varies with its mass (proportional to D^3) according to

$$LS \propto m^{1/4} \propto D^{3/4} \tag{2.47}$$

A cat weighs about 5 kg and a mouse weighs about 50 g. The cat, 100 times more massive, therefore lives about three times longer:

$$LS \propto 100^{1/4} = 3.16$$

The heart rate, HR, scales with mass as well

$$HR \propto m^{-1/4} \propto D^{-3/4} \tag{2.48}$$

We can see that the cat's heart beats one-third as often as the heart in the less-massive mouse:

$$HR \propto 100^{-1/4} = 0.316$$

An animal's metabolic rate also scales with its mass. The basal metabolic rate (measured in watts) of an animal has to do with its surface-to-volume ratio. As we saw in Equation 2.3, the surface-to-volume ratio, S/V, scales with the characteristic dimension, D, as

$$S/V = \frac{1}{D}$$

The bigger the creature, the less surface area it has to dissipate the heat generated by its mass. (By the way, your metabolic rate is probably close to about 100 W, similar to a light bulb.) A larger animal's metabolism must be slower than a smaller animal's metabolism to prevent overheating. This is not the only factor involved; other complex factors affect the metabolic scaling law. The metabolic rate, MR, scales with body mass as

$$MR \propto m^{3/4} \propto D^{9/4} \tag{2.49}$$

Bigger animals use energy more efficiently. Here we see that the cat—although 100 times more massive—has a metabolic rate only about 30 times that of the mouse:

$$MR \propto 100^{3/4} = 31.6$$

Amazingly enough, this relationship has been found to hold true across the breadth of the animal kingdom, from single-celled organisms to the blue whale!

2.3 ACCURACY OF THE SCALING LAWS

Scaling laws are a means of estimating phenomena at small scales. They provide useful information for engineers when designing small systems. Realize, however, that scaling law calculations can yield misleading or completely inaccurate results and therefore should be employed with discretion. Caution should be taken especially when considering a system reduced to the nanometer size scale. Mechanical systems tend to scale well even to the nanoscale, while electromagnetic scaling laws can fail dramatically at the nanoscale, and thermal scaling laws have variable accuracy.

The scaling laws treat matter as a continuum; for example, strength and modulus of materials are determined for the bulk. But at the nanoscale, the effects of atomic-scale structure, mean free path effects, thermal energy, and quantum mechanical uncertainties contradict this approximation for matter.

Table 2.1 summarizes the scaling relationships discussed in this chapter.

TABLE 2.1   Summarization of Scaling Laws Discussed in this Chapter

Physical Quantity	Scaling Law ($\propto D^X$)	Equation
Scaling in Geometry		
Volume	D^3	2.1
Surface area	D^2	2.2
Surface-to-volume ratio	D^{-1}	2.3
Scaling in Mechanics		
Strength (force)	D^2	2.4
Mass	D^3	2.7
Force of gravity	D^3	2.8
Striction force	D^2	2.9
Wear life	D	2.10
Power	D^2	2.11
Power density	D^{-1}	2.12
Force of a spring	D	2.13
Oscillating frequency of a spring	$D^{-3/2}$	2.14
Period of oscillation of a spring	$D^{3/2}$	2.15
Kinetic energy	D^3	2.16
Gravitational potential energy	D^3	2.17
Potential energy of a spring	D^2	2.18
Scaling in Electricity and Electromagnetism		
Electrostatic field strength	D^{-1}	2.19
Resistance	D^{-1}	2.20
Current	D	2.21
Power (dissipated)	D	2.22
Power (dissipated per unit area)	D^{-1}	2.24
Capacitance	D	2.25
Electrostatic force (parallel plates)	D^2	2.26
Magnetic force	D^2	2.28
Scaling in Optics		
Divergence angle	D^{-1}	2.30
Wavelength (photolithography)	D	2.32
Resolution	D^{-1}	2.33
Focal length	D	2.34
Scaling in Heat Transfer		
Energy to heat a system	D^3	2.35
Heat capacity	D^3	2.36
Heat conduction rate [$D > 1\ \mu m$]	D	2.37
Heat conduction rate [$D \leq 1\ \mu m$]	D^2	2.38
Thermal time constant	D^2	2.39
Heat convection rate [$D > 1\ \mu m$]	D^2	2.40
Fluids in Fluids		
Volumetric flow in a circular pipe	D^4	2.41
Pressure gradient in a circular pipe	D^{-2}	2.44
Terminal velocity of object through fluid	D^2	2.45
Time to terminal velocity	D^2	2.46

TABLE 2.1 **(continued)** Summarization of Scaling Laws Discussed in this Chapter

Physical Quantity	Scaling Law ($\propto D^X$)	Equation
Scaling in Biology		
Life span	$D^{3/4}$	2.47
Heart rate	$D^{-3/4}$	2.48
Metabolic rate	$D^{9/4}$	2.49

HOMEWORK EXERCISES

2.1 Give at least four advantages of miniaturization in machine design.

2.2 True or false? Miniaturization improves the factor of safety of a product.

2.3 Scaling laws are
   a. General engineering guidelines for miniaturization
   b. Useful for estimating how the characteristics of something will vary with changes in characteristic dimension
   c. Helpful estimates of a device's performance at the nanoscale
   d. Accurate predictors of physical characteristics at the macro-, micro-, and nanoscales

2.4 Scaling laws derive from
   a. Design engineers' experience
   b. Market demands
   c. Laws of physics
   d. The material used

2.5 The characteristic dimension is
   a. The dimension in which the object is largest in a three-dimensional representation of that object
   b. Metric units
   c. A representative measurement of something for comparison purposes
   d. A variable used to determine the surface-to-volume ratio of an object

2.6 True or false? The characteristic dimension, D, of an object is the average of that object's width, height, and length.

2.7 Based on the scaling laws, how many times greater is the strength-to-weight ratio of a nanotube ($D = 10$ nm) than the leg of a flea ($D = 100$ μm)? Than the leg of an elephant ($D = 2$ m)?

TABLE 2.2 Women's Weightlifting Gold Medalists

Weight Class (kg)	Lift (lb)
48	463.05
53	490.61
58	523.69
63	534.71
69	606.38
75	600.86

Source: Data from the 2004 Olympic Games in Athens, Greece.

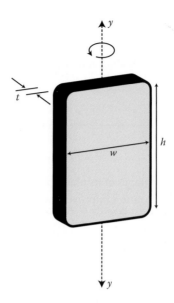

FIGURE 2.8   A micromirror. (For Homework Exercise 2.10.)

2.8   Data from the 2004 Olympic Games in Athens, Greece, are provided in Table 2.2. Plot the percentage body weight lifted versus weight class. Is the same trend evident in the women's weightlifting event as in the men's event shown in Figure 2.4? In what weight class is there a discrepancy, and how might this be explained?

2.9   Derive the mass-to-volume ratio as a function of characteristic dimension. Explain the result.

2.10   The micromirror shown in Figure 2.8 is used for redirecting light in an optical communication system. The torque needed to spin it on the y-axis is directly proportional to its mass moment of inertia, I_{yy}:

$$I_{yy} = \frac{1}{12}\rho htw^3$$

Here, ρ is the density of the mirror material.
   a. Derive the scaling law for I_{yy}.
   b. If the mirror's dimensions can be reduced to one-third the original size, what is the corresponding percent reduction in the torque required to turn the mirror?

2.11   What is the resistance of a cubic micrometer of copper ($\rho = 17.2 \times 10^{-9}\ \Omega$ m)?

2.12   How much more or less resistance does a cubic micrometer of copper have versus a cubic millimeter?

2.13   In designing an electrostatic actuator made from parallel plates, you can either double the plates' areas or halve the distance separating them. Which provides a greater improvement in the electrostatic force?

2.14 Consider two parallel wires 100 μm long, each carrying 20 μA of current, separated by 1 μm.

    a. What is the electromagnetic force between these wires?

    b. If the orientation is right, the electromagnetic force created by these two wires can be enough to lift one of the wires. The force of gravity opposing that motion, $F_g = mg$, where m is mass and $g = 9.8$ m/s². If the wire is made of copper (8.96 g/mL) with a diameter of 2 μm, what is the force of gravity holding it down?

    c. How many times greater or smaller is this force than the electromagnetic force being used to lift the wire?

    d. If the characteristic dimension of the wires was allowed to increase, and with it the current through the wires as governed by Equation 2.21, would the electromagnetic force ever overcome the gravitational force?

2.15 True or false? The smallest spot we can illuminate with a given lens depends on the wavelength of light we use.

2.16 Consider a compact disc on which the pits used to store the digital information measure 850 nm long by 500 nm wide. Visible light has the following wavelengths: 380–420 nm for violet; 420–440 nm for indigo; 440–500 nm for blue; 500–520 nm for cyan; 520–565 nm for green; 565–590 nm for yellow; 590–625 nm for orange; and 625–740 nm for red. Which of these colors, if any, could be used for building the pits on the disc?

2.17 Light passes through a lens with a refractive index of 1.6, and front and back surfaces with radii of 11 and 15 cm, respectively.

    a. What is the focal length of the lens?

    b. What would be the focal length of an identically proportioned lens 100 times larger?

2.18 The calorie is a unit of energy defined as the amount of energy needed to raise 1 g of water by 1°C.

    a. How many calories are required to bring a pot of water at 1°C to a boil? The pot is full to the brim, with diameter 20 cm and depth 20 cm. The density of water is 1000 kg/m³.

    b. If we consider that D for the pot is 20 cm, approximately how much more energy is needed to heat a hot tub with $D = 2$ m? How many calories is that?

    c. If energy costs $0.10 per kilowatt-hour (kW-h), how much does it cost to heat this hot tub?

    d. How does the price (in dollars) of heating the hot tub scale with the tub's characteristic dimension?

    e. What percentage of the cost of heating the hot tub would be saved by reducing the tub's characteristic dimension by 33%?

2.19 A can (355 mL) of Coke has 140 "food calories" (1 food calorie = 1 kcal). How many equivalently sized cans worth of water could be brought to a boil using the energy in a single Coke? Assume the water is initially at room temperature (20°C). The density of water is 1000 kg/m³.

2.20 True or false? At the microscale, reducing a pipe's radius causes an increase in the pressure drop per unit length.

2.21 A crude oil pipe's radius is reduced by 5%. What is the corresponding percentage change in the pressure drop per unit length?

2.22 A pipe's radius can be thought of as a characteristic dimension, D. What is the surface-to-volume ratio (S/V) of the inside of the pipe as a function of D?

2.23 True or false? The number of heartbeats during an animal's lifetime tends to be approximately one billion beats, no matter its size.

2.24 Plot mass versus metabolic rate (each as a function of characteristic dimension $1 \leq D \leq 1000$) on a log–log scale.

    a. Is the relationship linear or exponential on this graph?

    b. How much faster is the metabolic rate of an animal 10,000 times more massive than another?

Short Answers

2.25 Give two examples of products you use that have been miniaturized, and relate how these changes were improvements.

2.26 Write a paragraph supporting or refuting the idea that the scaling laws are truly *laws*.

RECOMMENDATIONS FOR FURTHER READING

1. T.-R. Hsu. 2002. *MEMS and Microsystems: Design and Manufacture.* McGraw-Hill.
2. S. E. Lyshevski. 2002. *MEMS and NEMS: Systems, Devices, and Structures.* CRC Press.
3. K. E. Drexler. 1992. *Nanosystems: Molecular Machinery, Manufacturing and Computation.* Wiley Interscience.
4. C. Phoenix. 2004. *Scaling Laws—Back to Basics,* Center for Responsible Nanotechnology, www.crnano.org
5. M. Wautelet. 2001. Scaling laws in the macro-, micro- and nanoworlds. *European Journal of Physics,* 22:601–611.

CHAPTER 3

Introduction to Nanoscale Physics

3.1 BACKGROUND: NEWTON NEVER SAW A NANOTUBE

On July 5, 1687, Isaac Newton published a three-volume work titled *Philosophiae Naturalis Principia Mathematica*. Students spending big money on physics textbooks this year might consider photocopying pages from Newton's book instead (and brushing up on their Latin), because most of what Newton figured out centuries ago is still just as true today. His three laws of motion are part of a broader field known as classical mechanics. Do not be fooled by the ancient-sounding name—classical mechanics works for apples or Apple computers. Even calculus, which Newton invented to help him develop his theories, is perfectly pertinent. So too are the approaches to classical mechanics developed in the ensuing centuries by the likes of Lagrange, Hamilton, d'Alembert, and Maxwell.

However, classical mechanics has limits. While it accurately describes the physical world as Newton saw it, Newton never saw a nanotube or an atom. If he had, chances are his book would have been a little longer, with more statistics.

And it might have included quantum mechanics. In order to understand nanotechnology, we will need to understand more about nanoscale objects such as atoms, electrons, and photons. We need to understand some quantum mechanics. In the next few sections, we will bring you and Isaac up to speed.

3.2 ONE HUNDRED HOURS AND EIGHT MINUTES OF NANOSCALE PHYSICS

Deep within the Sun, four hydrogen atoms are fused together by relentless pressure and heat. In the process, six photons are born. Three of these photons leave the Sun, hurl across the vacuum of space and arrive at the Earth, where they penetrate the atmosphere. One of them strikes an airborne particle of pollen and is reflected into the depths of the ocean. The other two photons continue on and slam head-on into a solar panel on the roof of a house. Like torch bearers handing their flaming batons to the next runners, each of these photons passes energy to an electron, and these two electrons carry on from there—joining a sea of electrons slowly drifting down a copper wire. One of these electrons bumps into a vibrating copper atom in the wire and releases its energy as heat. Without this extra energy, the electron loses its mobility and so remains behind, attached to the lattice of metal atoms. The remaining free electron eventually makes it all the way down the wire and into a tungsten light bulb filament 2 m in length but only 1/100 in. thick. The filament is wound up and coiled to fit inside the bulb. Because the filament is so much thinner than the wire, it is more resistant to electron flow. The electron (and some companions) bump into a tungsten atom, generating vibrations and heat. When this atom cools off, it releases a new photon. This one flies out of the bulb, reflects off the page of a book and into your eye.

By the end of this chapter you will understand how atoms, photons, and electrons make this everyday journey possible. Learning how it works may only make it seem more miraculous.

BACK-OF-THE-ENVELOPE 3.1

How much time does the entire epic relay race take—from the Sun to solar panel to eyeball?

The Earth's orbit around the Sun is elliptical, so the distance between the two varies but the average is about 150 million kilometers (1.5×10^8 km). Light travels at 299,972,458 m/s. To get from the Sun to the Earth it takes

$$\frac{150{,}000{,}000 \text{ km}}{299{,}972{,}458 \text{ km/s}} = 500 \text{ s}$$

That is about 8 min, 20 s. Electrons moving down a wire to a light bulb do not move with quite the same zip as photons, but slog their way along as a "sea" of charge. We learn more about this sea in Section 4.2.3 but for now, we can estimate their speed at about 10 cm/h. If the wire is 10 m long, it will be 100 h before that same electron makes it to the filament. If the photon emitted by the light bulb filament travels 3 m to your eyeball, then

$$\frac{3\text{m}}{299{,}972{,}458 \text{ m/s}} = 1 \times 10^{-8}\text{s}$$

So, that happens in 10 ns.
Total relay race time: 100 h, 8 min, 20.00000001 s.

3.3 THE BASICS OF QUANTUM MECHANICS

We know exactly where the planet Mars will be hundreds of years from now but to locate an electron, we can only guess. Such is quantum mechanics. It is a theory for correlating and predicting the behavior of molecules, atoms, and subatomic particles such as electrons, photons, and neutrons.

Quantum mechanics arose in the early part of the twentieth century out of difficulties reconciling Newtonian mechanics with experimental observations. Long-standing equations could not explain things that were happening in laboratories all over the world.

Scientists found that matter was not the comforting continuum they had for so long assumed it to be. Things such as energy and radiation did not increase or decrease smoothly but instead, skipped around. Imagine coming across an oven that could exist at 100°, 127°, or 350° and could not transition up or down through all the temperatures in between.

Armed with new experimental data, scientists rethought physics; and from 1900 to 1930, a theory of quantum mechanics was built from the bottom up. In some ways, the theory is still in its infancy. The principles have finally gained consensus among scientists but not complete agreement. Whereas classical mechanics is deterministic—one action equaling a reaction—the tools of quantum mechanics are probability and statistics, and educated guesses.

Quantum mechanics is broad in scope. It actually encompasses the once-universal classical mechanics as a subdiscipline, which is applicable under certain circumstances. In fact, the two disciplines can be interpreted such that there are no contradictions between the two. Simply stated: if you wish to understand the physics of a basketball, you can use classical mechanics, a buckyball, quantum mechanics.

3.3.1 Atomic Orbitals (Not Orbits)

The hydrogen atom is the simplest atom there is. It has one proton in its nucleus and one electron. The electron, as we learned in Chapter 1, is not embedded in the atom like a watermelon seed, and it is not orbiting the nucleus like a planet around the Sun. All we can say about the electron is that it inhabits a region of space enclosing the nucleus called an orbital. That is its territory but it does not really have a home. It does not rest. If you were to take a picture of a hydrogen atom—its nucleus and its electron—at a single instant, it might look similar to Figure 3.1a. Take another picture of it soon afterward, and the electron is somewhere else. We do not know how the electron got there, only that it *is* there. Take a picture of it again. And again. Pretty soon you have mapped out the electron's domain. It is a three-dimensional map of all the places we are likely to find the electron—its orbital. With hydrogen, this probability distribution is spherical. This is what we see in Figure 3.1b. There is a high probability of finding the electron within this sphere. But the probability is not 100%, meaning there is also a very slim chance that the electron is somewhere else, outside this sphere.

In general, orbitals represent a volume of space within which an electron is most likely to be, based on what kind of atom it is and how much energy the atom has. There can be one or, at the maximum, two electrons in any orbital. The shapes of these orbitals can be spherical or, at higher energy states, can assume a variety of intriguing lobe and ring shapes, as well as combinations thereof. Represented in Figure 3.2 are three such possibilities. The shapes reflect the repulsive nature of the negatively charged electrons, which distribute themselves so as to keep the atom as stable as possible. Electrons tend to fill the low-energy orbitals close to the nucleus before they fill those farther out. This maintains the atom at its lowest energy state—and when it is not in this state, an atom will radiate excess energy until it is. (Note that the orbitals in Figure 3.2 are for single atoms; the orbital configurations of molecules are significantly more complicated because probability distributions overlap and can be reinforced or canceled out.)

(a)

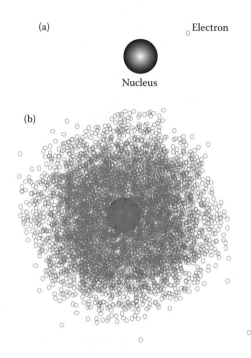

Nucleus

(b)

FIGURE 3.1   Electron orbital of hydrogen. The hydrogen atom is the simplest atom there is. It has a proton for a nucleus (most atoms also have neutrons in the nucleus) and an electron, as shown in (a). If you plot the location of this electron at various times, the picture would look something like the spherical probability distribution shown in (b). This is a kind of three-dimensional map of places one is likely to find the electron. (The scale of this figure is not at all proportional. The diameter of an electron is roughly 1000 times smaller than the diameter of a proton. The atom is mostly empty space. If the nucleus were a marble, the orbital in this case would be about as big as a football stadium.)

We organize orbitals into energy levels and sublevels. The levels are numbered 1, 2, 3, 4, and so on. These numbers are known as the quantum numbers, and we will learn about them in Section 3.3.3. The sublevels are designated by letters: s, p, d, f, g, h, ... and up through the alphabet from there. When writing the electron configuration for a particular atom, we specify how many electrons are in each occupied sublevel using a superscript number. For example, the electron configuration for magnesium is written $1s^2 2s^2 2p^6 3s^2$. This means it has electrons in the 1s, 2s, 2p, and 3s sublevels. We can add up the superscript numbers to find out that magnesium has $2 + 2 + 6 + 2 = 12$ electrons.

This naming convention is not exactly intuitive when it comes to determining which orbitals are filled for a given atom. Figure 3.3 can help. It shows the sublevels up to 6d in order of ascending energy. Electrons fill low-energy sublevels first, although in a few exceptional cases electrons begin to fill higher sublevels before a lower level is completely filled (see Homework Exercise 3.6). We can use this chart as a guide for determining how the electrons of an atom will be configured. Figure 3.4 shows the configuration for scandium as an example.

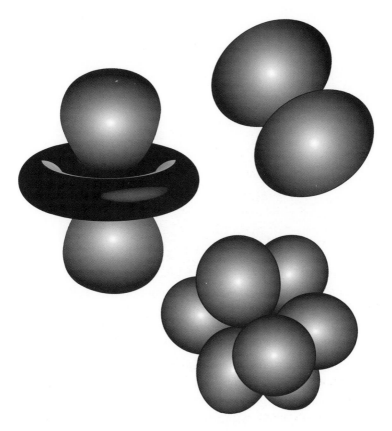

FIGURE 3.2   Schematic representations of electron orbital shapes.

BACK-OF-THE-ENVELOPE 3.2

Electrons in the highest occupied energy level of an atom are called valence electrons. We will discuss the role of valence electrons in later chapters. How many valence electrons does a sodium atom have?

Looking at the Periodic Table of Elements tells us that sodium (Na) has an atomic number of 11, which means it has 11 electrons. Using Figure 3.3 we can see that the first 10 electrons will completely fill energy levels 1 and 2. This leaves a single electron in the outermost occupied energy level, 3s. Sodium therefore has one valence electron.

3.3.2 Electromagnetic Waves

Look around you. The living things and inanimate objects you see are emitting radiation—thermal radiation, to be specific. Sometimes this radiation is visible to the human eye, such as when a toaster coil glows red, or when the tungsten filament of a light bulb gets white-hot. The higher an object's temperature, the higher the frequencies of its radiation; lower temperatures give lower frequencies. Thermal radiation is a type of electromagnetic (EM) wave.

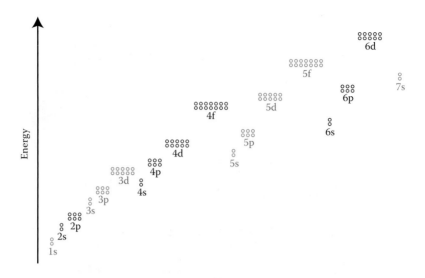

FIGURE 3.3 Filling the many levels and sublevels of atomic orbitals. Each orbital is represented by a vertical pair of circles. The sublevels contain varying numbers of orbitals: the s sublevels have one orbital each, p have 3, d have 5, and f have 7. With a few exceptions, electrons fill the sublevels in order of ascending energy: 1s, 2s, 2p, 3s, 3p, 4s, 3d, 4p, 5s, 4d, 5p, 6s, 4f, 5d, 6p, 7s, 5f, 6d.

The EM spectrum is shown in Figure 3.5. This spectrum encompasses the numerous wavelengths and frequencies of EM waves. The frequencies of thermal radiation are most often from infrared through the visible colors of the rainbow and up to ultraviolet. Nonthermal forms of EM radiation can be caused by magnetic fields; much of the radiation observed by radio astronomers is created this way.

Still, at the fundamental level, *all electromagnetic radiation is created the same way*: it all comes from changes made to the energy of electrons or other charged particles. We will discuss this in more detail in a moment.

What are EM waves? Let's start with what they are not. They are not mechanical waves like sound waves, water waves, or waves traveling down a rope. EM waves can move not only through air and water, but also through the vacuum of space—something mechanical

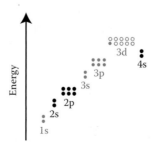

FIGURE 3.4 The electronic configuration of the scandium atom. The atomic number of Sc is 21, meaning it has 21 electrons. Orbitals containing electrons are shown as full circles; empty orbitals are open circles. The 4s sublevel has less energy than the 3d level and thus it fills up first. The electronic configuration is written as: $1s^2 2s^2 2p^6 3s^2 3p^6 3d^1 4s^2$.

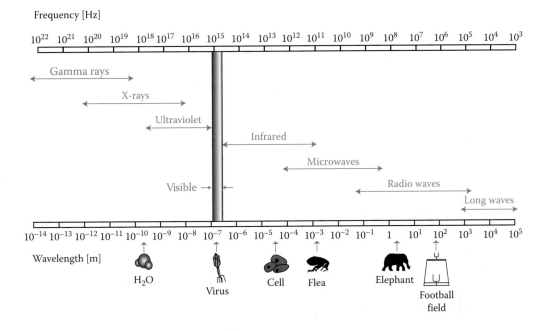

FIGURE 3.5 The EM spectrum. Everyday things of varying size aid in visualizing the size of wavelengths in different parts of the spectrum.

waves cannot do. No matter what their frequency, f, is, EM waves always travel through empty space at 299,792,458 m/s. (This is the speed of light, c; or, to be truly accurate, "the speed of electromagnetic radiation in a vacuum.") Because of this, we obtain the following relationship:

$$c = f\lambda \tag{3.1}$$

In this equation, λ is wavelength. Thermal radiation is emitted from substances at a continuous distribution of wavelengths but the frequency at which the radiation is most intense is a function of the temperature, T, and obeys the following relationship:

$$\lambda = \frac{0.2898 \times 10^{-2}\,\mathrm{m\,K}}{T} \tag{3.2}$$

BACK-OF-THE-ENVELOPE 3.3

A stoplight radiates visible red light, which has a frequency of about 430 Terahertz (THz). What is the wavelength of this light?

$$\lambda = \frac{c}{f} = \frac{3 \times 10^8\,\mathrm{m/s}}{430 \times 10^{12}\,\mathrm{s}^{-1}} = 700\,\mathrm{nm}$$

BACK-OF-THE-ENVELOPE 3.4

The temperature of your skin is about 95°F. At what wavelength is the radiation emitted by the atoms in your skin most intense?

Converting 95°F into 308 K and solving Equation 3.2, we obtain

$$\lambda = \frac{0.2898 \times 10^{-2}\,\text{mK}}{308\text{K}} = 9.4\,\mu\text{m}$$

This is infrared radiation, as we can see in Figure 3.6.

3.3.2.1 How EM Waves Are Made

EM waves are generated whenever a charged particle accelerates. This may seem counterintuitive: should not a particle *need* energy to accelerate, not *radiate* energy? And it is true: a particle *does* need energy to accelerate—it is, in fact, this energy that becomes the waves. It is simple to understand if we start by examining a single, stationary electron. The electric field around it is oriented in a static, radial pattern, as shown in Figure 3.7a. As long as the electron remains alone and does not move, the shape of the field does not change.

Next, an electron moving with a constant velocity (Figure 3.7b). At constant velocity, no force is acting on the electron; no energy is transferred to it. The electron carries the field with it, and if we somehow took a snapshot of the field at a particular instant it would look identical to the field of the stationary electron.

Finally, we examine an *accelerating* electron. Whether the electron had been at rest or coasting along at some constant velocity, a "kick" of energy was needed to make it accelerate (or a "kick" in the other direction to decelerate). When the electron accelerates, the field must accelerate along with it, continually updating so as to correspond to the new point of origin. Even if this updating happens at the speed of light, the field nearest the electron will react first, and the outer field will lag behind. This puts a bend in the field. The bend

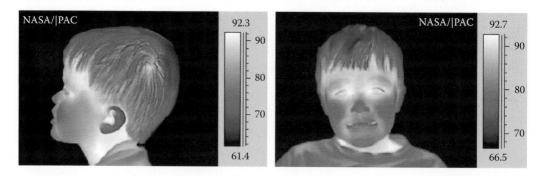

FIGURE 3.6 Warm objects emit EM waves that can be seen by an infrared camera. These images of a young boy were taken by a thermal infrared camera that translates infrared radiation into visible light. Different areas appear warm (bright) or cool (dark), depending on how close blood vessels are to the surface of the skin. The wavelength of this radiation is on the order of 10 μm. (Courtesy of NASA/JPL-Caltech.)

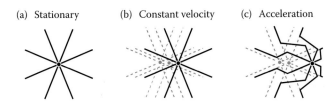

FIGURE 3.7   An accelerating electron creates EM radiation. In (a) the electron is stationary and the electric field lines remain straight. In (b) the electron has a constant velocity; the electric field is the same as the stationary case. In (c) the electron is suddenly accelerated, creating a bend in the field lines that propagates outward. This bend carries with it the energy of EM radiation. (Prior locations of the electron and its field lines are shown in gray.)

propagates away from the electron (see Figure 3.7c). Some of the energy originally exerted in the "kick" is expended in transmitting a bend throughout the field. *The bend carries energy. This energy is electromagnetic radiation.*

Most of the time, charged particles are made to accelerate and decelerate over and over again in an oscillatory manner. This makes for continuously varying, sinusoidal bends in the field—EM waves that radiate out sort of like ripples on a pond. Radio stations produce radio waves by quickly alternating the voltage applied to the wires of an antenna, making the electrons in the antenna oscillate.

EM waves are made up of two components: (1) an electric field and (2) a magnetic field. Because any time-varying electric field generates a magnetic field (and vice versa), the oscillating electric field creates an oscillating magnetic field, which in turn creates an oscillating electric field, and so on. These fluctuating electric and magnetic fields are at right angles to one another and to the direction of wave propagation. One type of EM wave, a plane-polarized wave, is graphically represented in Figure 3.8.

3.3.3 The Quantization of Energy

We learned in the previous section that the electrons of an atom exist for the most part inside orbitals, and that the shape of these orbitals and their distance from the nucleus are determined in large part by how much energy the atom has. In 1900, Max Planck's physical interpretation of thermal radiation sparked two wildly controversial ideas about the nature of atoms. Since then we have learned that both of these ideas were right.

The first idea was that atoms cannot have arbitrary amounts of energy—that their energy cannot start from zero and ramp up through all possible amounts to higher energy levels.

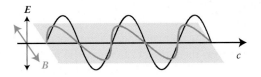

FIGURE 3.8   An EM wave. This is a plane-polarized wave, traveling at the speed of light, c. The oscillating electric field, E, is oriented on a plane perpendicular to the plane of the magnetic field, B.

They can only have discrete values, similar to the way money only comes in specific denominations (i.e., there are no 0.084-cent coins or $23 bills). The energy, E, an atom or molecule can have is governed by

$$E = nhf \tag{3.3}$$

Here, n is a positive integer (never a fraction) called a quantum number, and f is the frequency at which the atom vibrates. The h in the equation is a universally important constant named after Planck himself (see the sidebar "Quantum Mechanics' Constant Companion").

Planck's constant, $h = 6.626 \times 10^{-34}$ J s

Planck's work suggested that if atoms gain and lose energy in this quantized way, there must be specialized packets of energy floating around—a universal currency of energy exchange. And there is. It is the photon. Atoms emit or absorb photons when they hop up or down between quantum states. The energy of an emitted photon is easy to determine: it is the energy difference between these two quantum states. If the atom's energy hops up, for example, from the $n = 3$ state to the $n = 4$ state, it has *absorbed* a photon with energy, E_{photon}. In this case,

$$E_{\text{photon}} = 4hf - 3hf = hf$$

The same is true when the atom's energy hops down, for example, from the $n = 2$ to the ground state ($n = 1$). The atom *emits* a photon, the energy of which is

$$E_{\text{photon}} = 2hf - 1hf = hf$$

This leads us to conclude

$$E_{\text{photon}} = hf \tag{3.4}$$

BACK-OF-THE-ENVELOPE 3.5

A photon has no mass, but it can have momentum. How? To find out, we will use Albert Einstein's legendary mass–energy equivalence equation.

$$E = mc^2 \tag{3.5}$$

E is energy, m is the mass of the particle, and c is the speed of light. This energy is actually the "rest energy" of a particle—the amount of energy the particle represents when it is not moving at all. The equation in its entirety, known as the energy–momentum relationship, is

$$E^2 = p^2c^2 + (mc^2)^2 \tag{3.6}$$

Here, p is the relativistic momentum of the particle, a modified definition of momentum developed to account for travel velocities approaching the speed of light—which is where the classical conservation of momentum (think colliding billiard balls) is not quite accurate. For one thing, the photon has zero rest mass, and the classical definition of momentum is mass times velocity. The momentum, p, of a photon can be determined by setting $m = 0$ in Equation 3.6, which gives

$$p = \frac{E}{c} \tag{3.7}$$

We know that $E = hf = hc/\lambda$. Thus, the momentum of a photon can also be expressed as

$$p = \frac{E}{c} = \frac{hc}{c\lambda} = \frac{h}{\lambda} \tag{3.8}$$

This relationship holds true for particles in general, including electrons and neutrons and baseballs.

Atoms radiate or absorb energy only by changing quantum states—which entails changing sublevels. If the atom remains in a particular sublevel, it has neither absorbed nor emitted any photons.

3.3.4 Atomic Spectra and Discreteness

In the last section, we discussed how objects emit thermal radiation over a continuous spectrum of frequencies. There is an exception, however. Gases, when excited by electricity or heat, can give off multiple, *single-frequency* bands of radiation instead of a continuous spectrum. The frequency bands such gases emit represent discrete atomic energy levels. As an electron in a high-energy state skips down to a lower-energy state, the atom emits a photon. The energy of this photon equals the exact energy difference between the two energy levels. If electrons were not restricted to discrete energy levels, we would see a continuous spectrum emitted. Instead, we see only specific wavelengths. In the case of hydrogen, some of the wavelengths emitted correspond to visible light: 656 nm (red), 486 nm (blue), 434 nm (indigo), and 410 nm (violet). If we were to examine the light given off by sodium vapor street lamps and mercury vapor lights, we would see specific wavelengths as well.

Not only do atoms *emit* specific wavelengths of light, they also *absorb* specific wavelengths. This makes sense: since the electrons can only skip up to specific states, a photon with exactly that energy difference can be absorbed. This behavior can be used in identifying unknown substances. The absorption bands observed in the Earth's atmosphere helped us figure out what gases it consisted of. Also, in early studies of the solar spectrum, the bands did not correlate with *any* element. Why? Because it was a new one—helium, derived after the Greek word *helios*, meaning "sun."

QUANTUM MECHANICS' CONSTANT COMPANION:
PLANCK'S CONSTANT (INTRODUCED IN 1900)

Geometry has pi ($\pi = 3.15149$), chemistry has Avogadro's number ($N_a = 6.022 \times 10^{23}$), and quantum mechanics has Planck's constant ($h = 6.626 \times 10^{-34}$ J s). This lowercase letter stands for something fundamental to all nanoscale things from atoms to light, and appears in any mathematical result truly reflecting quantum mechanical behavior.

We get it from Max Planck (1858–1947) (see Figure 3.9), a German physicist whose work in 1900 spawned a pair of radical new ideas: (1) energy is quantized, and (2) atoms trade energy around in little packets, later dubbed "photons." In fact, this concept of a particle of light was so strange at the time that Planck himself resisted it. Nonetheless, the concept took hold.

The significance of h is its simplicity: a photon's energy is h multiplied by its frequency. Planck hit upon the constant while trying to solve "the ultraviolet catastrophe." This problem is about the intensity of radiation emitted by an ideal emitter, which varies with the body's temperature and the frequency of the radiation. (The closest thing to an ideal emitter is a black body, which can be made in a lab using a large cavity with a tiny hole in the side. Radiation that enters the hole has a very slim chance of coming back out because it is absorbed by the cavity walls. Any radiation that comes *out* of the hole, therefore, is presumed to consist entirely of thermal radiation from the inner cavity walls and contain no reflected radiation.) Classical physics could not explain why experimental data showed that the intensity of the thermal radiation did not go to infinity at high frequencies. Planck thought the equation he came up with (which used h because it just so happened fit the data best) was merely a mathematical trick. When he presented his theory about energy packets, most scientists (himself included) did not think it was realistic. But it was, and Planck was honored for his work in 1918 with the Nobel Prize. In accepting the prize, he called his work "a happily chosen interpolation formula." After its discovery, he said he had busied himself "with the task of elucidating a true physical character for the formula … until after some weeks of the most strenuous work of my life, light came into the darkness, and a new undreamed-of perspective opened up before me" (Planck's complete Nobel Lecture can be found at www.nobel.org).

Although scientifically fruitful, Planck's life was beset with tragedy. During World War I, he lost a son in action and two daughters died during childbirth. Planck refused to leave Germany during Hitler's regime, although his home was destroyed by bombs and his son was executed by the Nazis, who accused him of plotting to assassinate Adolf Hitler. For his outspoken disapproval of the way the Nazis treated his Jewish colleagues, Planck was forced to resign as president of the Kaiser Wilhelm Institute of Berlin in 1937. After the war, he was renamed president. The institute was renamed as well: today it is known as the Max Planck Institute.

3.3.5 The Photoelectric Effect

Shine high-frequency light on certain metals and electrons will fly off the surface. This phenomenon, called the photoelectric effect, baffled scientists in the late 1800s but it also provided the proof Albert Einstein needed to extend Planck's ideas about quantization. The problem was that classical physics could not explain the following observations about the photoelectric effect:

1. There is always some frequency of light below which electrons are not emitted from the metal surface. Scientists call this the cutoff frequency, and it varies with the metal.

FIGURE 3.9   Max Planck. (Photo courtesy of the Clendening History of Medicine Library, University of Kansas Medical Center.)

However, should not electrons be emitted no matter what, as long as the frequency of the light is intense enough (i.e., more energy delivered per surface area)?

2. No matter how high or low the light's intensity, the maximum kinetic energy of the ejected electrons is the same (see Figure 3.10a). The only thing that raises the electrons' maximum kinetic energy is the frequency of the light used. But should not more intense light make for more energetic electrons?

3. Electrons are emitted within nanoseconds of the light hitting the surface, even when the light's intensity is low. But, is not the light's energy distributed over the surface? Will it not take time for the surface to build up energy and start emitting electrons?

In 1905, Einstein cut through such confusion (a service to science meriting the 1921 Nobel Prize). Here is the gist of his realization: "Light is nothing but photons." This is a *monumental* discovery. It means that EM radiation is like a wave *and* like a particle. A photon, as currently understood, is a massless, chargeless particle of EM radiation that is never at rest but always traveling like a wave. If the photon is in empty space, it travels at the speed of light. This is not something that can be easily visualized or sketched. Neither is it *necessary* to see a picture; what is necessary is to accept that the concept of a particle and the concept of a wave are not mutually exclusive. EM radiation somehow behaves like *both* simultaneously. We will discuss this wave/particle duality in the next section.

Recall Equation 3.4, which says photons have energy, $E_{photon} = hf$. When light shines on a metal surface, the photon, as a little packet of energy, is not divided among the atoms on the metal's surface. The photon transfers *all* of its energy to a single electron in the metal. This one electron is emitted from the surface with a maximum kinetic energy, KE_{max}:

$$KE_{max} = hf - \phi \qquad (3.9)$$

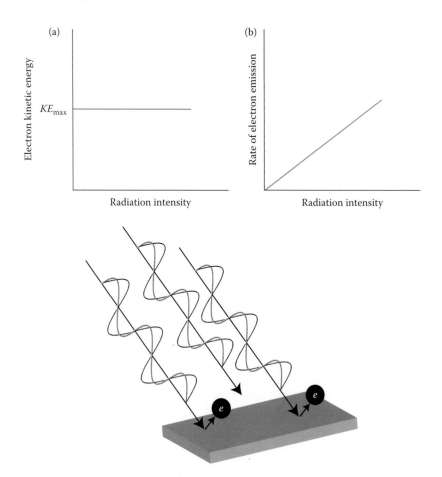

FIGURE 3.10 The photoelectric emission of electrons from a metal surface. The maximum kinetic energy, KE_{max}, of electrons emitted by incoming photons at a given frequency is constant, no matter what the radiation intensity (a), and the rate of electron emission is directly proportional to radiation intensity (b).

In this equation, ϕ is the work function of the metal—the minimum energy binding the electron to the metal. It is typically a few electron-volts (eV), such as 4.08 eV for aluminum. This explained what was confusing about the photoelectric effect:

1. The reason for the cutoff frequency is that a photon must have more energy than the work function of the metal to free up an electron. If the frequency of the photon is not high enough, no electrons are emitted—no matter how many photons hit the surface. If the photon has more than the work function worth of energy, the extra energy is converted into kinetic energy in the electron. The cutoff frequency, f_c, is the frequency at which the maximum kinetic energy of the emitted electrons goes to zero using Equation 3.9, or

$$f_c = \frac{\phi}{h} \tag{3.10}$$

2. Increasing the light's intensity (i.e., increasing the number of photons per unit of time) does not mean that the electrons will have higher kinetic energy; it just means there will be more electrons emitted. Their kinetic energy, as we see in Equation 3.9, depends on the frequency of the light and the work function—not on the light intensity (see Figure 3.10b).

3. Electrons are emitted instantaneously because even the very first photon to reach the surface can transfer its packet of energy to a surface electron, causing it to fly off the surface. It is a matter of individual photons interacting with individual electrons—not some kind of uniform absorption or buildup of energy to a critical point.

The photoelectric effect can be used to make useful devices. Consider the increasingly ubiquitous digital camera. Photons reflected from your subject enter the camera's lens, where they strike a photoelectric surface. Electrons are emitted, collected, and measured. The amount of electrons emitted depends on how much light struck a particular location (pixel) on the photoelectric surface. Each pixel location is covered by a red, green, or blue filter, and the amount of light through each type is used to determine what colors will appear in the rendered image. Another example is a burglar alarm. Ultraviolet light, invisible to the human eye, shines across a doorway to a photoelectric detector on the other side. When the rate of electron emission drops due to an interrupted light beam, an alarm or lock can be triggered.

QUANTUM MECHANICS OF THE SUN

A gigantic burning ball that powers photosynthesis, that drives the seasons and ocean currents, that alters your mood, that enables all food and fossil fuel on Earth—the Sun is also a quantum mechanical marvel (see Figure 3.11). Orbiting it are eight planets and their moons, tens of thousands of asteroids, and trillions of comets, and yet all these things make up less than 1% of the mass in our solar system. The remaining 99.86% of the mass is the Sun, most of which is hydrogen plasma. Plasma is the fourth state of matter—a high-temperature gas made of broken-up, charged atoms called ions and independent electrons that move about freely. The Sun's energy is generated by nuclear fusion, made possible in part by the Sun's 16×10^6 K core temperature. Fusion is when two atomic nuclei combine to create a single nucleus.

The hydrogen atom is the simplest atom and has a single proton nucleus. Inside the Sun, hydrogen atoms are jammed together in an iterative sequence called the proton–proton chain. In this process, the nuclei of four hydrogen atoms are successively forced together until a single nucleus is formed. The new nucleus consists of two protons and two neutrons. This new nucleus is the same as that of another element, helium. Four hydrogen atoms have fused into a single helium atom. The mass of the new nucleus is less than the mass of the four proton nuclei that made it. This leftover mass, m, becomes energy, E, according to Einstein's famous relation, $E = mc^2$, where c is the speed of light. This amounts to about 25 MeV of energy and takes the form of two neutrinos and six photons:

$$4H \, \text{protons} \rightarrow He + 2 \, \text{neutrinos} + 6 \, \text{photons}$$

(Neutrinos are tiny, elusive particles that travel at the speed of light and lack electric charge. Very little is known about them.)

Every second, approximately 700 million metric tons of the Sun's hydrogen becomes helium through fusion. The Sun sheds about 5×10^6 metric tons of mass per second in the form of energy. Some of this energy fuels the Sun's endless fusion reactions and some is emitted into space as EM radiation. The Sun emits EM radiation in all parts of the spectrum. It also emits protons and electrons, known as solar wind. Solar wind is redirected away from Earth by the planet's magnetic field. On the other hand, much of the Sun's EM radiation reaches the Earth's surface as visible and infrared radiation, and even as radio waves. The Earth's atmosphere shields us from almost all the ultraviolet radiation, x-rays, and gamma rays.

It is mind-boggling to consider the life of a single photon in the Sun. Born of the extraneous mass of the Sun's ongoing fusion reaction, it is jettisoned across the 92,960,000-mile void of space, through Earth's atmosphere, and into your window. The same photon that was 8 min ago inside the Sun is now inside your eyeball.

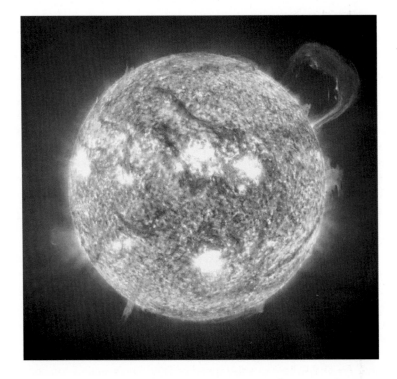

FIGURE 3.11   Our quantum mechanical Sun. This image was taken using the Extreme Ultraviolet Imaging Telescope (EIT) on September 14, 1999, measuring radiation in the 30.4 nm wavelength. The handle-shaped feature in the upper right is an eruption of (relatively) cool, dense plasma in the Sun's atmosphere, the temperature of which is about 60,000 K. (Image courtesy of NASA.)

3.3.6 Wave–Particle Duality: The Double-Slit Experiment

A particle is an object with a definite position. It acts as an individual thing. On the other hand, a wave is a periodic pattern, lacking a definite position. It can interact with other waves constructively or destructively. When something behaves as both a wave and a particle, this constitutes wave–particle duality. Photons and electrons are such things. This phenomenon is thus far absolutely impossible to explain in any classical manner and is at the heart of quantum mechanics.

The double-slit experiment is one of the better ways to observe the quantum behavior of electrons in action. It is helpful to first understand the experiment in simpler ways. Since we are trying to understand how something can behave like a particle or like a wave, let us first conduct the experiment using bullets and then using water.

3.3.6.1 Bullets

Take a look at the experimental setup shown in Figure 3.12. There is a machine gun that sprays bullets randomly in the directions shown. In front of the gun is a wall with two slits (slits 1 and 2) just big enough for the bullets to pass through. The bullets can ricochet off the slit's walls as they pass through the slit. Behind the wall is a backstop. Bullets always arrive one at a time and always as whole bullets (they cannot break into fragments along the way). If we let the gun fire away for a specified time and measure the *number* of bullets arriving at particular locations along the backstop (as shown by the stacks behind the wall), we can divide by the total number of bullets that hit the backstop and get a probability of a

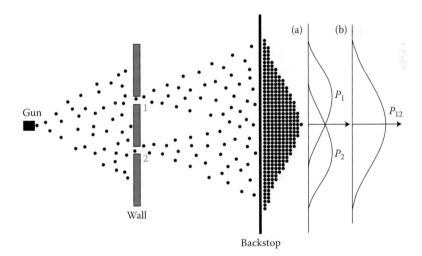

FIGURE 3.12 The double-slit experiment with bullets. A machine gun sprays bullets toward a wall with two slits. Behind the wall is a backstop. The stacks of bullets behind the backstop show how many bullets hit at particular locations during a given time. If we divide by the total number of bullets that hit the backstop during that time, we obtain the probability of a bullet arriving at a particular location. If we do the experiment with slit 2 covered, we obtain the curve P_1 shown in (a). If we cover slit 1, we obtain the curve P_2. With both slits open, we get the sum of these probabilities: $P_1 + P_2 = P_{12}$. This probability distribution is shown in (b).

bullet arriving at a particular location. This probability distribution is shown in Figure 3.12b. Why is the maximum probability right in the middle, between the two slits? Well, if we do the experiment again, this time covering slit 2 so that bullets can only pass through slit 1, we will get the curve P_1 shown in Figure 3.12a. If we cover slit 1, we obtain curve P_2. When both slits are open, the probabilities simply add together, giving us the curve in Figure 3.12b and

$$P_{12} = P_1 + P_2 \tag{3.11}$$

3.3.6.2 Water Waves

We will now perform a similar experiment with water waves, as diagrammed in Figure 3.13. We have a shallow pool of water and on one side there is a circular wave source. Like the prior experiment, we have a wall with two slits and behind it a backstop. We assume that this backstop absorbs all waves so that there is no reflection. This time, instead of counting the number of bullets, we measure the height of waves. This is a way to measure wave *intensity*, or the energy being carried by the waves as they arrive at various locations along the backstop. This intensity can be of any value, small for tiny waves hitting the detector, and gradually increasing to higher values with taller waves.

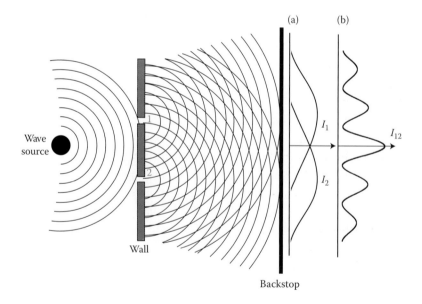

FIGURE 3.13 The double-slit experiment with water waves. A circular wave source creates waves in a shallow pool. The waves pass through slits in a wall and continue on to a backstop that prevents reflection. We measure wave height at various locations along the backstop. The height, or *intensity*, tells us how much energy a given wave has. If we cover slit 2, the intensity at the backstop is the curve I_1 shown in (a). If we cover slit 1, the intensity is the curve I_2. When both slits are open, the waves propagating from slit 1 interfere with those from slit 2. The interference can be constructive or destructive and leads to the shape of the curve we measure when both slits are open (I_{12}).

If we cover slit 2, the intensity at the backstop is the curve *I*1 shown in Figure 3.13a. If we cover slit 1, the intensity is the curve *I*2. However, when both slits are open, the intensity is not simply the sum of *I*1 and *I*2. The waves propagating from slit 1 interfere with those propagating from slit 2.

Sometimes, the waves add to one another to create a larger amplitude (a phenomenon unique to waves known as constructive interference) and sometimes they subtract from one another for a smaller amplitude (destructive interference). For example, the interference is constructive if the distance from slit 1 to a single detection point along the backstop is exactly one wavelength smaller than the distance from slit 2 to the same detection point because two wave peaks can arrive at the detection point simultaneously. Or, the interference will be destructive if the distance from slit 1 to the detection point is exactly half a wavelength smaller than the distance from slit 2 to the detection point because one wave's peak can arrive as the other wave's trough arrives, canceling out one another. We see then that although the two waves start out in-phase at the slits, in most cases they travel different distances to get to each point along the backstop and therefore have a relative phase difference, Φ, upon arriving at most points. This interference is the reason for the shape of the curve of *I*12, which can be expressed as

$$I_{12} = I_1 + I_2 + 2\sqrt{I_1 I_2}\, \cos \Phi \qquad (3.12)$$

3.3.6.3 *Electrons*

Now we are ready to conduct the double-slit experiment with electrons. In 2002, a *Physics World* magazine reader poll named this particular experiment the all-time, "most beautiful experiment in physics." It was first performed by English scientist Thomas Young some time around 1801, using light instead of electrons.

The setup is diagrammed in Figure 3.14. An electron "gun" emits electrons, each with nearly the same energy and in random directions toward the wall with the two, closely spaced slits. Beyond the wall is the backstop. As we did with the bullets, we count the number and location of electrons hitting the backstop. And like the bullets, the electrons always arrive whole at the backstop—never as partial pieces of electrons. As for the probability of an electron arriving at various locations along the backstop, the shape of the curve does not look like the probability did for bullets. It looks like the curve labeled P_{12} in Figure 3.13b. We are counting the electrons like bullets, but they are behaving like waves! We can see the interference pattern very clearly in Figure 3.15, which shows data collected during an actual dual-slit experiment with electrons.

How is this behavior possible?

Let us try to analyze it. We do the experiment again, this time closing slit 2. This gives us the curve P_1 shown Figure 3.14a. And when we cover slit 1, we obtain P_2. However, the result we observe when both slits are open is not the sum of probabilities from each slit being open. In the case of electrons:

$$P_{12} \neq P_1 + P_2 \qquad (3.13)$$

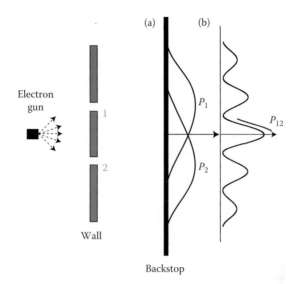

FIGURE 3.14 "The most beautiful experiment in physics." The double-slit experiment was first performed by Thomas Young, using light. Here we use electrons, which are fired from a gun toward the two slits. We count the number and location of electrons hitting the backstop, just like we did with the bullets. If we do the experiment with slit 2 closed, the probability distribution looks like curve P_1 in (a). Covering slit 1, we observe P_2. However, the probability curve when both slits are open does not look like it did with bullets, as we might expect. The curve of P_{12} for electrons is the same as the curve of I_{12} for waves. While electrons arrive one at a time at the backstop, like particles, their probability of arrival is subject to interference, like waves. This is wave–particle duality.

There is interference at work here. In fact, mathematically, the curve of P_{12} for electrons is the same as the curve of I_{12} for waves. This means it is not true to say that the electrons go either through slit 1 or slit 2. The only way to explain this curve is to say that—somehow—the electron is simultaneously present at both slits and is simultaneously interacting with both slits.

When Young first performed this experiment, he shined a beam of light through an opaque screen with a pair of parallel slits and observed a pattern of alternating dark and light bands on a white surface on the other side of the screen. He reasoned that light consisted of waves.

So this is what we can say about electrons (the same being true of photons): they arrive one at a time, like particles, and their probability of arrival is subject to interference, like waves. The results of this experiment are not contradictory since we never see an electron behave like a particle and like a wave at the same time. Only one or the other. Niels Bohr's principle of complementarity says that neither the wave model nor the particle model adequately describes matter and radiation. In fact, the models complement one another and must be used in tandem.

This is *wave–particle duality*, and it is the best explanation we have—thus far.

Consider this: in 1906 the Nobel Prize for physics was awarded to J. J. Thomson for showing that electrons are particles; and then in 1937, J. J.'s son, George Thomson, was

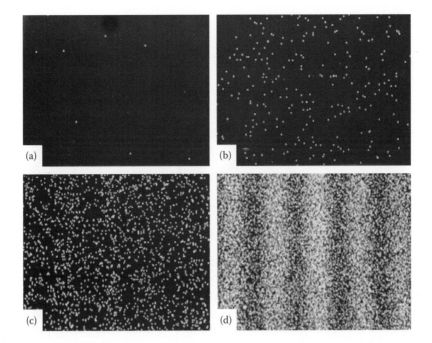

FIGURE 3.15   Experimental data from the electron dual-slit experiment. Each white dot depicts the arrival of a single electron at the backstop detector during an actual dual-slit experiment. Both slits were open. The data shown were captured at four times during the 20-min experiment. The number of electrons in each frame is (a) 8 electrons, (b) 270 electrons, (c) 2000 electrons, and (d) 6000 electrons. We can see the interference pattern in (d), similar to the probability curve P_{12} shown in Figure 3.14. (Reprinted courtesy of Dr. Akira TONOMURA, Hitachi, Ltd., Japan.)

awarded the same prize for showing that electrons are waves. Perhaps one of George's two sons or his two daughters will provide the next major insight!

3.3.7 The Uncertainty Principle

In order to find out if electrons are really passing through one slit or the other, we can do one more hypothetical experiment—a subtle variation of the double-slit electron experiment we just discussed. All we do differently this time is shine lights at the back sides of the two slits, as shown in Figure 3.16. In this way we can tell which slit a given electron takes to get through the wall. The lights enable us to watch an electron because as it passes through the light, we see a tiny flash (using a microscope), and then the electron hits the backstop. If we see a flash near slit 1, the electron went that way; if the flash is near slit 2, the electron went that way. When we do the experiment, we find that electrons go through either slit 1 or slit 2.

We can keep track of which slit a given electron goes through and also the corresponding probability curve at the backstop. This is analogous to prior experiments where we closed one slit or the other—only this time, instead of knowing that an electron came through one slit because the other one was closed, we know *because we watched it* come through the slit. The strange thing is that the total probability curve no longer exhibits the

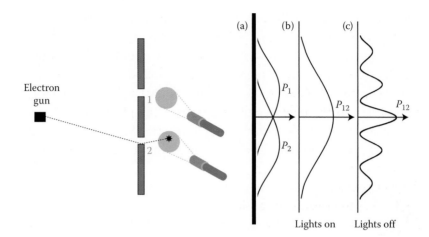

FIGURE 3.16 One more time, under the lights. Shining light at the back sides of the slits enables us to see tiny flashes when electrons pass through the slits. Just as before, we count the number and location of electrons when they hit the backstop, and we observe that all the electrons that pass through slit 1 give the probability curve P_1. Those passing through slit 2 give the curve P_2. However, when we determine the overall probability distribution, we must add P_1 and P_2 together, which gives the curve P_{12} shown in (b). The interference is gone. By observing the electrons' paths, we have somehow altered their behavior. If we turn off the lights again, we get the usual wave interference pattern shown in (c).

wave behavior we saw in P_{12} in Figure 3.14. Ever since we started observing the electrons with light, the curve started exhibiting particle behavior (like P_{12} in Figure 3.12). The interference pattern is gone. When we turn off the light and repeat the experiment, the curve goes back to looking like the wave's curve again.

The only thing we can conclude about this is that the way we are observing the electrons is changing their behavior. In 1927, Werner Heisenberg decided that it was not sloppiness, or improper experimental procedures, or inadequate equipment that were making it so difficult to watch the electrons without disturbing them. It was the way of nature, now known as the Uncertainty Principle. In order to measure something, you have to interact with it. In relation to the double-slit experiment, the Uncertainty Principle means that no device can ever be built to tell us which slit the electrons go through without also disturbing the electrons and ruining their interference pattern.

More generally, the Uncertainty Principle states the following: If we measure the x-component of an object's momentum with an uncertainty Δp_x, and at the same time measure its x-position, we can only know this x-position with limited accuracy. (We picked the x direction but it also applies for the y or z directions.) Specifically, the uncertainty in an object's position and the uncertainty in its momentum at any moment in time must be such that

$$\Delta x \Delta p_x \geq \frac{h}{4\pi} \tag{3.14}$$

The more precisely something's position is determined, the less precisely its momentum is known. If Δx is very small, Δp_x must be large, and vice versa.

When we saw an electron pass through one of the two slits in our hypothetical experiment, we saw it because a photon of light had bounced off of it and into our eye (through a microscope). We learned in Back-of-the-Envelope 3.5 that a photon can have momentum. During the collision of the photon and the electron, the photon transfers an unknown amount of momentum to the electron. Although we were able to pinpoint quite accurately where the electron was (meaning Δx is very small), we simultaneously changed the electron's momentum, but by how much we cannot know very well (because Δp_x is large).

Heisenberg's Uncertainty Principle is almost like a dare, a challenge to people to just try and figure out a way to measure the position and momentum of something, anything—an electron, a basketball, a comet—with greater accuracy. And no one has figured out a way to do it yet.

The idea of uncertainty fits well within the framework of quantum mechanics—a tenet of which is that we never know for sure what is going to happen under a given set of conditions, only the probability of specific outcomes. Accepting quantum mechanics means feeling certain that you are uncertain.

3.3.8 Particle in a Well

To speak qualitatively, electrons and photons are particles that behave like waves, or waves that behave like particles. Next, we conquer a quantitative mathematical description of their motion and their energy. One simple way we can approximate a particle/wave is to consider a particle at the bottom of an infinitely deep, square well. That is, the walls are so high that the potential energy at the top of each one is infinity, so the particle will never have enough energy to get out. The well is therefore a "potential well."

We consider the particle's motion only in one dimension. If this were an ordinary particle, we would treat it like a speck of mass, and its motion ricocheting back and forth between a pair of parallel walls would be straightforward to describe—its velocity, its acceleration. However, with this unique type of particle, it is more appropriate to draw an analogy with a unique kind of wave—the standing wave.

Standing waves can be found on plucked guitar strings, or on a resonating beam—any place that a wave is confined to a given space. If we stretch a string tight and attach it to parallel walls of the well, and then vibrate it at one of its resonance frequencies, a wave pattern arises. Three such standing waves on stretched strings are shown in Figure 3.17. The strings take on a sinusoidal shape in which certain points actually stand still. Although rather simple, the mathematical model that describes this type of standing wave also suitably elucidates the behavior of confined quantum particles like electrons and photons.

The strings are fixed to the walls on each end, giving our model boundary conditions. In order for the wave on the strings to have zero amplitude at the fixed ends, the wavelengths can only be certain lengths. This constrains the wavelength to a set of values:

$$\lambda = \frac{2L}{n}, \quad n = 1, 2, 3, \ldots$$

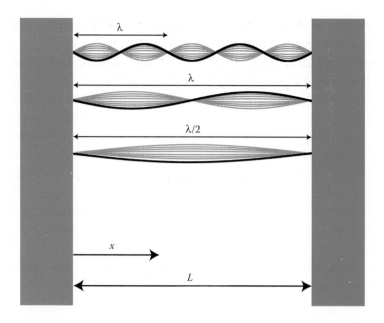

FIGURE 3.17 Standing waves on strings. The shape of the waves is sinusoidal, and certain points along the strings stand still. The only permissible wavelengths are those that make the waves' amplitudes equal to zero at the walls.

Here, L is the distance between the opposite walls. We can see by this equation that a standing wave confined this way must have a quantized wavelength. The equation governing the vertical displacement, y, at a given location, x, along the string is

$$y(x) = A\sin\left(\frac{2\pi}{\lambda}x\right)$$

Here, A is the amplitude. By combining these two equations, we get an equation describing the motion of a standing wave on a string:

$$y(x) = A\sin\left(\frac{n\pi}{L}x\right), \quad n = 1, 2, 3, \ldots$$

Particles such as electrons and photons are typically described using a more generalized wave function, denoted simply by ψ. While ψ itself is not a measurable quantity (it does not have units), all the measurable quantities of particles—such as energy and momentum—can be determined using ψ. In fact, the wave function's absolute square, $|\psi|^2$, is proportional to the probability that the particle occupies a given point at a given time. The wave function for a particle in a box is exactly the same as for the vertical displacement of the string:

$$\psi(x) = A\sin\left(\frac{n\pi}{L}x\right), \quad n = 1, 2, 3, \ldots \tag{3.15}$$

Here, A is the maximum value for this wave function, L is the width of the well, and x is the distance from one wall. This equation gives the allowed wave functions for a particle in an infinitely deep potential well. Figure 3.18 shows graphs of ψ and of $|\psi|^2$ as functions of x.

But what does all this really mean? Well, first, we know that the wave function must be zero at the walls of the well, just as the string must have zero amplitude at the wall. This is the way to mathematically express the fact that the particle fills the well but cannot exist outside of it. For this to happen, only certain wavelengths are allowed—namely, those for which $\lambda = 2L/n$.

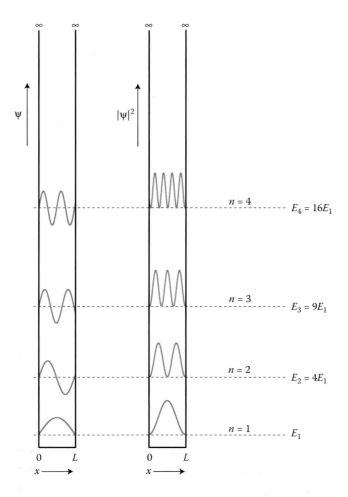

FIGURE 3.18   The quantized states of a particle in a potential well. Graphed here are the wave function and the probability distribution, $|\psi|^2$, versus x for $n = 1, 2, 3,$ and 4. These correspond to the lowest four energies the particle is allowed to have. We can see that both ψ and $|\psi|^2$ are zero at the boundaries of the well—this constraint is what forces the wavelengths to take on quantized values. Also note that $|\psi|^2$, which shows us the probability of finding the particle at a given value of x, is always positive because a negative probability is meaningless. The well's width is L and its depth is considered infinite.

Knowing this, we can substitute what we learned about the momentum, p, of a particle from Equation 3.8:

$$p = \frac{h}{\lambda} = \frac{h}{(2L/n)} = \frac{nh}{2L}, \quad n = 1, 2, 3, \ldots$$

Here, h is Planck's constant. Since the particle's momentum, $p = mv$ (m being the particle's mass and v being its velocity), we can next develop a relationship for the particle's allowed values of kinetic energy, KE:

$$KE = \frac{1}{2}mv^2 = \frac{p^2}{2m} = \frac{(nh/2L)^2}{2m}, \quad n = 1, 2, 3, \ldots$$

This is what we have been working toward and now we have arrived at an expression for the energy of the particle in the potential well:

$$E = \left(\frac{h^2}{8mL^2}\right)n^2, \quad n = 1, 2, 3, \ldots \tag{3.16}$$

The energy of the particle is quantized. Knowing what we know now about quantum mechanics, this is what we expect. The lowest energy state the particle can have occurs when $n = 1$, which gives $E_1 = h^2/8\,mL^2$. The next energy level up is $n = 2$, which gives $E_2 = 4h^2/8\,mL^2 = 4E_1$. We can see in Figure 3.18 the four lowest energy levels for a particle confined in a potential well. The spacing between energy levels gets larger as n increases. Note that $n = 0$ is not allowed. This is in contradiction to classical mechanics, where zero energy is acceptable and the energy of a particle can be any value.

The particle in a well model, while oversimplified, happens to be quite accurate for the lower energy levels; it becomes progressively less accurate at higher levels.

How does this energy quantization affect a real particle? Consider an electron in an atom. When this electron drops from one energy state to a lower state, a photon is emitted. Or, the electron can become more energized if it absorbs a photon. Either way, the energy of the photon must be equal to the difference in energy between the two states. Atoms radiate or absorb energy in the form of photons. This is how they change quantum states.

BACK-OF-THE-ENVELOPE 3.6

Using the particle in a well model, estimate how much energy is given off (in the form of a photon) when an electron in an atom jumps from the seventh to the sixth energy level. What is the wavelength of this photon? Assume the atom has a radius of 100 pm.

If we use Equation 3.16 and take the width of the well (in this case the diameter of the atom) to be $L = 200$ pm, we determine that

$$E = \frac{(6.626 \times 10^{-34} \text{ Js})^2}{8(9.11 \times 10^{-31} \text{kg})(200 \times 10^{-12} \text{m})^2} n^2 = 1.5 \times 10^{-18} n^2 \text{ J}$$

From this result we can easily find the difference in energy between the $n = 6$ and $n = 7$ states:

$$\Delta E = 1.5 \times 10^{-18} (7)^2 - 1.5 \times 10^{-18} (6)^2 = 1.95 \times 10^{-18} \text{ J}$$

Having determined how much energy is given off, we can now calculate the wavelength of the emitted photon. We know that $\Delta E = hf = hc/\lambda$, so the wavelength is

$$\lambda = \frac{hc}{\Delta E} = \frac{(6.626 \times 10^{-34} \text{ Js})(3.0 \times 10^{8} \text{ m/s})}{1.95 \times 10^{-18} \text{ J}} = 100 \text{ nm}$$

Referring to the EM spectrum, we can see that this wavelength corresponds to ultraviolet light.

3.4 SUMMARY

Classical physics perfectly describes many natural phenomena. Just not many nanoscale phenomena. Understanding nanotechnology necessitates an understanding of atoms, electrons, and photons—all of which are governed by quantum mechanics.

- Electrons inhabit regions of space enclosing an atom's nucleus called orbitals. These are organized into levels and sublevels, depending on how much energy the electrons have, with lower energy levels nearer the nucleus. Like standing waves, the electrons of an atom can take on only very specific wavelengths (energies). Electrons in the highest occupied energy level are called valence electrons.

- Radiation is made out of fluctuating electric and magnetic fields called EM waves that propagate through empty space at 299,792,458 m/s. These waves are generated whenever a charged particle (like an electron) accelerates. The EM spectrum encompasses numerous wavelengths, including those of visible light.

- A photon is a massless, chargeless "packet" of EM radiation. It behaves like a wave and like a particle at the same time. So do electrons. This is known as wave–particle duality. We can observe wave–particle duality at work in the dual-slit experiment. Atoms emit or absorb photons when their electrons hop up or down between quantum energy levels.

- The Uncertainty Principle dictates that the more precisely something's position is determined, the less precisely its momentum is known.

HOMEWORK EXERCISES

3.1 When were the ideas of quantum mechanics first introduced?

3.2 True or false? The field of quantum mechanics falls under the larger field of classical mechanics.

3.3 Give the electronic configuration of the following atoms:
   a. Hydrogen
   b. Silicon
   c. Silver

3.4 Which of the following lists of sublevels is given in ascending energy level?
   a. 3s, 3p, 3d, 4s
   b. 1s, 2s, 3s, 2p
   c. 2p, 3s, 3p, 4s
   d. 3p, 3d, 4s, 4p

3.5 Look up the number of valence electrons for the following atoms:
   a. Silicon
   b. Beryllium
   c. Oxygen
   d. Argon

3.6 An oddity in the order of filling orbitals is that an atom can be more stable if it has either a full or a half-filled sublevel. This leads to an exception in the filling rules when dealing with the 3d sublevel. When the 3d would typically have four electrons in it and the 4s a full pair, the 3d instead becomes exactly half filled by "stealing" one of the electrons from 4s—leading to a configuration that ends with ... $3d^54s^1$. This same thing happens when the 3d has nine electrons so that it can be more stable as a full sublevel. Name the atom in each of these cases.

3.7 At what speed do the following waves travel?
   a. Gamma rays
   b. Microwaves
   c. Radio waves

3.8 A stoplight radiates visible green light that has a wavelength of about 545 nm. What is the frequency of this light?

3.9 A popular FM radio station, KNNO 100.9 (MHz)—"All Nanotechnology, All the Time"—broadcasts using a transmitter with a power output of 200 kW. Roughly how many photons does this transmitter emit into the air per hour?

3.10 Calculate the wavelength for the following "particles":
   a. An electron (9.11×10^{-31} kg) moving at 100,000 m/s
   b. A proton (1.67×10^{-27} kg) moving at one-tenth the speed of light
   c. A mercury atom (201 g/mol, Avogadro's number $= 6.022 \times 10^{23}$) moving at 10 m/s
   d. A baseball (145 g) moving at 40 m/s

3.11 The peak wavelength of light coming from a giant red star is 660 nm. Calculate the approximate surface temperature of this star in degrees Fahrenheit.

3.12 Various sources give the temperature of the Sun's surface as 5500°C–6000°C. What range of wavelengths is emitted by this range of temperatures?

3.13 The human eye is most sensitive to light with a frequency of 5.36×10^{14} Hz. An object at what temperature (Celsius) radiates at this frequency?

3.14 True or false? Atoms typically absorb light at all of the same wavelengths they emit light.

3.15 What was the "ultraviolet catastrophe"?

3.16 The metals listed in Table 3.1 are illuminated with light having a wavelength of 180 nm.

  a. Determine the maximum kinetic energy of the electrons emitted by these metals due to the photoelectric effect.

  b. Determine the maximum wavelength of light to achieve electron emission with the photoelectric effect for each of these metals.

  c. Sketch a general plot of KE_{max} vs. frequency of light for the photoelectric effect. The plot should indicate f_c.

3.17 The Sun generates an average 3.74×10^{26} W of power. Assume the wavelength of the Sun's radiation is 500 nm.

  a. Estimate how many photons the Sun emits every second.

  b. Estimate the temperature of the Sun.

  c. If 4.45×10^{-8}% of the Sun's photons reach Earth, estimate the average intensity of this radiation on the Earth's surface in W/m^2 (radius of the Earth: 6.37×10^6 m).

3.18 Electrons can behave like which two dissimilar things?

3.19 In the double-slit experiment with bullets, what is the detector measuring? What does the detector measure in the water wave experiment?

3.20 In the electron double-slit experiment, what is the detector measuring? Is this similar to the bullet or water wave experiment? Does the probability curve resemble the curve in the bullet or the wave experiment?

3.21 What wave phenomenon explains the shape of the probability curve for electrons in the double-slit experiment?

3.22 According to the Heisenberg Uncertainty Principle, the more precisely you know the position of an electron, the less precisely the _____is known.

3.23 If an experimenter is able to measure the location of an atom's electron to a precision of 0.06 nm, what is the uncertainty in the velocity of the electron? (Mass of an electron: 9.11×10^{-31} kg.)

3.24 Consider a football field. Standing along the sideline are 100 people, one person at each 1-yard hash mark. The first person weighs 30 kg, the next 31 kg, and so on up to 130 kg. The people can only move in straight lines toward

TABLE 3.1 Photoelectric Effect Data for Various Metals

Metal	Work Function (eV)
Cesium	1.90
Sodium	2.46
Copper	4.70
Zinc	4.31
Silver	4.73
Lead	4.14
Iron	4.50

the opposite sideline. At a particular moment we are able to determine the distance of each of these people from the sideline to within 0.5 yards.

a. Create a plot that shows the minimum uncertainty of the people's speed (m/s) at this moment vs. their mass (kg).

b. If these uncertainties in velocity prevail for 1 min, whose position is more uncertain after this time, the lightest person or the heaviest person?

c. If instead of having a person at one end of the field we had an electron (mass: 9.11×10^{-31} kg), what would be the initial uncertainty of its momentum?

3.25 We can use the mathematical model of a particle trapped in a potential well to approximate the behavior of electrons in an atom.

a. Plot the wave function, Ψ, as a function of x for an electron in the sixth energy level in an atom that is 200 pm in diameter. Set the maximum value of the wave function to 1.

b. On the same graph, plot the probability of finding the electron as a function of x.

c. At what value(s) of x is the probability of finding the electron highest?

RECOMMENDATIONS FOR FURTHER READING

1. R. Serway and J. Jewett. 2009. *Physics for Scientists and Engineers.* Brooks Cole.
2. Roger S. Jones. 1992. *Physics for the Rest of Us: Ten Basic Ideas of 20th Century Physics That Everyone Should Know … and How They Have Shaped Our Culture and Consciousness.* McGraw-Hill.
3. David K. Ferry. 2001. *Quantum Mechanics: An Introduction for Devices Physicists and Electrical Engineers.* Taylor & Francis.
4. A. Fromhold. 1991. *Quantum Mechanics for Applied Physics and Engineering.* Dover Publications.
5. R. Feynman. 1995. *Six Easy Pieces.* Basic Books.
6. R. Schmitt. 2000. Understanding electromagnetic fields and antenna radiation takes (almost) no math. *EDN Magazine* March 2, p. 77.

CHAPTER 4

Nanomaterials

4.1 BACKGROUND: MATTER *MATTERS*

It matters to nanotechnology for the same reasons it matters to every other field of engineering: people make stuff out of matter. We use it. Matter that can be used for something has its own name: material. Gold veins buried inside a mountain and carbon floating around in the air are not immediately useful. To be useful, material must be collected. Then we bang on it, or melt it, or attach it to other material, or rearrange it until it meets our needs. The study of materials is the study of matter *for the purpose of using it*.

As human beings, we cannot help but be awed by the stunning symmetry of atoms in a crystal. But as engineers and scientists we are interested in matter for more pragmatic reasons. We have to ask how it might be useful as an ingredient in something we make. Can we use materials in new ways if they are in smaller pieces? The short answer is yes, and hence a chapter on nanomaterials. The longer answer follows.

4.2 BONDING ATOMS TO MAKE MOLECULES AND SOLIDS

Put a few atoms together and they become a molecule. Pack a lot of atoms together in a pattern and you get a solid. All atoms can form solids. Interatomic forces serve as the mortar, holding the atoms together. This is the case at the nanoscale, and all the way on up to the macroscale. The reason your body will not just vaporize and drift away into the ether is that the atoms that comprise you are stable where they are, held together by interatomic forces.

When two atoms are "infinitely" far apart (which to an atom might just be a micron), there is no force between them. When they are brought closer together, attractive and repulsive electrostatic forces (caused by electric charge held by the atoms) act on the atoms.

The interaction of two atoms to form a molecule, and later a solid, is a matter of conserving energy—a fundamental principle of the universe. Shown in Figure 4.1 is a general plot of the potential energy of two atoms separated by a distance, x. On the plot, negative energy corresponds to a net attractive force that pulls the atoms together. Positive energy corresponds to a net repulsive force that pushes the atoms apart. We can see that the net

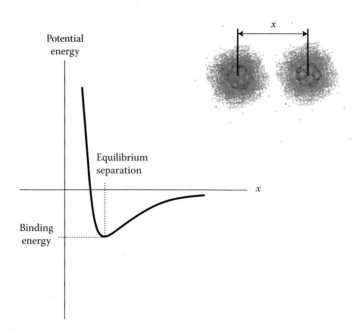

FIGURE 4.1 Total potential energy of a pair of atoms as a function of separation distance, x. At large separation distances, attractive forces pull the atoms toward one another, toward the equilibrium separation distance. At small separations, repulsive forces dominate.

force is attractive when the atoms first begin to interact, pulling them toward one another. This net attraction increases until the atoms reach an equilibrium separation—the lowest energy level on the plot. This is the distance at which the attractive and repulsive forces balance for a net force of zero.

When atoms are separated by this equilibrium distance, energy is required to move them closer together or to move them further apart. Unless energy is added, the atoms are content to exist where they are. When enough energy is added to drive them together, they become molecules. Repeat the process a few thousand times, and you get a nanoparticle. Repeat it a trillion *trillion* times (10^{24}) and you will get a baseball, a calculator, or even your hand.

The mechanisms by which these forces bond molecules and solids together are complex. There are three main kinds of bonds that hold together molecules: (1) ionic bonds, (2) covalent bonds, and (3) van der Waals bonds (Table 4.1). These three kinds of bonds are also what hold together the more complex, large-scale crystalline formations of atoms within solids, as well as nanoscale materials. Metals are one exception. A special arrangement of atoms and electrons known as the metallic bond holds solid metal together. We will discuss these bonding types now.

4.2.1 Ionic Bonding

When an atom loses or gains extra electrons, it becomes either positively or negatively charged. The sodium atom, Na, tends to give up its highest energy electron to become the

TABLE 4.1   Main Types of Bonding

Bonding	Among Atoms and Molecules	Within Solids
Ionic	Oppositely charged atoms are attracted to one another to make molecules. Example: Na$^+$ and Cl$^-$ make the NaCl molecule	Crystal structure is formed by an array of atoms held together by opposing charges. Example: Magnesium oxide, a crystal formed by O^{2-} and Mg^{2+} molecules
Covalent	Two atoms share electrons; their atomic orbitals overlap to make a molecular orbital. Example: Two hydrogen atoms make the H$_2$ molecule	Crystal structure is formed by an array of atoms sharing electron orbitals. Example: Diamond, made from C atoms each sharing electrons with four other C atoms
van der Waals (three types)	Two atoms are attracted to one another by weak electrostatic forces. Example: Polar H$_2$O molecules cling to one another, endowing water with unique characteristics	Atoms already organized as a solid are still subject to weak interaction forces with other atoms. Example: Covalently bonded layers of carbon in graphite stack-like paper with weak van der Waals forces; layers can slide, making graphite soft (see Figure 4.8)
Metallic	The same as "Within Solids"	Metal atoms are held together by a "sea" or "gas" of electrons; these electrons can move freely among the atoms, making metals conductive. Example: Copper atoms in a telephone wire

Note: These bonding mechanisms are those that enable atoms to form molecules as well as solids. Much about the way matter behaves can be explained by the way in which the atoms are held together.

positively charged ion, Na$^+$; whereas the chlorine atom, Cl, tends to pick up a spare electron to fill out its electronic configuration, thereby forming Cl$^-$. Negative and positive ions are drawn together (by Coulomb attraction) and we obtain NaCl (table salt).

The Coulomb attraction energy, E, that drives ionic bonding is given by

$$E = \frac{z_1 z_2 e^2}{4 x \pi \varepsilon_0 \varepsilon} \tag{4.1}$$

Here, x is the distance between the ions, ε_0 is the permittivity of free space, and ε is the dielectric constant of the medium between the ions (or the relative permittivity). The magnitude and sign of the two ions are given in terms of e, the elementary charge ($e = 1.602 \times 10^{-19}$ C), multiplied by the valence state of the ions, z_1 and z_2. In the case of Na$^+$, for example, $z = +1$; for Cl$^-$, $z = -1$. An ion such as Ca^{2+} has $z = +2$, and so on.

As a quick review, energy is a force applied over a distance. Taking the negative derivative of the energy equation with respect to distance, or $-dE/dx$, gives us the force, F, between the atoms. Here, negative forces and energies are considered attractive and positive forces and energies are considered repulsive. The Coulomb force, F, is therefore given by

$$F = -\frac{dE}{dx} = \frac{z_1 z_2 e^2}{4 x^2 \pi \varepsilon_0 \varepsilon} \tag{4.2}$$

BACK-OF-THE-ENVELOPE 4.1

How much energy and how big a force hold the NaCl molecule together?

First, we need to know the separation distance, x. We know Na has an atomic radius of 186 pm. However, when the Na atom loses an electron, it shrinks. Na^+ has a radius of 97 pm. The opposite is true of Cl, which upon acquiring an extra electron grows in radius from 99 to 181 pm. The sum of these two ionic radii is 0.278 nm. The dielectric constant of air is 1. The binding energy is therefore

$$E = \frac{(1)(-1)(1.602 \times 10^{-19} \text{ C})^2}{4\pi(0.278 \times 10^{-9} \text{ m})(8.854 \times 10^{-12} \text{ C}^2\text{N}^{-1}\text{m}^{-2})(1)} = -8.3 \times 10^{-19} \text{ J}$$

This energy is attractive. The binding force is

$$F = \frac{(1)(-1)(1.602 \times 10^{-19} \text{ C})^2}{4\pi(0.278 \times 10^{-9} \text{ m})^2(8.854 \times 10^{-12} \text{ C}^2\text{N}^{-1}\text{m}^{-2})(1)} = -3 \times 10^{-9} \text{ N}$$

When *numerous* ions come together to form a solid, the ionic interactions become more complex. In the case of NaCl, the positive sodium ions self-arrange so as to minimize the repulsion between one another and maximize the attraction to the negative chlorine atoms, thereby achieving an equilibrium where breaking apart the crystal structure thus formed would require more energy than maintaining it. Each sodium ion is surrounded by six chlorine ions, and each chlorine ion is surrounded by six sodium ions, as shown in Figure 4.2.

More than 200 compounds assume the same crystal structure as NaCl, one of many ionic crystal structures. Ionic crystals tend to be stable, hard crystals with poor electrical conductivity because they lack free electrons. They have high vaporization temperatures and are especially soluble in water (or any polar liquid) because the permanently dipolar H_2O molecules tug at the charged ions in the crystal, breaking the bonds and dissolving the solid. Typically, ionic crystals are transparent to radiation in the visible range, but absorb radiation in the infrared range. The heavy ions in the crystal have low natural

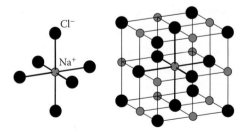

FIGURE 4.2 Crystal structure of table salt. Each sodium atom is attracted to six neighboring chlorine atoms, and vice versa. NaCl is an example of ionic bonding. The length of the cube edge is 563 pm.

frequencies and therefore tend to absorb low-energy (lower frequency) photons from the infrared region of the spectrum.

4.2.2 Covalent Bonding

Two atoms share electrons in a covalent bond. The atoms' individual orbitals overlap to form a mutual molecular orbital. This happens with hydrogen: two hydrogen atoms bond covalently as H_2, each atom sharing its one electron so that the molecule has two shared electrons. This effectively elongates the spherical orbital shape of a single hydrogen atom into more of a pill shape where the probability of finding one of the shared electrons is highest in the region between the two nuclei. The H_2 molecule is represented in Figure 4.3. Covalent bonds are very stable and hold together such molecules as F_2, CO_2, CO, H_2O, and CH_4.

Solids held together by covalent bonding are, like covalent molecules, highly stable. Covalent forces are of very short range, on the order of interatomic separations (100–200 pm), with typical binding energies from 4×10^{-19} to 1×10^{-18} J per bond. For example, the oxygen–hydrogen bond in a water molecule has a strength of 7.6×10^{-19} J.

Diamond is formed entirely from covalent carbon atoms—each sharing electrons with four neighbors, all 0.154 nm away, in what is known as a "giant molecule" throughout the solid crystal. Silicon and germanium both crystallize in the same pattern as diamond. Covalent crystals can be extremely hard due to the cohesiveness of covalent bonds, which can also make for very high melting temperatures (4000 K in the case of diamond) and poor conductivity.

4.2.3 Metallic Bonding

About two-thirds of the elements in the Periodic Table are metals. Many of the physical properties we associate with metals, including their luster, malleability, and high electrical

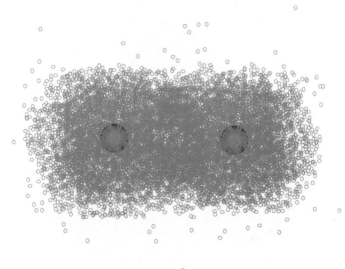

FIGURE 4.3   A covalent bond between hydrogen atoms. The probability of finding an electron in the H_2 molecule is highest in the region between the two nuclei.

and thermal conductivity, result from metallic bonding. When metal atoms organize into solid form, the electrons in their outermost orbitals (known as valence electrons) disassociate from the atoms. These newly free electrons form a "sea" that can flow in and around the lattice of positive ions. This is similar to covalent bonding in that the electrons in both cases are shared among two or more atoms; the discreteness of the atoms is lost.

The metallic bond is nondirectional. It does not lead to structures quite like those in ionic and covalent crystals. Although it is difficult to separate metal atoms in a solid, it is somewhat easier to move them around, as long as they stay in contact with each other. Because of this, metal atoms can be dissolved in other types of metals to form alloys, with customizable metal-to-metal ratios. One particular ratio might improve tensile strength, another corrosion resistance.

The energy binding the "sea" of shared electrons to the atoms in a metal (1–3 eV) is about the same amount of energy carried by photons of visible radiation. Thus, light interacts strongly with metal electrons and is absorbed and reemitted near a metal's surface instead of being completely absorbed; hence, metals' shiny appearance. The free "sea" of electrons is what conducts heat and electricity so well.

4.2.4 Walking through Waals: van der Waals Forces

Dutch scientist Johannes Diderik van der Waals (Figure 4.4) did not like the fact that real gases did not obey the ideal gas law. So he rewrote it. He pursued the idea that there were attractive forces between the molecules in a gas, forces the ideal gas law ignored, and added terms to the equation to account for them. Granted, the very concept of a "molecule" was

FIGURE 4.4 Johannes Diderik van der Waals (1837–1923). (Photo courtesy of Smithsonian Institution Libraries, Washington, DC.)

still in its infancy, a topic of hot debate. And van der Waals could not quite explain how these attractive forces worked, but his pioneering studies were nonetheless worthy of the 1910 Nobel Prize in Physics.

The unique set of forces that bear his name are better understood now. They are short-range in nature and lead to weak bonds among atoms and molecules. Although weaker than covalent, metallic, and ionic forces, van der Waals forces have a significant impact on the physical properties and behavior of all matter, and are especially relevant at the nanoscale.

Some people simply say "van der Waals forces" when referring to *all* the weak interaction forces between nonbonded molecules. There are three types of van der Waals forces, and, as you will see below, the many names given over the years to these three types seem to justify the practice of just calling them "van der Waals forces" for simplicity's sake. Keep in mind that, at any moment, more than one type of van der Waals force can be at work between atoms.

First, there is the *dipole–dipole force* (sometimes called the *orientation* or *Keesom* force). This force occurs between polar molecules such as HCl and H_2O. Polar molecules have an unequal distribution of electrons, making one side of the molecule more positive, the other more negative. When polar molecules are near one another, they favor an orientation where the positive portion of one molecule aligns with the negative portion of its neighbor, like a pair of bar magnets.

Such is the case for a particularly important type of dipole–dipole interaction known as hydrogen bonding. In the case of water, the attraction created by hydrogen bonding is what keeps H_2O in liquid form at more temperatures than any other molecule its size (instead of separating into individual vapor molecules). Since the hydrogen bonds require extra energy to break, water is a good place to store heat. Animals store excess heat in the water in their bodies. When the water has absorbed as much heat as possible, it evaporates away from the body and takes heat with it. Hydrogen bonding is what endows water with a high surface tension—the reason you can fill a glass with water slightly above the rim and it does not spill out. A network of hydrogen bonds enables water molecules to bind together as an ice crystal and is the reason why water is less dense as a solid than as a liquid, a rarity among substances. A hydrogen bond has only about 1/20 the strength of a covalent bond.

The next van der Waals force is the *dipole-induced–dipole force* (sometimes called the *induction* or *Debye* force). This weak force arises when a polar molecule brings about the polarization of a nonpolar molecule nearby. An example of a nonpolar molecule is the noble gas argon, which has a homogeneous distribution of electrons around its nucleus. When in close proximity to a polar molecule such as HCl, argon's electron distribution is distorted to one side of the nucleus. This is an induced dipole and it causes the molecules to be attracted toward each other.

Imagine two molecules, in this case two identical atoms, moving toward one another. One of them has a stronger dipolarity than the other at the moment they begin to interact. This is because the dipole is always fluctuating so its strength oscillates, even becoming neutral (with zero strength) once per oscillation. The atom with the stronger pole when the two atoms "meet" induces a synchronous, and opposite, fluctuating dipole in the other. The molecules are attracted by the dipole of their neighbor. *This tug is the dispersive force.*

4.2.4.1 The Dispersion Force

Like the force of gravity, the dispersion force (sometimes called the London dispersion force) acts on *all* atoms and molecules. This sets the dispersion force apart from solely electrostatic forces, which require charged or dipolar molecules. In fact, even nonpolar molecules can be made into dipoles, and this is the basis of the dispersion force. This force can act as a "long-range" force between molecules separated by up to 10 nm, but it also acts at interatomic spacings (~200 pm).

To understand how the dispersion force works, we will use the simplest kind of molecule there is: a single atom. We have learned that as electrons move around the atom's nucleus, they are like particles appearing here and there—on one side of the nucleus or another. Think about taking a snapshot of hydrogen: wherever the single electron is at that instant is, in essence, a very short-lived pole. Take another picture, the electron has moved and the pole has moved with it. The same is true of larger atoms and larger molecules. This rapidly fluctuating dipole, created by electrons as they move about the atom, creates momentary pockets of charge density called finite dipole moments.

When two molecules, in this case two atoms, come near one other, the one with the stronger fluctuating pole induces a synchronous, although opposite, fluctuating pole in the other. The molecules are attracted by the pole of their neighbor. *This tug is the dispersive force.* The larger the molecule, the larger the possible dipoles, and the larger the dispersive force. This is why larger molecules tend to have higher boiling points: because they can have larger fluctuating dipoles, larger molecules can be "stickier." In other words, they are more difficult to pry away from neighboring molecules into a vapor form.

4.2.4.2 Repulsive Forces

There is one final component of the van der Waals interaction between molecules; this component is repulsive. Repulsion dominates the interaction of molecules as soon as they are too close together—when their electronic orbitals begin to overlap. These repulsive forces are quantum mechanical in nature and referred to by many names—*exchange repulsion, hard core repulsion* (Figure 4.5), *steric repulsion,* and, in the case of ions, *born repulsion.*

FIGURE 4.5   Some "hard core" repulsion.

The Pauli Exclusion Principle governs this phenomenon. According to this principle, there can never be more than two electrons at a time in any one orbital. Put simplistically, this means that if two atoms or molecules get close enough to one another, their electron distributions can overlap only if the electrons are promoted to higher, unoccupied energy levels. Therefore, extra energy is necessary if two atoms or molecules are to be pushed any closer to each other than the equilibrium distance.

Repulsive forces play a role in how molecules pack together in solids—how closely and "comfortably" they can remain in proximity. If the molecules cannot pack together in a stable lattice, the melting point of the solid will be low. Meanwhile, higher melting points tend to correspond to better packing.

4.2.4.3 *The van der Waals Force versus Gravity*

Gravity makes home runs and slam dunks difficult. Gravitational forces give us the ocean tides and the orbits of the planets. However, the force of gravity at the nanoscale is negligible. It equals the acceleration of gravity ($g = 9.8$ m/s² on Earth) multiplied by the object's mass. At the nanoscale, the masses are so minute that other forces such as van der Waals forces become the dominant forces. This is among the more difficult concepts for us to grasp, subject as we are to Earth's constant tug on our bodies.

To get a feel for the relative magnitude of these forces, we start with the Lennard–Jones potential and model the interaction energy, E, of a pair of atoms separated by a distance x. This is a good way to approximate the van der Waals interaction. The Lennard–Jones potential equation is used often in simulating molecular dynamics. In the case of two of the same type of atom, it can be written as

$$E(x) = \varepsilon \left[\left(\frac{2r_{vdw}}{x} \right)^{12} - 2 \left(\frac{2r_{vdw}}{x} \right)^{6} \right] \tag{4.3}$$

In this equation, ε is known as the well depth and r_{vdw} as the van der Waals radius (or hard sphere radius). We can see both the attractive van der Waals interaction (which varies with the inverse-sixth power of the separation distance) and the repulsive interaction (which tends to vary by roughly the inverse-12th power of the distance). The parameters ε and r_{vdw} vary by atom. This relation is shown graphically in Figure 4.6. At large separation distances (not shown), the interaction forces fall off as the 7th (instead of the 6th) power of the distance.

BACK-OF-THE-ENVELOPE 4.2

How does the force of gravity on an atom compare to the van der Waals force?

To find out, let us evaluate a hydrogen atom in the presence of another hydrogen atom. For the hydrogen–hydrogen interaction, the well depth = 1.4×10^{-22} J, and the van der Waals radius $r_{vdw} = 0.12$ nm. Using these parameters in Equations 4.3 and 4.4, we can graph the interaction energy and the interaction force as shown in Figure 4.7. At the equilibrium separation distance $x = x_e$, the energy is at a minimum and the interaction force is zero. Unless additional energy is added to the system, the atoms will remain at this separation distance. Because force is the derivative of energy, we can determine where $E(x)$ is at a minimum by setting $F(x) = 0$ and solving for x_e. This calculation yields $x_e = 2r_{vdw} = 0.24$ nm. If we substitute this value into the equation for $E(x)$, we obtain $E_{min} = -1.4 \times 10^{-22}$ J, or an attractive energy of 0.14 zJ.

The maximum force between the atoms occurs when they are separated by $x = x_s$. To determine the value of x_s we first take the derivative of the force equation with respect to x:

$$\frac{dF}{dx} = 12\varepsilon \left[\frac{13(2r_{vdw})^{12}}{x^{14}} + \frac{7(2r_{vdw})^6}{x^8} \right]$$

Now, setting $dF/dx = 0$ and solving for x gives $x = x_s = 0.27$ nm. Substituting this value into the equation for $F(x_s)$, we get the maximum adhesion force between the two atoms:

$$F_{max} = -1.6 \times 10^{-12} \text{ N, or an attractive force of 1.6 pN}$$

How does this compare with the force of gravity, $F_g = mg$, where $g = 9.8$ m/s^2. The mass of hydrogen is 1 g/mol. Dividing this by Avogadro's number gives us the mass per atom: 1.66×10^{-24} g/atom. We can now calculate the force of gravity on a single hydrogen atom:

$$F_g = (1.66 \times 10^{-27} \text{ kg})(9.8 \text{ m/s}^2) = 1.6 \times 10^{-26} \text{ N}$$

The difference is tremendous. The van der Waals force tugging one hydrogen atom toward its neighbor is roughly *100 trillion times stronger* than the gravitational force tugging it toward the ground. Rollercoaster rides can put g-forces of 2–3 g's on riders; fighter pilots can handle up to about 9 g's for very short periods of time. Try and imagine 100,000,000,000,000 g's!

While this equation is an approximation, it correlates well with experimental measurements and can teach us a great deal about the forces acting between a pair of atoms. If we wish to talk in terms of interaction force, we take the negative derivative of Equation 4.3—just as we did with the Coulomb force in deriving Equation 4.2 earlier in the chapter. The force between the atoms is

$$F(x) = 12\varepsilon \left[\frac{(2r_{vdw})^{12}}{x^{13}} - \frac{(2r_{vdw})^6}{x^7} \right] \tag{4.4}$$

The van der Waals forces enable creatures such as geckos and spiders to cling to walls and ceilings. Such creatures have nanoscale hairs on their feet. The size and abundance of these hairs mean that a great deal of surface area of their bodies can make contact with a surface. The van der Waals forces from so many molecular interactions between the wall and the hairs hold the creature to the wall with orders of magnitude more force than that of gravity's relentless downward pull.

This is not to say that van der Waals forces do not affect us at larger scales also. It is this force that keeps us from falling through the floor or walking through walls. The atoms cannot go through one another; the only way to move through the wall or the floor is to move the atoms out of the way. This lends richer meaning to the act of touching something

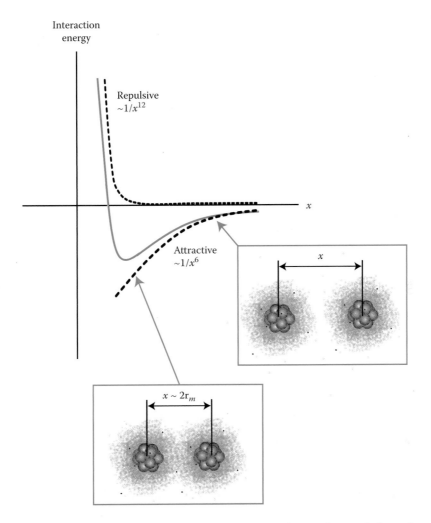

FIGURE 4.6 The van der Waals interaction. The graph shows the van der Waals forces between two molecules separated by a distance, x. The attractive forces, proportional to $1/x^6$, dominate at larger separation distances, but the repulsive force, which scales as $1/x^{12}$, dominates at closer range—separation distances of about twice the molecular radius, r_m, if both molecules are the same. The combination of these repulsive and attractive forces (shown as dotted lines) governs the van der Waals force interaction (thick gray line).

with your hand. Each van der Waals interaction starts off attractive, but then like a pair of extremely powerful opposing magnets cannot be brought completely together. *So, are you really touching this book, or are the atoms in your hand just very, very close to the atoms in the paper?*

4.3 CRYSTAL STRUCTURES

All substances form solids. And just about every solid has a crystal structure. (Plastic and glass are examples of noncrystalline solids.) Crystals are formed because of nature's intrinsic efficiency: energy is never wasted when a more efficient option exists. Atoms organize

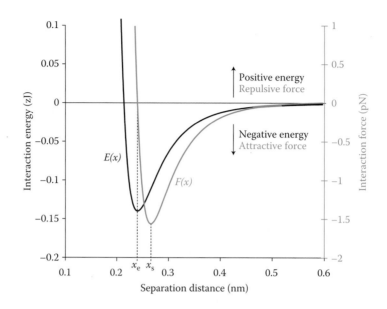

FIGURE 4.7 The interaction of a pair of hydrogen atoms. The interaction energy as a function of separation distance is graphed in black; the interaction force as a function of separation distance is in blue. The energy is at a minimum and the interaction force is zero at the equilibrium separation distance, $x = x_e$. The force curve is the derivative (slope) of the energy curve. See Back-of-the-Envelope 4.2.

themselves into crystals because energy can be minimized that way. In ionic crystals, the electrostatic energy of the ions is minimized. In covalent and metallic crystals, the act of sharing electrons minimizes their kinetic energy.

Crystals have pattern structures that are repeated over and over in all directions. The regular, periodic configuration of a crystal is represented by a lattice, which is a three-dimensional array of points. These "points" are simply representations of atoms, ions, or molecules. The smallest repeating unit of a lattice—the lowest common denominator of a crystal, so to speak—is the unit cell.

Unit cells fall into seven classes, which are the seven shapes a cell can have. These shapes are cubic, tetragonal, orthorhombic, monoclinic, hexagonal, triclinic, and trigonal. The atoms, ions, and molecules can be located on the corners of these shapes, in the centers, and on the faces. This makes for a total of four types of unit cell:

1. *Primitive* has a lattice point at each corner.

2. *Body-centered* has a lattice point at each corner plus one at the center of the cell.

3. *Face-centered* has a lattice point at each corner plus one at the center of each face.

4. *Face-centered (two faces)* has a lattice point at each corner, as well as in the centers of two opposing faces.

As observed in Table 4.2, there are a total of 14 possible combinations of the seven crystal classes and the four types of unit cell. These 14 types of lattice are called the

TABLE 4.2 The Bravais Lattices

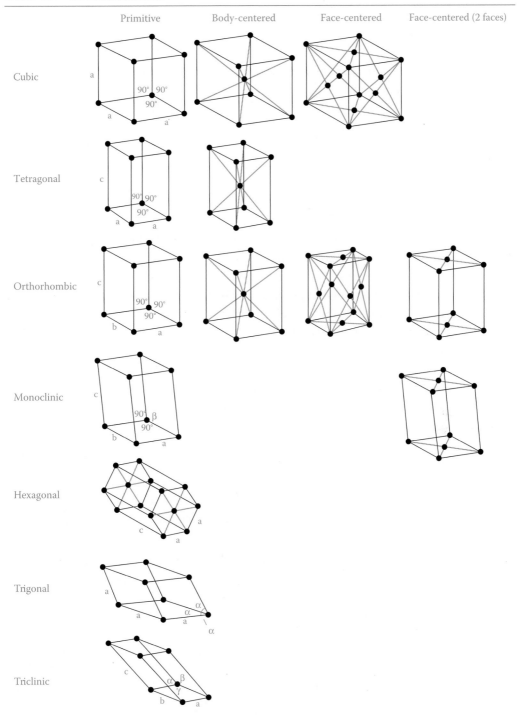

Note: All crystalline materials assume one of these 14 arrangements.

Bravais lattices. All crystalline materials, including nanomaterials, assume one of these 14 arrangements. Take, for example, the NaCl crystal from Figure 4.2. The NaCl unit cell is cubic. There is one face-centered array of Na^+ ions, and filling in all the gaps is another face-centered array of Cl^- ions.

Most bulk materials are not made entirely of a single crystal, but are actually many single crystals packed together. This makes most materials polycrystalline—an aggregate of single-crystal grains oriented in many directions. The size of these grains can be as small as a few nanometers wide or large enough to see with the naked eye. The regions between the grains are known as grain boundaries. Also keep in mind that most crystals are not perfectly complete; the lattice usually contains imperfections, defects, and vacancies.

4.4 STRUCTURES SMALL ENOUGH TO BE DIFFERENT (AND USEFUL)

What do we know about materials at this point? We know how they are held together in molecular form and in larger, more organized solid crystals. We know about van der Waals forces and their drastically heightened importance at the atomic and molecular scale. We know about the various forms crystals can assume. We have spent the time to gain an understanding of these fundamental properties of materials for one reason: to understand how these properties change as the amount of material diminishes, even all the way down to individual atoms, and how very small pieces of material acquire unique properties and, therefore, unique uses.

Our discussion will focus on stable structures with features so small they give materials unique and useful physical properties.

A lot of information is contained within this statement. Let us examine it piece by piece. *Stable structures* are not gases or liquids, lacking consistent organization. A stable structure is organized, typically as a solid or as a specialized type of molecule. Or even as an ordered array of just a few atoms. A stable structure can be synthetic or naturally occurring. The important thing is that the features of the structure are small (typically ranging from micrometers to picometers), and the very size of these features endows the structure with *unique and useful physical properties*—properties that do not exist at larger (or smaller) sizes.

A basketball-, a marble-, and a pollen-sized ball of gold vary only in size. They all melt at 1064°C. They are all of the color gold. However, a ball of gold just a few nanometers in diameter melts at about 750°C. And a liquid filled with these nanometer-sized balls turns the liquid red, not gold. Gold particles are just one of the many structures having unique and usable properties owing solely to their small size.

No matter how small a structure is, some of the things we have learned about materials are still the same. Covalent, ionic, metallic, and van der Waals bonds still hold everything together. (In both a small and a large piece of graphite, covalently bonded layers of carbon are held in a stack by weak van der Waals forces. See Figure 4.8.) Crystals still form with the usual 14 lattice structures, even when there are limited numbers of atoms; however, the lattice structure of a small piece of material can be different from the lattice structure of the bulk material.

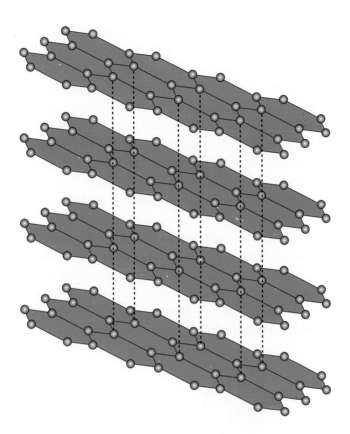

FIGURE 4.8 The structure of graphite. Weak van der Waals forces, represented by dashed lines, hold covalently bonded layers of carbon together.

4.4.1 Particles

Tiny particles of matter with diameters typically ranging from a few nanometers to a few hundred nanometers have distinctive properties. Such a particle can be made from a single crystal, an aggregate of crystals, or be completely noncrystalline; it can be a metal, a semiconductor, or an insulator. Particles with sizes larger than typical molecules but too small to be considered bulk solids can exhibit hybrid physical and chemical properties. The smaller a particle, the higher is its surface-to-volume ratio. In Chapter 2 we learned that the surface-to-volume ratio scales as $1/D$, where D is the characteristic dimension of the object (see Equation 2.3). This physical characteristic of particles can affect the way they behave with one another and with other substances.

In chemical and physical processes, only the surface of an object is exposed to the reaction and participates in the process. In a lump of sugar, a small percentage of the sugar molecules are on the surface, whereas powdered sugar has a high percentage. This is why a 1-g lump of sugar takes a long time to dissolve in water compared to the same mass of powdered sugar, which has more molecules in direct contact with the water.

While a bulk solid material will typically have less than 1% of its atoms on the surface, a particle can be so small as to have over 90% of its atoms on the surface. As more atoms are added, this percentage drops: a particle with a 10-nm diameter has about 15% surface atoms; a 50-nm particle has about 6% surface atoms. A few examples of particles with high surface-to-volume ratios are shown in Figure 4.9. The close packing of the atoms in these structures can be the result of either hexagonal or cubic crystal structures. As we see in Figure 4.10, there is a sharp decline in the percentage of surface atoms with increasing particle size.

For a given material, the crystal structure of a particle is not necessarily the same as the structure found in the bulk. Ruthenium particles 2–3 nm in diameter can have BCC and FCC structures not found in bulk ruthenium. Platinum and indium can also take on different crystal structures in particle form.

van der Waals forces dominate the interaction of particles, which behave almost like atoms.

The high surface-to-volume ratio of particles also makes them inherently more reactive as catalysts in chemical reactions, a property exploited by engineers to quicken commercial chemical production. Atoms at the corners and edges of the particles are even more reactive than those along surface planes. As we can see in Figure 4.9, smaller particles have more of their atoms at edges and corners than larger particles do. Gold is a useful catalyst in numerous chemical reactions, including oxidation processes crucial to production of

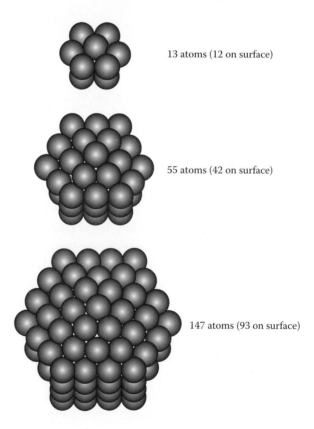

13 atoms (12 on surface)

55 atoms (42 on surface)

147 atoms (93 on surface)

FIGURE 4.9 Particles. The fewer the atoms in the particle, the higher the percentage of atoms on the surface, endowing such particles with heightened chemical reactivity and size-dependent melting properties.

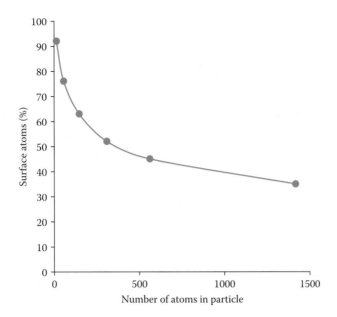

FIGURE 4.10 The percentage of surface atoms on a particle as a function of the total number of atoms. While not inclusive of *all* lattice types or particle sizes, the points plotted here are representative of particles in general and show the overall trend toward a higher percentage of surface atoms with smaller particles.

agrochemicals, pharmaceuticals, and other products. As part of one such process, gold particles 2–15 nm in diameter are small enough to squeeze into specific locations along certain long-chain hydrocarbon molecules and insert a single oxygen atom—making the process more efficient and eco-friendly because the harsh chemicals traditionally used, such as chlorine and organic peroxides, are no longer necessary.

BACK-OF-THE-ENVELOPE 4.3

One property significantly affected by decreasing the characteristic dimension, D, of a particle is the melting temperature, T_m. We can estimate the effect of particle size (assuming the particle is a sphere) using this scaling law equation:

$$T_m \simeq T_{m,bulk}\left(1 - \frac{1}{D}\right) \tag{4.5}$$

Here, $T_{m,bulk}$ is the melting temperature of the bulk material and D is the particle diameter in nanometers.

How does the melting temperature of a particle vary with its size? With its surface-to-volume ratio?

To find out, let us take a look at a gold particle. Bulk gold melts at 1337 K. If we use Equation 4.5 and graph the approximate melting temperature T_m versus particle diameter D with 1 nm $\leq D \leq$ 20 nm, we obtain the curve in Figure 4.11a. The surface-to-volume ratio of a spherical particle, where surface area is $4\pi(D/2)^2$ and volume is $(4/3)\pi(D/2)^3$, is graphed versus melting temperature in Figure 4.11b. As we can see, there is an inverse linear relationship in this case between the melting temperature and the surface-to-volume ratio. Atoms at the surface are more easily accessed and rearranged than atoms in the center of the particle and the melting process starts earlier. The melting temperature of particles is always lower than the bulk, as much as 1000 K in some cases.

There are hundreds if not thousands of methods used to make particles. For the most part, particles are created from vapor or from liquid. In the first case, atoms in vapor form are made to condensate and group together into particles. One simple way to do this is to evaporate a material at very high temperatures in a furnace until the vapor is supersaturated, and then cool it off. Particles form by condensation. In laser pyrolysis (from the Greek words for "fire" and "separate"), lasers tuned to the specific absorption energy of the vapor material cause the vapor to decompose into its constituents, which then condense into particles—such as when a laser breaks apart the $(Fe(CO)_5)$ molecule, causing a supersaturation of iron vapor that condenses to form tiny iron particles. Or, plasmas with temperatures of about 10,000°C can break materials into individual atoms, which upon removal from the plasma can be cooled until they agglomerate into particles. And yet another method is to bombard a solid with highly energetic, charged atoms of an inert gas like argon. This causes whole particles of atoms to eject from the solid's surface—a process known as sputtering. In spray pyrolysis, the desired material is dissolved into a solvent liquid and then sprayed as micron-sized aerosol droplets; processing steps cause the droplets to evaporate and leave behind a solid particle.

4.4.1.1 Colloidal Particles
Butter, milk, cream, blood, toothpaste, fog, smoke, smog, asphalt, ink, paint, glue, sea foam—all of these are types of colloids, which are mixtures of two phases of matter,

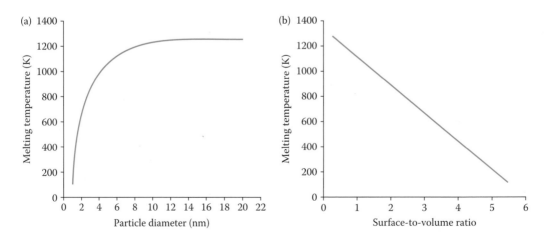

FIGURE 4.11 Melting temperatures of particles. (a) Estimated melting temperature, T_m, as a function of particle diameter, D; and (b) T_m as a function of a particle's surface-to-volume ratio.

typically solid particles dispersed in a liquid or a gas. The particles are known as colloidal particles; and when suspended in a liquid, the resultant mixture is known as a "sol." The particles tend to adsorb and scatter visible radiation (light), making the liquid appear turbid. They also impede the liquid's flow, making it more viscous. Through a series of steps, a sol can be aggregated from a liquid into a networked structure, or "gel." This sol–gel process is widely used to make small particles and materials containing them. Techniques for controlling the size and shape of colloidal particles, as well as preventing the particles from clumping into larger chunks, are centuries old and continue to be honed.

4.4.2 Wires

Wires are ubiquitous. They are the arteries of electricity, taking it across the country, throughout your city, within the walls of your house, into the labyrinth of electronic components in your computer. While getting electricity across great distances will remain a job for big wires, small wires enable the intricate network of transistors, diodes, logic gates, and interconnects necessary for faster, smaller, more powerful computers.

Wires can serve purposes beyond mere electricity transport—if small enough, they acquire new properties, turning ordinary wires from passive components into active ones—as in sensors, transistors, or optical devices. Wires with nanometer-scale diameters are ideal systems for investigating how size affects the electrical transport and mechanical properties of materials. Such wires are crystalline in structure and can be made from metals, semiconductors, superconductors, polymers, and insulating materials. They have large aspect ratios (length/diameter). Wires with 1–200-nm diameters and lengths of several tens of microns are common. Other wire-type structures such as belts and rods with different aspect ratios and cross-sections can also be made at this scale (see Figure 4.12).

Because wires, like particles, can achieve high surface-to-volume ratios at small dimensions, the physical properties of wires can vary, depending on their surface condition and geometrical configuration—so much so that two wires made of the same material can behave in completely different ways if the crystal structures, surface conditions, or aspect ratios are different. For example, a pair of gallium nitride wires, both less than 100 nm in diameter, will each emit radiation at different wavelengths if the crystal structures of the wires are different.

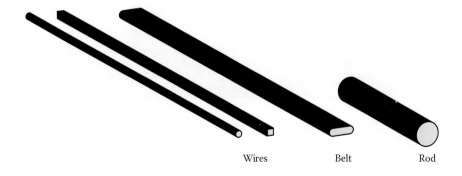

Wires   Belt   Rod

FIGURE 4.12   Three high-aspect-ratio structures.

To make wires this size, numerous methods have been used. One intuitive way is to use a template that has tiny holes in it. Anodic alumina (Al_2O_3) is commonly used as a template and has pore diameters of 10–200 nm. The holes are filled with the wire material either (1) in liquid form at high temperature and pressure, (2) in vapor form, or (3) by electrochemical deposition, where ions are pulled into the template using a voltage bias. The material then adopts the shape of the cavity in which it is deposited (see Figure 4.13). Often, additional wire material is added; as it rises (or "grows") out from the template, it maintains the templated shape, forming a wire.

Another way to make small wires is known as the vapor–liquid–solid technique. Silicon wires, among others, can be made using this process. Here is how a droplet of catalyst is put on a surface. In the case of silicon wires, the catalyst can be a gold nanoparticle or a thin gold film that then aggregates into nanoparticles upon heating. Silicon gas is introduced and it absorbs into the gold until the gold becomes supersaturated with silicon. At this point, a speck of silicon solidifies. This speck serves as a seed, the preferred site for continued deposition of silicon. The wire begins to grow. As it does, the molten gold bead stays on the top of the wire like a flower on a stem, collecting silicon until the desired wire length is achieved (see Figure 4.14). The longer the reaction is allowed to go on, the longer the wire.

There are many other techniques for making wires, but the process is usually similar to the two described here: material collects and "grows" as a stem from a seed, one atom or molecule at a time, into a wire. Wires can also be grown inside a flowing liquid in order to orient them in a particular direction.

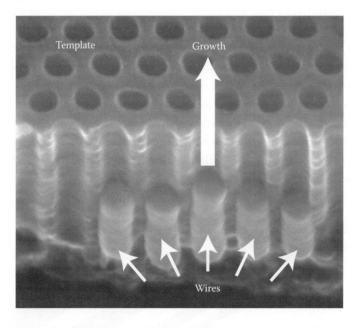

FIGURE 4.13   Cross-sectional view of a porous anodic aluminum oxide template used to grow wires. The template and substrate have been cleaved to view the cross-section of the pores from the side using a SEM. Each pore is approximately 50 nm in diameter and the five metallic cobalt wires shown are approximately 100 nm long (an aspect ratio of 2:1). Wires like this can be grown with aspect ratios exceeding 250:1 using thicker templates. (Photo courtesy of J. C. LaCombe and G. P. Sklar.)

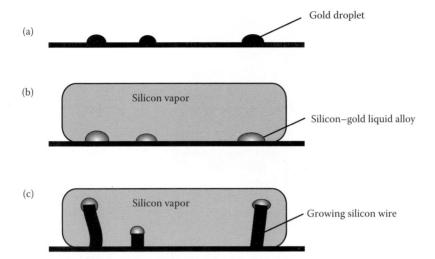

FIGURE 4.14 The vapor–liquid–solid method makes a silicon wire. (a) Beads of gold are deposited on a surface. (b) Silicon vapor absorbs into the beads, forming a liquid alloy. Once the bead is super-saturated, a speck of solid silicon forms inside it. (c) As more silicon vapor is absorbed, this seed grows into a wire, taking the alloy bead up with it.

4.4.3 Films, Layers, and Coatings

Beauty may or may not be skin deep. But usefulness can definitely be skin deep—even if the skin is only a few nanometers thick. Because so many physical interactions are really just surface interactions, the type of surface an object has can make it more useful or completely useless. Depositing a thin layer of material onto a product can modify and improve the product's performance—making it more efficient, less prone to friction, or more resilient to wear, as examples.

One novel, bottom-up approach to surface coating is self-assembly. In self-assembly, the individual units of the coating material—typically molecules or particles—are designed such that they template themselves and form a densely packed layer. This layer is often less than 1 μm thick. The process is practically autonomous—a matter of merely providing the right conditions for the units to organize and assemble on their own. Sometimes the coating consists of only a single layer of highly organized molecules. This type of coating is known as a self-assembled monolayer (SAM).

A SAM has three components, as shown in Figure 4.15. The head group forms a chemical bond that holds the molecules on the substrate; the backbone chain group serves as a spacer, defining the thickness of the layer; the tail group (sometimes called the surface terminal group), is the functional, outermost part of the layer. This makes the tail group the crucial component since it is what physically interacts with the environment and defines the surface's properties. Thiols, which are compounds of sulfur and hydrogen atoms (–SH), are frequently used as head groups on gold-coated surfaces because sulfur bonds so readily to gold; oxygen atoms can serve as the head group on silicon surfaces. Other head groups include silane (such as trichlorosilane, $-SiCl_3$) and carboxyl (–COOH). Backbones are often formed from an alkyl chain of carbon and hydrogen atoms (in the ratio C_nH_{2n+1}).

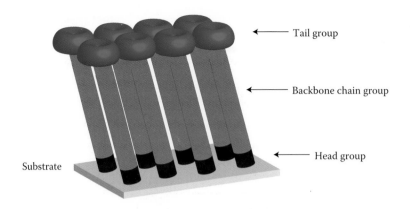

FIGURE 4.15   A self-assembled monolayer.

SAMs can "functionalize" a surface—making it chemically receptive to a specific type of molecule or reaction, or render the surface chemically inert, or alter the surface energy. They can make a surface hydrophobic, hydrophilic, adhesive, slippery, rough, electrically positive, or electrically negative.

Particles, too, can be dispersed as coatings using several techniques, including self-assembly, as well as chemical vapor deposition and electrodeposition. A thin layer of particles can significantly enhance an object's surface area, boosting the number of available interaction sites. A spray-on coating containing a photocatalyst and nanometer-sized titanium dioxide (TiO_2) particles can make a pane of glass "self-cleaning." The particles are so small they are transparent to visible radiation, and chemically fused to the glass. The coating changes the glass from hydrophobic to hydrophilic, so instead of beading up and running off in dirty streaks, water spreads out evenly in a sheet and washes away the dirt, which has been broken down by the photocatalyst using energy from the ultraviolet radiation. Your own skin can even benefit from a thin layer of TiO_2 particles. Although transparent, the particles absorb cancer-causing ultraviolet radiation, making them a great ingredient in sunscreens and cosmetics. (This will be discussed in Chapter 8, "Nanophotonics.")

Nanometer-thick metal layers are used in computer hard drives. These ubiquitous data storage devices take advantage of a phenomenon known as giant magnetoresistance (GMR), evident in alternating layers of magnetic and nonmagnetic metals. The application of a magnetic field causes a localized change in the electrical resistance that can be coded as a 1 or a 0. Another device for storing information is the digital video disc (DVD). While compact discs (CDs) use infrared lasers, DVDs use blue lasers since blue is a smaller wavelength for higher resolution and therefore higher density data storage. Newer Blu-ray discs are blue–violet lasers. Blue lasers can be made using layers of indium-doped gallium nitride just 3–4 nm thick. (These devices are known as quantum well lasers and are discussed in greater detail in Chapter 8.)

Flat-screen displays for computers, cell phones, cameras, and televisions also use thin films. Liquid–crystal displays can be made from an array of transistors printed on a film of organic material just 20 nm thick. Older displays used silicon layers 2 mm thick and were not as bright or as flexible. Another type of display uses organic light-emitting diodes

(OLEDs), where the light-emitting layer is a film of organic polymer flexible enough to be woven into fabrics and clothing. Imagine watching the news on your buddy's T-shirt!

As with particles, the crystal structure of a thin film can be different from the bulk material, often due to the added surface energy from so much extra surface area. Copper is one example. In bulk form, it is FCC, like the sodium chloride in Figure 4.2, but thin films of copper can be BCC.

4.4.4 Porous Materials

Sometimes it is the *absence* of material that gives a solid special uses—as is the case with materials having very small pores. Zeolites are the best known of these porous materials, with rigid crystal frameworks of pores and cavities among the lattice of atoms, like tunnels intersecting among a network of caves. The cavities range in width from 200 to 2000 pm and the pores from 300 to 1000 pm. The cavities can hold loose atoms or "guest molecules" such as water. Heating can remove the water; in fact, the Swedish mineralogist who first described zeolites as a mineral group in 1756 noticed that when he took a blowtorch to chunks of the material they hissed and boiled—he named them after the Greek words for boil (*zeo*) and stone (*lithos*).

There are 56 naturally occurring zeolites, all of which belong to a larger class of crystal structures known as aluminosilicates. Zeolite A is depicted in Figure 4.16. More than 150 other forms of zeolite have been man-made. Other aluminosilicates have also

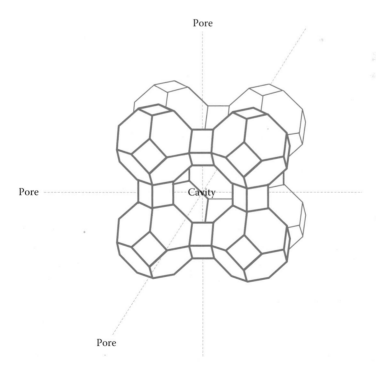

FIGURE 4.16   The zeolite A framework. The schematic shows the enclosed cavity and the pores leading into it. The formula of zeolite A is $Na_{12}[(SiO_2)_{12}(AlO_2)_{12}]27H_2O$ and the pores are roughly 400 pm wide.

been synthesized, most importantly those with larger pore sizes (1.5–10 nm), as well as other framework structures based on zeolites known as zeotypes. Certain types of naturally occurring clay have pore structures like zeolites, as does a recently discovered class of porous materials known as metal-organic frameworks (MOFs). The structures of MOFs are similar to zeolites, with interconnected networks of channels and large cavities that can create the largest known surface areas, by mass, of all ordered materials—up to 4500 m^2/g. That is close to the area of a football field in 1 g.

The uses of such high-surface-area materials range from improving sensors and catalysts to high-density storage of hydrogen. Take for example a square patch of silicon. If the area of the patch is 100 m^2, that is all the active area we have to work with—to capture specific molecules or catalyze reactions or any other application. But if we spread a fraction of a gram of highly porous material such as a metal-organic framework (MOF) onto the patch, we boost the active surface area to 1000 m^2. That is a 10 million-fold improvement.

BACK-OF-THE-ENVELOPE 4.4

A single gram of highly porous, low-density MOF material can have a surface area of 4500 m^2. How does this value compare with the surface area of a 1-g cube of gold? If you could flatten out the gold into a sheet, how thin must it be to achieve the same surface area as the MOF?

The density of gold is 19300 kg/m^3, so a 1-g piece has a volume of 5.18×10^{-8} m^3. A cube with this volume has sides measuring $(5.18 \times 10^{-8})^{1/3} = 0.0038$ m. The surface area of this cube is therefore $6 \times 0.0038^2 = 8.7 \times 10^{-5}$ m^2. That is about 51 million times less surface area.

If we flatten out the gram of gold into a sheet so that the area of both sides together is 4500 m^2, the thickness, t, would have to be such that $(2250$ $m^2)t = 5.18 \times 10^{-8}$ m^3. (We assume that the surface areas of the sides of the sheet are negligible.) Therefore $t = 230$ picometers (pm). That is practically atomic thinness, as the diameter of a single gold atom is approximately 180 pm.

The unique porosity of these materials can make them useful in water softening. Hard water is full of positively charged magnesium and calcium atoms. These atoms can collect on the inside of pipes, clogging them and reducing heat conduction through them, as well as react with soap to form a sticky scum instead of a lather. Zeolites used in detergents can soften water by capturing the magnesium and calcium atoms in their pores. It is this application that has driven the development of the majority of synthetic zeolites.

Zeolites can also serve as sieves on a molecular scale, the ultrafine mesh created by the many channels granting passage to only those molecules having specific sizes or shapes. As such, chemical processes requiring separation or purification often employ zeolites. Similar to particles with high surface areas, zeolites make efficient catalysts, their porous structure creating direct surface-to-surface contact with up to 100 times more molecules than a traditional "chunk" of catalyst material.

When used in tandem, the *filtering* and *catalysis* functions of zeolites enable specialized processes not possible with conventional materials. In *reactant shape-selective catalysis*, only molecules of given dimensions enter the pores and reach the catalytic sites; in *product shape-selective catalysis*, only products of given dimensions can squeeze out of the catalysis sites and move on to the next process step; and in *transition-state shape-selective catalysis*, certain reactions can be prevented inside the cavities simply because the products of such reactions would be too large to fit.

4.4.5 Small-Grained Materials

As we learned in Section 4.3, bulk materials are rarely one gigantic, single crystal. They are more often polycrystalline, made from *grains*, where each grain is a separate crystal. Grains are oriented in all directions, and the regions between them are known as grain boundaries. Materials with small grain sizes appear in nature, and we are also able to make them artificially. One method involves compacting and consolidating particles at high pressures and temperatures until they have melded together as a solid. In this case, the particles become the grains. If the particles are 100 nm in diameter, the grains of the new material will be roughly that size also.

Grain size makes a big difference in the hardness and stretchiness of a material and the amount of stress it can withstand. Why is this? And do smaller grains make for stronger materials? First, let us remind ourselves how materials behave when we pull on them.

A rubber band, a copper rod, or a bone in your body—no matter what it is, when we tug on it, it stretches. Just about everything is somewhat elastic. We can actually see the rubber band stretching, but the copper rod also stretches, infinitesimally, as the atomic bonds stretch. Once the force is removed, the material springs back to its original size and shape. But if we do not remove the force and instead keep pulling, the material eventually exceeds its elastic limit. It starts to permanently elongate, a phenomena known as plastic deformation. Plastic deformation involves the breaking of atomic bonds and the making of new bonds in new locations, which is why the material does not spring back to its original shape. If the material is brittle, it will not elongate much at all. Instead, it cracks. If ductile, the material will stretch out and "neck" like a piece of gum that gets skinnier and skinnier as you pull it apart until finally it breaks.

During plastic deformation, dislocations occur within the crystal lattice of atoms. The dislocations are places where one section of the lattice is able to slip across another section. Figure 4.17 illustrates a dislocation. Grain 1 and Grain 2 are adjacent grains with different orientations. As the applied force leads to dislocation in Grain 1, we see the bonds break and the top section of the lattice start to slide across the bottom section. When this slipping section of lattice reaches the grain boundary, it halts. Because the alignment of the atoms is different in Grain 2, the slip plane is not continuous.

As the grains in a material get smaller, so too does the ease with which the atoms can slip. There are more grain boundaries to prevent slippage. Yield stress, σ_y, is the amount of stress a material can withstand before plastic (permanent) deformation begins. The Hall–Petch equation relates a material's yield stress to its grain diameter:

$$\sigma_y = \sigma_o + \frac{k}{\sqrt{d}} \tag{4.6}$$

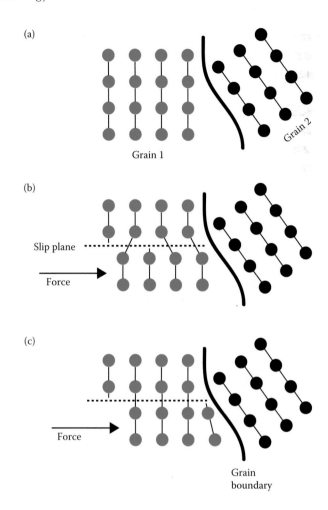

FIGURE 4.17 An edge dislocation. During plastic deformation, sections of the crystal lattice slip across one another. In (a), Grain 1 and Grain 2 are adjacent grains with different orientations. In (b), an applied force causes a dislocation in Grain 1. Bonds rupture as the top section of the lattice starts to slide. In (c), the slipping section of lattice reaches the grain boundary; but because the slip plane does not continue into the differently oriented Grain 2, the slipping stops.

Here, σ_o is the frictional stress (N/m²) that opposes dislocation, k is a constant, and d is the grain diameter in micrometers. Small grains make for harder, stronger materials that can withstand higher forces before plastic deformation begins. We can see the exponential increase in yield strength at small grain sizes in the graph in Figure 4.18.

The Hall–Petch equation holds true for grain sizes ranging from 100 nm to 10 μm. A material's elastic modulus (Young's modulus), E, which is the force needed to elongate it, does not vary much over this range of grain sizes.

However, at the smallest of grain sizes—less than about 30 nm—the Hall–Petch equation is no longer accurate. Materials with grains this minute lose yield strength. Plastic deformation starts to occur at lower and lower stresses as the grains shrink. The elastic modulus of materials with grains smaller than 5 nm also suddenly drops off. When the

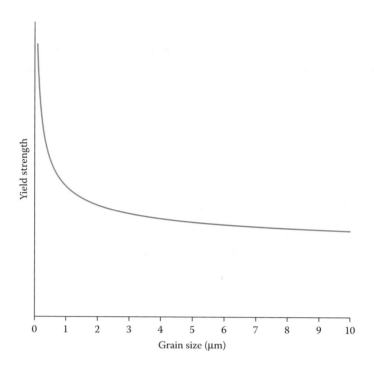

FIGURE 4.18 The Hall–Petch equation relates a material's grain size to its yield strength. The relation holds true for grain sizes from 10 μm down to 100 nm.

grains are that small, dislocations like the one shown in Figure 4.17 are not very likely to occur at all. Instead of behaving like an agglomeration of *grains*, materials behave more like a dense web of *grain boundaries*. There simply is no room for the lattices to slip across one another because they collide with a grain boundary as soon as they budge. Lattice dislocation therefore cannot be the way very small-grained materials deform. Instead, it is believed that individual atoms migrate, diffuse, and slide along grain boundaries and through triple junctions (Y-shaped grain boundary intersections).

We learned earlier in this chapter how the melting temperature of particles decreases when the particles are very small. Well, grains are analogous to particles (especially when we fabricate solids out of particles). The melting temperature of a solid decreases with grain size. While the temperature dependence of σ_o and k in the Hall–Petch equation can be neglected for conventional grain sizes, the effect of melting temperature can be considered when the grains are smaller than 30 nm. Materials can soften up, in effect, when their melting temperature drops due to smaller grain size.

It has proven difficult to draw conclusions about just how ductile small-grained materials are because testing methods vary and experimental samples are prone to miniscule flaws and pores that can alter experimental results. Still, it is generally agreed that metals with grain sizes less than 30 nm are brittle and do not elongate much during plastic deformation. Consider copper. The conventional coarse-grained, polycrystalline variety will elongate up to 60% before it cracks, whereas copper with grain sizes less than 30 nm elongates 5% at most.

4.4.6 Molecules

"Today's manufacturing methods are very crude at the molecular level," says Ralph Merkle, a longtime and outspoken proponent of nanotechnology. "Casting, grinding, milling and even lithography move atoms in great thundering statistical herds. It's like trying to make things out of LEGO® blocks with boxing gloves on your hands. Yes, you can push the LEGO blocks into great heaps and pile them up, but you can't really snap them together the way you'd like" (http://www.foresight.org/GrandPrize.2.html). Merkle and others see a day in the future when we might gain unprecedented positional control over atoms such that we can chemically or mechanically, or even biologically, assemble them as we please. This paradigm shift from top-down to bottom-up engineering is a central tenet of what is sometimes referred to as "molecular nanotechnology."

The material properties of certain specialized molecules are useful to us and justify their inclusion in our discussion of nanomaterials. In molecular machines and molecular electronics, an individual molecule *is* a device. A single molecule is the material used. The molecules covered here can be incorporated into larger devices but what sets them apart is that they can be useful on an individual basis.

4.4.6.1 Carbon Fullerenes and Nanotubes

Carbon is found in every corner of the universe. Here on Earth, every single living thing contains carbon atoms. This particular atom forms a staggering variety of complex molecules, especially with hydrogen, oxygen, and nitrogen—millions and millions of organic compounds. There are only three crystalline forms of pure carbon. Two are well known: (1) diamond, one of the hardest known materials, and (2) graphite, one of the softest known materials. Carbon atoms form very strong covalent bonds with other carbon atoms, which explains diamond's strength. As we saw in Figure 4.8, graphite has covalently bonded carbon also, in layers, but it is the weak van der Waals bonds holding the layers together that make graphite soft.

4.4.6.1.1 Carbon Fullerenes Fullerenes are the third crystalline form of carbon. Although fullerenes exist in interstellar dust and even can be found in geological formations on Earth, no one had ever seen one until 1985, when they were finally found, in Texas of all places. (See "The Discovery of C_{60}.")

THE DISCOVERY OF C_{60}: HOW ONE BEER AND 32 PIECES OF PAPER HELPED WIN THE NOBEL PRIZE

On its way toward Earth, some of the light from the stars is absorbed by interstellar dust. By examining the light that *does* make it to Earth, we have clues about what the dust is made of because different materials absorb different wavelengths. One of the absorbed wavelengths is 220 nm, in the ultraviolet region of the electromagnetic spectrum. This wavelength piqued scientists' interest. It might have been due to graphite particles—or maybe some other type of carbon.

It was known already that there were long linear carbon chains floating around in space between the stars. Harold W. Kroto, an English chemist from the University of Sussex, thought these snake-like carbon molecules might be born of carbon-rich red giant stars. In 1985, he contacted Richard Smalley at Rice University in Texas. Smalley had a machine. It used an intense laser pulse to generate temperatures of tens of thousands of degrees—hot enough to vaporize anything. The machine could simulate the conditions in red giants and maybe recreate the illusive carbon snakes. Smalley and his colleague from Rice, Robert Curl, decided to help Kroto give it a try.

With help from students, these men superheated carbon vapor and cooled it into small clusters that were sent into a mass spectrometer for analysis. The measured masses showed clusters made of between two and 30 carbon atoms—strong evidence for long carbon chains with unpaired bonds at either end. These were what Kroto wanted to see. However, there was another peak on the graph that demanded curiosity: a 60-carbon cluster that kept showing up, again and again.

Why 60? If bonded together in the most energetically favorable way, what form would 60 carbon atoms take? They imagined two possibilities: (1) a flat sheet of graphite like chicken wire cut from a sheet (see Figure 4.8) or (2) some type of spherical, dome shape.

Smalley checked out a book about architect Buckminster Fuller, the architect who had popularized the geodesic dome, in order to see a photograph of one. It was a clue. Then later that night, Smalley tried modeling such a structure on his computer. No luck. So he went low-tech. He cut hexagons out of a pad of legal paper. Each of the six edges represented carbon bonds and each of the six points a carbon atom. He connected the edges of the hexagons in an effort to build a large sphere. He could not. At midnight, he considered going to bed but had a beer instead. He recalled Kroto saying he had made a geodesic dome with his kids, and that it might have included pentagons as well as hexagons. Maybe.

Smalley cut out a pentagon and gave it a try. He attached hexagons to the five edges of the pentagon. It formed a bowl. He kept adding hexagons and pentagons until he had made a hemisphere, then a complete sphere. Theoretical chemists had as early as 1966 postulated a third form of carbon, other than graphite and diamond, that might just be spherical. But there was no real evidence. Now Smalley had a sphere in his hands. He had made it using 12 pentagons and 20 hexagons, simulating realistic carbon bonds. In 1991, after he shared the Nobel Prize with Kroto and Curl for the discovery of C_{60}, Smalley wrote about his late-night epiphany:

> "Even though I had put the thing together with flimsy paper and transparent tape, it seemed beautifully symmetrical and quite sturdy. When I dropped it on the floor, it actually bounced. Surely this had to be the correct shape. Here was a way...that explained the magic of sixty." (Great Balls of Carbon: The Story of Buckminsterfullerene, by Richard Smalley, *The Sciences*, March/April 1991, pp. 22–28.)

They named the molecule the buckminsterfullerene in honor of the imaginative architect. Today, C_{60} is often called a "buckyball." The various other geodesic carbon clusters are known as "fullerenes"—the third form of carbon. Smalley liked the form so much that he renovated his house to include a skylight shaped like a half-buckyball, the glass framed in steel struts representing the atomic bonds.

Fullerenes are hollow and made entirely of carbon. They can be spherical, elliptical, or tubular. (The tubes are probably the most widely known and intensely studied material in nanotechnology and are discussed in the next section.) A feature common to fullerenes' numerous intriguing molecular shapes is that every atom is a surface atom. This makes

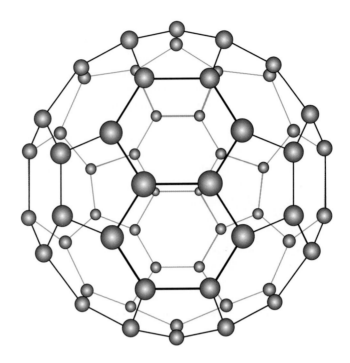

FIGURE 4.19 The C_{60} fullerene. A sphere, this unique molecule resembles a soccer ball in the arrangement of hexagonal and pentagonal carbon bonds.

fullerenes unique. The most common form is called C_{60}. It has 60 carbon atoms bonded together just like the hexagons and pentagons on a soccer ball. It is roughly 700 pm in diameter (see Figure 4.19). Larger fullerenes have been found or synthesized, including C_{70}, C_{76}, C_{80}, C_{84}, C_{90}, C_{96}, ... even up to C_{540}.

The "cage" structure of fullerenes has potential for use as a type of molecular storage enclosure for other types of substances, for example, drugs that could be injected into the body inside fullerenes and later released at a specific site and time. Groups of the C_{60} type can, like gigantic atoms, take on FCC crystal structures held together by van der Waals forces. If potassium vapor is diffused into such a crystal, the potassium atoms take up the empty spaces between the fullerenes and form a compound, K_3C_{60}. Although C_{60} itself is an insulator, K_3C_{60} is conductive—in fact, at very low temperatures (below 18 K), the electrical resistance across the crystal becomes zero: it is a superconductor. Also, due to its bond structure, C_{60} is unlikely to react with other substances, making it a possible lubricant or the smallest ball bearing ever made.

4.4.6.1.2 Carbon Nanotubes   As molecules go, nanotubes are rock stars. Or maybe super-heroes. It is difficult to find an article about nanotechnology that does not tout these simple tubes of carbon and all their mighty powers. Technically, nanotubes are cylindrical fuller-enes, but the words "nano" and "tube" do well to sum them up (see Figure 4.20). They are hollow, roughly 1.5 nm in diameter, and typically a few hundred nanometers long. However, nanotubes can also be hundreds of microns long—an astounding aspect ratio. Because of

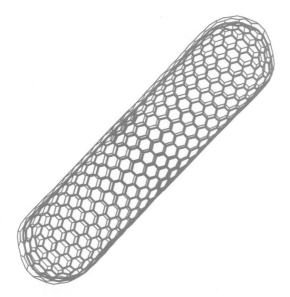

FIGURE 4.20 A single wall nanotube. Each hexagon represents the bonds among six carbon atoms.

the strong carbon bonding that holds them together, these tubes tend not to bind with other atoms: they are chemically stable and inert.

Nanotubes resemble rolled-up chicken wire. Imagine a sheet of graphite like the one in Figure 4.8. While tubes are not actually formed by rolling up graphite sheets, their structures can be more readily understood if we think of it that way. This idea is shown in Figure 4.21. Depending on how the sheet is rolled, the tube can take on three possible structures, or chiralities: (1) armchair, (2) zigzag, or (3) chiral. The chirality of a tube determines some of its properties—most notably, whether it will conduct electricity like a metal or like a semiconductor. The armchair type behaves like a metal, while certain varieties of the zigzag and chiral types can behave like semiconductors.

The conductivity of metallic tubes can be more than 50 times higher than that of copper, making them good nanoscale wires, and the semiconducting type may soon find use as transistors in the ever-shrinking circuitry of computers. Some of the other astounding and useful properties of nanotubes are their high thermal conductivity—similar to that of diamond, one of the best known thermal conductors—and their strength. The elastic modulus of single wall tubes is about 1000–3000 GPa. These values are the highest ever measured for any material and are about 10 times higher than that of steel, although the tubes weigh six times less. Tubes are also incredibly flexible and resilient, allowing them to bend under high stress and spring back to their original shape. Tennis rackets, bicycle components, golf clubs, and car bumpers are among the first products to contain nanotubes. They have also been used in the fuel lines of cars to carry away spare electrons in the line and prevent the buildup of static electricity, which might ignite the fuel.

Although chemically inert, nanotubes can adsorb up to 100 times their own volume in hydrogen atoms and thus could be developed as a safe place to store hydrogen for fuel cells.

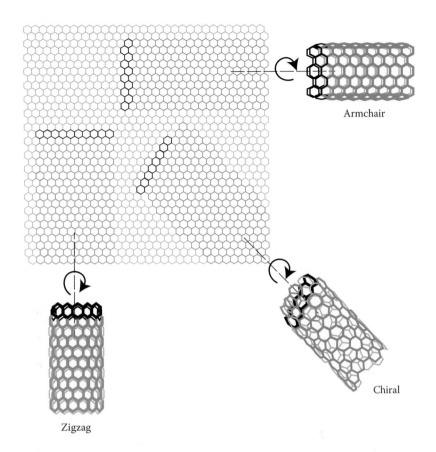

Armchair

Chiral

Zigzag

FIGURE 4.21 Nanotube chirality. The three types of tube can be understood by visualizing a rolled-up sheet of carbon atoms. Depending on how the sheet is rolled up, the tube will be an armchair type, zigzag type, or chiral type, as shown.

The tubes can also form coaxially, one inside another, like Russian dolls. This type of tube is called a multiwall nanotube (as opposed to a single wall nanotube) and can be tens of nanometers in diameter, made from numerous tubes. Both the single wall and multiwall types occur commonly.

How do we make nanotubes? Usually in one of four ways: (1) arc discharge, (2) laser ablation, (3) catalyzed decomposition, or (4) chemical vapor deposition. In arc discharge, a DC voltage is applied between a pair of electrodes made of graphite. High voltages cause an arc of electricity between the electrodes, which begins to vaporize the graphite. The evaporated atoms redeposit elsewhere—as plain carbon, as particles, as fullerenes, and as tubes. In laser ablation, a graphite target is vaporized by a laser instead, and the evaporated carbon reforms in tubular form. Catalyzed decomposition is essentially a matter of heating a carbon-containing gas to 900–1200°C, causing it to decompose and reform as tubes. The same is true of chemical vapor deposition, although this process often entails the growth of tubes on a surface. Of course, there are innumerable subtleties and complexities to these growth methods, more so than can be related here. It suffices to say, for our purposes, that no matter which growth method is used, there are two essential components: (1) vaporized

carbon, and (2) some type of catalyst site. Particles 1–5 nm in diameter are usually the catalyst, providing a location where tube nucleation can begin. The manner in which the carbon atoms *use* the catalyst and bond together into a tubular shape is still a matter of intense speculation and study.

Nanotubes have also been made from other materials, including metals such as nickel, copper, and iron, oxides such as TiO_2 and ZnO, nitrides such as BN and GaN, and numerous other inorganic molecules.

4.4.6.2 Dendrimers

Dendrimers have a branched structure that earned them their name, taken from the Greek word *dendra*, meaning "tree." These networked molecules are constructed step-by-step, with each step known as a "generation." Two strategies can be used, as shown in Figure 4.22. In divergent synthesis, molecular branches are added outward from a core molecule.

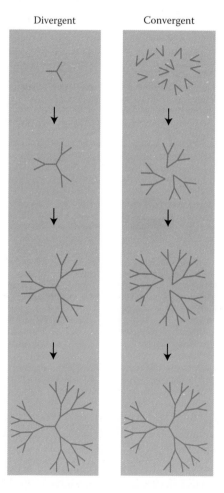

FIGURE 4.22 Divergent and convergent synthesis of a dendrimer. In this example, both techniques require three generations. The resulting dendrimer has trifunctional branching—three branches coming together at every branching point.

In convergent synthesis, premade molecular "wedges" are coupled together and eventually bonded to a common core. Typical dendrimers are held together by covalent bonds. They usually measure less than 5 nm across, although bigger, heavier varieties measuring hundreds of nanometers have also be made. These larger dendrimers are significantly more complex since they are the result of numerous generations and tend to be spherical.

We learned earlier about how the outermost layer of atoms in a SAM is the functional group that gives the layer its properties. In dendrimers, the outer branches are called end groups and the nature of the atoms and molecules within these end groups give the dendrimer its "personality." Characteristics such as chemical reactivity, stability, solubility, and toxicity are determined by the end groups. This means that by selecting the right end groups, dendrimers can be tailored for specific uses. Often, dendrimers trap other "guest" molecules within the gaps between their many branches, which is useful for storage of unstable molecules or for trapping and removing harmful molecules from contaminated water. The potential for dendrimers in commercial applications is broad and includes uses in coatings, adhesives, catalysts, and sensors.

4.4.6.3 Micelles

Micelles are self-assembling structures made from a certain type of molecule called a surfactant. Surfactants are characterized by a hydrophobic ("water fearing") tail and a hydrophilic ("water loving") head, as depicted in Figure 4.23a. The hydrophilic part is polar and the hydrophobic part is a hydrocarbon chain. In water, the tails of such molecules bunch together so as to minimize contact with the water, leaving their heads facing the water. This self-assembling behavior gives rise to a number of forms, three of which are

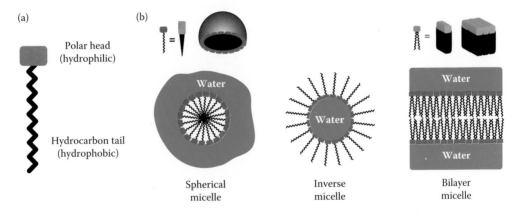

FIGURE 4.23 Micelle varieties. A micelle is made from a certain type of molecule, called a surfactant, shown in (a), which has a hydrophobic tail attached to a polar head. When a number of such molecules are in close proximity, they arrange themselves so as to keep their "heads in the water." Three varieties of micelle are shown in (b). With a single hydrocarbon tail, the surfactant is wedge shaped and tends to form spherical shapes with other wedge-shaped surfactants, similar to pieces of a pie. With a double hydrocarbon tail, the surfactant is similar in girth from head to tail, and forms layers. The bilayer variety is a critical component of biological membranes, such as cell membranes (discussed in Chapter 10).

shown in Figure 4.23b. Cylindrical micelles are also possible. You may notice a similarity between the spherical micelle and the dendrimers discussed in the prior section; indeed, some micelles are dendrimers.

When surfactant molecules are present in water above a certain concentration, known as the critical micelle concentration (CMC), micelles 2–10 nm in diameter spontaneously take form. The critical concentration is quite low; as an example, a surfactant with a hydrocarbon chain with 16 carbon atoms in it will have a CMC in the picomolar range (one part surfactant to one trillion parts water). Depending on the chemical properties of the micelle and the concentration of the solution, a given micelle can remain stable for as long as a few months or as short as a few microseconds.

One medical application that has been proposed for micelles involves attaching medicine to the hydrophobic cores of spherical micelles and then injecting them into the bloodstream, where the micelles would later dissociate and release their therapeutic cargo. The bilayer variety already serves a very important and ubiquitous role as the main component of biological membranes, like cell membranes. This is discussed in greater detail in Chapter 10, "Nanobiotechnology." Because micelles can incorporate otherwise insoluble molecules into their "dry" core, they help us digest fats, cholesterol, and many vitamins—all of which are hydrophobic—by making them soluble in water and therefore able to move through our digestive system. Similarly, micelles can be used in soaps to capture insoluble organic compounds from surfaces and allow them to be washed away in water.

4.5 SUMMARY

Be it a particle, a wire, a porous mesh, or a film; be it manmade or biological; be it a single molecule—a nanomaterial differs from a conventional polycrystalline material not only because of the size of its structures, but also in the way we can use it. The electronic, optical, magnetic, chemical, and mechanical properties are substantially affected by the scale of a material's features. Consider the nanotube. It is made from all carbon atoms, just like graphite. In fact, it is essentially a rolled-up tube of graphite, yet it behaves almost nothing like graphite. This is because of its unique, nanometer-scale structure.

Often, smaller grain sizes (and the correspondingly larger fraction of grain boundaries) endow nanomaterials with new mechanical characteristics. Also, nanomaterials usually have more surface area by volume than bulk materials. With more atoms exposed, surface effects tend to dominate nanomaterials' interactions.

Fundamental material properties that were long considered constants (melting temperature, electrical conductivity, ductility, etc.) are suddenly subject to manipulation. We can tailor these properties by engineering the dimensions of a material's features.

Among the more important developments in our growing dominion over materials is our ability to confine electrons such that their intrinsic, quantum properties manifest, and become useable. By confining electrons inside a nanometer-sized structure, we have access to the quantum traits discussed in Chapter 3—and powerful means of controlling the electrical, optical, magnetic, and thermodynamic behaviors of solid materials. The usefulness of quantum mechanics in some of the nanomaterials discussed in this chapter will remain a focus as we continue our study of nanotechnology.

HOMEWORK EXERCISES

4.1 How is "material" different from "matter?"

4.2 True or false? All substances form solids.

4.3 Interatomic forces are usually which type of force?

4.4 True or false? On a plot of the potential energy of two atoms versus their separation distance, the lowest energy point on the curve is the separation distance where attractive and repulsive forces are equal.

4.5 True or false? The radius of an ion is roughly equivalent to that of the uncharged atom. Explain your answer.

4.6 The forces of covalent bonds usually act over what range of interatomic separations?

4.7 True or false? A single water molecule is formed by van der Waals bonding, while covalent bonding holds numerous water molecules together.

4.8 True or false? The dispersion force can cause a nonpolar molecule to behave like a dipolar molecule.

4.9 Explain why smaller molecules tend to have lower boiling points.

4.10 Rank the following bonds from strongest to weakest and provide the bond energy: the bond between hydrogen and oxygen in a water molecule; the bond between sodium and chloride in the NaCl molecule; the bond between atoms in a metal; the van der Waals bond between adjacent hydrogen atoms.

4.11 In Back-of-the-Envelope 4.2, we determined the van der Waals attractive force between a pair of hydrogen atoms. Of course, in practice we more often deal with devices and objects, and van der Waals forces often lead to parts sticking together. The attractive energy between a spherical body of radius R and a flat surface, separated by a distance x (see Figure 4.24) is given by

$$E(x)_{\text{sphere-surface}} = -\frac{HR}{6x} \qquad (4.7)$$

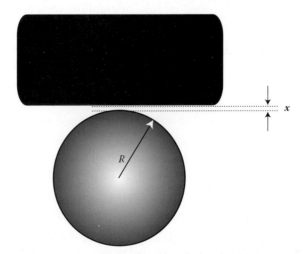

FIGURE 4.24 A spherical body of radius, R, separated from a surface by a distance, x. (See Homework Exercise 4.11.)

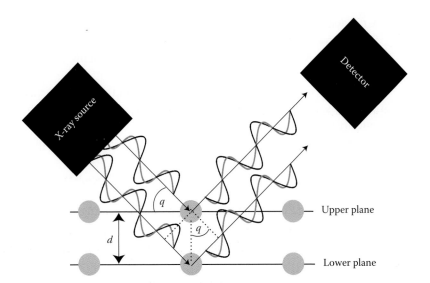

FIGURE 4.25 X-rays reflected by the top two planes of a crystal. The gray dots represent atoms in the crystal lattice. The planes are parallel. (See Homework Exercise 4.13.)

In this equation, H is the so-called Hamaker constant. In the case of solids separated by air, $H \approx 10^{-19}$ J.

a. Determine the equation for the attractive force with respect to x.

b. Now determine the force of gravity pulling the sphere down, away from the surface, as a function of R. Assume that the particle is made of silicon, with a density of 2330 kg/m³.

c. How large can the sphere be before gravity pulls it away from the surface if they are separated by 10 nm of air? By 1 nm?

4.12 The spacing between atoms in a crystal is about 100 pm. What forms of electromagnetic radiation have wavelengths short enough to fit between the atoms in the crystal?

4.13 A given crystal structure (such as NaCl) can be represented as consisting of planes of atoms, as shown in Figure 4.25. A beam of x-rays can be reflected off the crystal, where some of the beam penetrates through the atoms of the upper layer and strikes the atoms in the lower plane. A pair of incident x-rays from an x-ray source is reflected from the crystal as shown and into an x-ray detector.

a. How much farther does the beam reflected from the lower plane travel from the source to the detector than the one reflected from the upper plane?

b. If the beams have wavelength, λ, under what conditions will the two reflected beams constructively interfere with each other (i.e., have maxima at the same points)?

c. The relationship you derived in part (b) was first derived by W.L. Bragg (1890–1971). If we can experimentally measure the diffraction angle and we know the wavelength used, what feature of the crystal can we use this relationship to determine?

d. X-rays with a wavelength of 140 pm are reflected from an NaCl crystal and are found to constructively interfere at an angle of incidence of 14.4°. Calculate d.

4.14 True or false? Adjacent atoms within a solid each contain electrons with identical amounts of energy.

4.15 The properties of particles are largely dependent on their surface-to-volume ratio. With so many of their atoms on their surface, particles can be quite chemically reactive.

 a. Derive the equation for surface-to-volume ratio as a function of particle radius (treating the particle as a sphere).

 b. Graph the surface-to-volume ratio for radii of 1–100 nm. Use a logarithmic scale on the y-axis.

4.16 The following equation has been used to predict the melting temperature, T_m, of a spherical particle of radius, r:

$$\left(1 - \frac{T_m}{T_{m,\text{bulk}}}\right) = \frac{2}{\rho_s L r}\left[\gamma_s - \gamma_l\left(\frac{\rho_s}{\rho_l}\right)^{2/3}\right]$$ (4.8)

Here, $T_{m,\text{bulk}}$ is the bulk melting temperature, ρ_s and ρ_l are the densities of the solid and liquid phases, respectively; γ_s and γ_l are the surface tensions of the solid and liquid phases, respectively; and L is the latent heat of the material.

 a. On a single graph, plot the melting temperature of gold particles according to the scaling law in Equation 4.5 and according to Equation 4.8 for particle radii from 1 to 10 nm. ($T_{m,\text{bulk}} = 1336$ K, $\rho_s = 1.84 \times 10^4$ kg/m³, $\rho_l = 1.728 \times 10^4$ kg/m³, $\gamma_s = 1.42$ J/m² and $\gamma_l = 1.135$ J/m², and $L = 6.276 \times 10^4$ J/kg.)

 b. Which equation predicts the higher melting temperatures over this range?

 c. At which particle radius does Equation 4.8 begin to vary more than 20 K from the scaling law in Equation 4.5?

4.17 Name two pyrolysis processes used to make particles with nanometer diameters.

4.18 What is the aspect ratio of

 a. A tree branch 5 cm thick and 2 m long?

 b. A silicon wire 2 nm thick and 50 μm long?

4.19 Draw a schematic of a SAM, labeling the three main building groups (head group, tail group, backbone chain). Describe the function of each group.

4.20 Name at least five properties of a surface that a SAM can be used to tailor.

4.21 Name a common SAM head group.

4.22 Determine the missing atom (designated by "?") in the SAM structures shown in Figure 4.26.

4.23 Most synthetic zeolites have been developed for what purpose?

4.24 Redraw the graph in Figure 4.27.

 a. Label the two curves either brittle or ductile.

 b. Label the yield stress on each curve.

 c. Assume curve (a) was measured during a test of a material with an average grain size of 8 μm. Using a dashed line, redraw the curve on the same graph, this time assuming the test was performed using a specimen of the same material, but with a grain size of 2 μm.

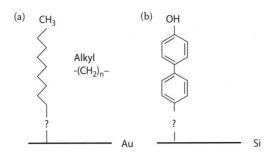

FIGURE 4.26   SAM head groups. (See Homework Exercise 4.22.)

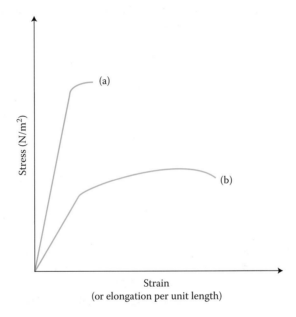

FIGURE 4.27   Stress–strain curves. (See Homework Exercise 4.24.)

4.25 True or false? All living things on Earth contain carbon atoms.

4.26 Name the three crystalline forms of pure carbon.

4.27 The unique structure of a fullerene molecule makes every atom.

4.28 Estimate the surface-to-volume ratio of a C_{60} fullerene by treating the molecule as a hollow sphere and using 77 pm for the atomic radius of carbon.

4.29 What are the three types of nanotubes?

4.30 What key property does the chirality of a nanotube determine?

RECOMMENDATIONS FOR FURTHER READING

1. R. Smalley. 1991. Great balls of carbon: The story of buckminsterfullerene. *The Sciences* March/ April, 22–28.

2. J. Israelachvili. 1992. *Intermolecular and Surface Forces.* Academic Press.

3. G. Schmid. 2005. *Nanoparticles: From Theory to Application.* Wiley-VCH.
4. K. Klabunde. 2001. *Nanoscale Materials in Chemistry.* Wiley-Interscience.
5. T. Tsakalakos, I. Ovid'ko, and A. Vasudevan. 2001. *Nanostructures: Synthesis, Functional Properties and Applications.* Kluwer Academic Publishers.
6. C. N. R. Rao, A. Müller, and A. K. Cheetham. 2004. *The Chemistry of Nanomaterials: Synthesis, Properties and Applications.* Wiley-VCH.
7. L. Smart and E. Moore. 2005. *Solid State Chemistry: An Introduction.* CRC Press.

CHAPTER 5

Nanomechanics

5.1 BACKGROUND: THE UNIVERSE MECHANISM

In 1900, a Greek man diving for sponges discovered a shipwreck near the island of Antikythera. Among the statues, jewels, pottery, furniture, and bronze artwork recovered from the wreckage, there was a corroded green object about the size of a shoebox. It had dials and an intricate configuration of 30 bronze gears, very much like a clock. Later analyses showed that it had been under the sea for more than 2000 years and most likely came from ancient Greece.

The intended use for the "Antikythera mechanism," as it later became known, remains enigmatic despite intense study. Experts on ancient Greece and scientists who have built working replicas of the mechanism generally come to the same conclusion: it was used as a kind of astronomical clock to model the motions of celestial bodies.

For as long as man has been building machines, there have been those who viewed the universe as one.

Mechanics deals with the motions of objects and the forces that cause these motions. Galileo and Descartes both believed mechanics could reveal the nature of everything—be it a flower, a river, or a person. Their mechanistic viewpoint left little room for spontaneity, randomness, or purpose. The universe, in their minds, was like one big clock. Later on, Isaac Newton built on the ideas of these two thinkers by developing a completely mathematical theory for the motion of the universe. Newton's 1687 scientific treatise, *Philosophiae Naturalis Principia Mathematica*, gave us the famous three laws of motion and the foundation for classical mechanics. The importance of Newton's work and its impact on the world since its publication cannot be overstated—because he was onto something: the universe is very much a mechanical thing.

Classical mechanics views the world in a deterministic way, where every motion is the inevitable consequence of the conditions prior to the motion. According to classical mechanics, everything operates with machine-like precision. Every motion can be predicted *exactly*. In the years since Newton's revelations, we have found classical physics to be immensely useful, but occasionally insufficient. Here are some of the main questions that classical mechanics cannot correctly answer:

- Why do some fundamental physical quantities, such as the energy of an atom, only assume specific values?

- Why do some objects, such as electrons and photons, act like both a wave and a particle?

- Why cannot we know an object's exact momentum and its exact position at the same time?

We learned the answers to these and other related questions in Chapter 3 "Introduction to Nanoscale Physics." There we began our discussion on the behavior of objects at the nanoscale—including the weird and wonderful phenomena of quantum mechanics. What classical mechanics sees as continuous, quantum mechanics sees as jumpy. Classical mechanics says, "I know exactly where that thing will go next," to which quantum mechanics replies, "But there is a chance that it will not go there at all."

Galaxies, planets, and cannonballs move according to classical mechanics. Yet look deeper into any of these objects and there is even more motion: the atoms themselves are a hive of activity. As it turns out, both approaches to mechanics come in handy when discussing *nano*mechanics. This chapter relies on classical *and* quantum mechanics to explain the forces that move objects at the nanoscale.

5.1.1 Nanomechanics: Which Motions and Forces Make the Cut?

If mechanics deals with the motions of objects and the forces that cause these motions, then what forces and motions qualify as "nano?"

The motions that we are most concerned with are oscillations. An oscillation, or vibration, is any motion that repeats itself in time. All objects tend to oscillate at a particular frequency, which depends on properties such as the object's mass and stiffness. This frequency is known as the object's resonance (or natural) frequency. The resonance frequency of a nanoscale object or system is characteristically high, simply because its mass is so characteristically small. Higher frequencies open doors for new applications, such as extremely sensitive mass detectors and novel ways to store information in a computer.

The forces that arise when an atom is close to another atom are exclusive to the nanoscale: on any larger scale, we can only observe the average of a multitude of such forces. Take for example what happens when you put your hand on a wall. The sheer number of atoms involved makes it impossible to distinguish the intimate details of what is really happening at the nanoscale. You cannot feel the tickle of van der Waals forces between the atoms in your skin and the atoms in the wall as these forces switch from attractive to repulsive. You cannot feel electron orbitals as if they were Braille dots.

However, the technology of nanomechanics enables us to *observe*, *measure*, and even *exert* forces on the molecular atomic level. These three crucial capabilities provide new ways to interact with our world. This chapter discusses how we can push and pull on a single molecule, "feel" the atoms on the surface of an object, and conduct a type of mechanical chemistry where we measure the forces between molecules during a chemical reaction.

5.2 A HIGH-SPEED REVIEW OF MOTION: DISPLACEMENT, VELOCITY, ACCELERATION, AND FORCE

Motion is inherent to mechanics. Therefore, we begin our discussion with a quick review of a few of the most basic principles of motion: displacement, velocity, acceleration, and force.

Displacement, say Δx, is an object's change in position—the distance it moves in the x direction between its initial and final positions:

$$\Delta x = x_{final} - x_{initial} \tag{5.1}$$

We can measure displacement in meters (m).

Velocity, v, is the displacement of an object over a given time, Δt:

$$v = \frac{\Delta x}{\Delta t} \tag{5.2}$$

Note that this velocity is an average velocity during $\Delta t = t_{final} - t_{initial}$. The "instantaneous" velocity would be the velocity given by this formula if Δt went to zero. Of course, it is impossible to measure something in zero time, and thus all measured velocities are truly average velocities. We can express velocity in units of meters per second (m/s).

Acceleration, a, is the change in velocity over a given time:

$$a = \frac{\Delta v}{\Delta t} \tag{5.3}$$

Again, this is an average value—because Δt would have to go to zero to measure a truly instantaneous acceleration. We can measure acceleration in meters per second per second (m/s²).

From these concepts we have four very useful equations for motion in a straight line under constant acceleration. Velocity as a function of time, where v_0 is the initial velocity, is

$$v = v_0 + at \quad \text{(for constant } a\text{)} \tag{5.4}$$

Displacement as a function of velocity and time, where x_0 is the initial displacement, is

$$x = x_0 + \frac{1}{2}\left(v + v_0\right)t \quad \text{(for constant } a\text{)} \tag{5.5}$$

Loading to displacement as a function of time:

$$x = x_0 + v_0 t + \frac{1}{2}at^2 \tag{5.6}$$

Therefore, combining Equations 5.4 and 5.5, we see that velocity as a function of displacement is

$$v^2 = v_0^2 + 2a(x - x_0)$$ (5.7)

Newton's Second Law (of his three famous laws) relates to force, F. It explains what will happen to an object when a force is applied to it. The answer is that the object will accelerate. It is considered positive acceleration if the force causes the object to speed up in the x direction, or negative acceleration—sometimes called deceleration—if the force causes the object to slow down (accelerate in the $-x$ direction). According to Newton's law, the acceleration is directly proportional to the net force acting on the object and inversely proportional to the object's mass:

$$\sum F = ma$$ (5.8)

The net force equals the sum of all the forces, $\sum F$, acting on the object. To honor the man who taught us the most about them, we measure forces using Newtons (1 N = 1 kg m/s²).

Figure 5.1 shows the displacement, velocity, and acceleration of a particle. At $t = 0$, a force begins pushing the particle and it accelerates in the x direction, gaining velocity every second for 5 s. The graph of velocity can be determined by the derivative (slope) of the displacement graph at each moment in time; in turn, the acceleration is equal to the derivative of the velocity. In this example, the displacement increases exponentially, so the velocity increases linearly and the acceleration is constant.

BACK-OF-THE-ENVELOPE 5.1

A metal particle with a mass of 5 femtograms (fg) is at rest when suddenly acted upon by opposing magnetic fields. One field exerts 7 attonewtons (aN) of force in the positive x-direction, and the other exerts 10 aN in the negative x-direction. What will be the direction and magnitude of the particle's velocity after 10 s?

The net force on the particle is simply

$$\sum F = -10 + 7 = -3\,aN$$

Thus, its acceleration is

$$a = \frac{\sum F}{m} = \frac{-3 \times 10^{-18}\,N}{5 \times 10^{-18}\,kg} = -0.6\,m/s^2$$

Velocity, v, expressed as a function of time and acceleration is

$$v = v_0 + at$$

Here, the initial velocity is $v_0 = 0$. So the final velocity, v, after 10 s is

$$v = at = (-0.6\,m/s^2)(10\,s) = -6\,m/s$$

That is, after 10 s, the particle is traveling at 6 m/s in the negative x-direction.

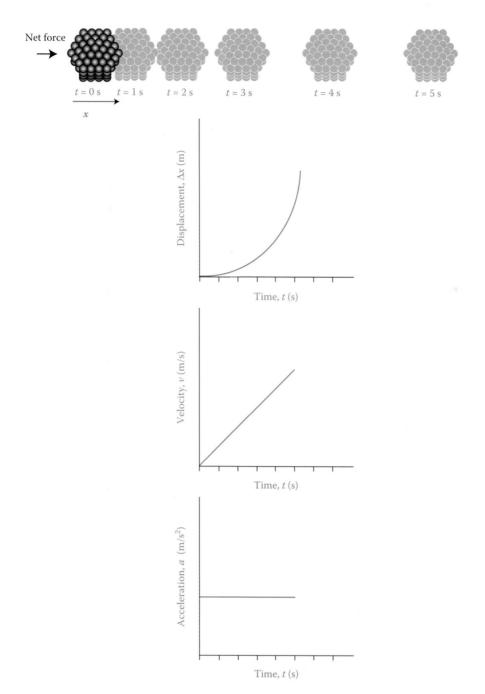

FIGURE 5.1   A speedy review of displacement, velocity, and acceleration. A particle is forced from rest at $t = 0$ and accelerates in the x direction, gaining velocity every second for 5 s. Velocity is the derivative (slope) of the displacement graph at each moment in time; in turn, the acceleration is the derivative of the velocity. In this example, the displacement increases exponentially, so the velocity increases linearly and the acceleration is constant.

5.3 NANOMECHANICAL OSCILLATORS: A TALE OF BEAMS AND ATOMS

When you hum a tune to yourself, your vocal chords oscillate, creating sounds that in turn cause your eardrums to oscillate. However, if you could listen really well, you could hear the humming of the entire universe. That is because everything in it is oscillating—everything is "humming."

Constant oscillatory motion is common to all matter. Atoms are oscillating back and forth at all times. Oscillation is also of fundamental importance to many nanomechanical systems.

The simplest oscillating system consists of a mass coupled to something elastic. A ball connected to a spring is a perfect example. This simple system can be used to represent the two nanoscale systems that we will study in detail in this chapter: (1) beams and (2) atoms. A beam has an inherent elasticity and can be deflected; atomic bonds are also inherently elastic. We can therefore represent each of these systems quite well using a ball-and-spring model, as shown in Figure 5.2.

The story of oscillatory motion at the nanoscale begins with classical mechanics and an understanding of beams. Then the plot thickens as we map what we learn to quantum mechanics so as to better understand atoms. In the end, we see that these two approaches actually overlap to yield similar results.

5.3.1 Beams

Our study of nanomechanical oscillators begins with one of the world's simplest and most common structures—the beam. A beam can be free at one end (like a diving board) or fixed on both ends (like a bridge). Numerous nanoscale structures qualify as beams, including wires, nanotubes, rods, belts, columns, and cantilevers—indeed, any flexible object that extends outward from a fixed base can be approximated as a beam. Likewise,

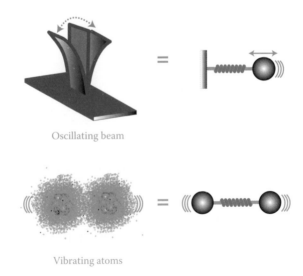

Oscillating beam

Vibrating atoms

FIGURE 5.2   Beams and atoms oscillate like balls on springs. This is how we approximate these two important systems throughout the discussion of nanomechanics.

the back-and-forth motion of a beam as it oscillates can be approximated by the motion of a ball on a spring, as shown in Figure 5.2. Let us now take a closer look at the details of this oscillation.

5.3.1.1 Free Oscillation

In the absence of any external forces, a ball on a spring undergoes free oscillation. The only force exerted on the ball is the force of the spring, F_{spring}. This force is proportional to the distance, x, that the spring is displaced from its equilibrium position:

$$F_{spring} = -kx \qquad (5.9)$$

The stiffness, k, of the spring is measured in Newtons per meter (N/m). In the case of a rectangular cantilever beam (free at one end), the spring constant for vertical deflection is given by

$$k_{beam} = \frac{E_M w t^3}{4L^3} \qquad (5.10)$$

Here, E_M is the modulus of elasticity (a material constant), w is the beam's width, t its thickness, and L its length.

The spring force is always directed toward the equilibrium position, opposite to the displacement. Figure 5.3a shows the motion of a ball on a spring. All that is needed to make this simple system oscillate is an initial disturbance. In the first instant (top), the spring is elongated and thus exerts a force on the ball that tugs it back toward equilibrium. This is the only force on the ball and it results in an acceleration, a. Using Equation 5.8

$$\sum F = F_{spring} = -kx = ma$$

Here, m is the mass of the ball. (Note that we are assuming that the spring's mass is negligible.)

When the spring is elongated, the displacement, x, is positive. This means that when the ball is released at time, $t = 0$, the acceleration will be in the negative ($-x$) direction, toward the equilibrium point. The velocity of the ball increases as it shoots toward the equilibrium position of the spring. For all positive values of x, the spring will continue to pull the ball to the left. At the equilibrium point ($x = 0$), the spring exerts no force, although the ball is traveling fast enough to continue beyond equilibrium and begin to compress the spring, moving the ball into the negative displacement ($-x$) region. Now the spring will exert a force on the ball in the positive x direction, so the acceleration will be in the positive x direction. The force will slow the speeding ball to a stop, reverse it, and push it back toward equilibrium again. This cycle repeats indefinitely. As it oscillates, the ball traces a sinusoidal pattern as a function of time, and the amplitude of the sine wave never diminishes. This is known as *simple harmonic motion*.

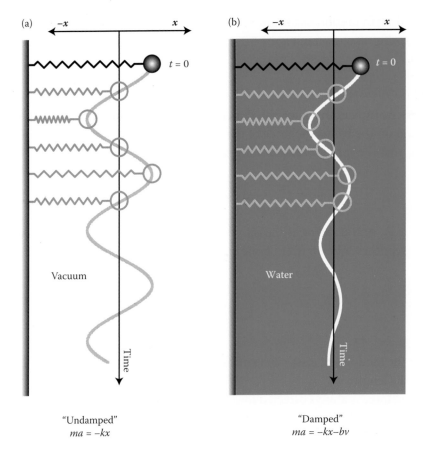

FIGURE 5.3 The displacement of a ball on a spring as a function of time. In (a), the only force on the ball is the spring force, $F_{spring} = -kx$, where k is the spring constant and x is the displacement of the ball. In this undamped system, the ball oscillates indefinitely in a vacuum environment (considering the ideal case where there is) no energy loss in the spring, creating a sinusoidal pattern on the graph of displacement versus time. In (b), the system is in water, giving rise to a damping force, $F_{damp} = -bv$, where b is the damping coefficient and v is the ball's velocity. This damped system still oscillates, although the amplitude diminishes over time, and the oscillation frequency is lower than that of the undamped system.

A simple harmonic oscillator is the prototype of any system with a mass that makes small vibrations about an equilibrium point. This includes the vibrations of beams and atoms; the acoustic, thermal, and magnetic properties of solids; and the oscillation of electromagnetic radiation. The frequency at which the simple harmonic system oscillates is an important quantity mentioned earlier, known as the natural (or resonance) frequency, f_n, and determined by

$$f_n = \frac{1}{2\pi}\sqrt{\frac{k}{m}}$$

(5.11)

Here, k is the stiffness of the spring and m is the mass of the ball (the spring's mass being negligible). In the case of a rectangular beam, we use an effective mass, $m_{eff} = 0.24\,m$, where m is the total mass of the beam. This accounts for the fact that not all of the beam's mass is localized at its end, as it is with a ball at the end of a spring. Frequency in this formula is measured in cycles per second, or Hertz (Hz). A cycle is one complete oscillation. We can also express the natural frequency as an angular frequency, ω_n, with units of radians per second (rad/s):

$$\omega_n = 2\pi f_n \tag{5.12}$$

The time each cycle takes is the period. At a frequency, f, the period is

$$\tau = \frac{1}{f} \tag{5.13}$$

The maximum displacement of the system is the amplitude, A (see Figure 5.4).

This is obviously an idealized system. In reality, there are always dissipative forces, such as friction. These forces not only retard the motion of the object, but also rob the system of energy. We can lump these dissipative forces together into a single force, called the damping force, F_{damp}. Unlike the spring force, which is proportional to displacement, the damping force is proportional to the velocity, v, of the object, and can be expressed as follows:

$$F_{damp} = -bv \tag{5.14}$$

Here, b is known as the damping coefficient, which varies from system to system. The negative sign indicates that this force always acts in the direction opposite to the object's motion.

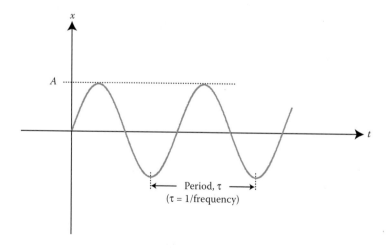

FIGURE 5.4   The amplitude, A and period, τ of oscillation shown on a graph of displacement versus time.

Damping occurs anytime an object moves through a fluid such as air or water. The mechanical energy is gradually lost to the fluid in the form of heat; this loss of energy is known as viscous damping. Vibrational energy can also be transmitted from an oscillating object via the support structure. This support-related loss (sometimes called damping loss) occurs at the base of beams. Additional energy is stolen by the internal friction within an elastic element, due in part to imperfections in the atomic lattice. In most cases, this internal friction occurs in the bulk (volume) of the spring or the beam. However, in the case of beams with nanometer thicknesses, there is a high surface-to-volume ratio and surface effects such as imperfections, residues, and coatings can be significant sources of damping.

In Figure 5.3b we see a spring–ball system operating in water. The spring force still works the same way, either pushing or pulling the ball back toward equilibrium. However, whenever the ball is moving, the damping force resists it. (And the only time the ball is not moving is when it turns around at the points of maximum displacement.) Although the system still oscillates, it oscillates at a frequency lower than the undamped natural frequency, and the amplitude decreases over time until the motion ultimately ceases. Notice, however, that the frequency does not change as the amplitude decreases. In simple harmonic motion, frequency does not depend on amplitude.

Without any external force to push the ball, its acceleration is determined by the spring and damping forces only:

$$\sum F = F_{\text{spring}} + F_{\text{damp}} = -kx - bv = ma$$

5.3.1.2 Free Oscillation from the Perspective of Energy (and Probability)

Thus far, we have examined the mass–spring system from the perspective of force. Let us take another look at this system to understand how it uses energy.

We can think of energy as a force applied through a distance. Energy is measured in joules (J), where $1\,\text{J} = 1\,\text{N}\,\text{m}$. So any time the spring is forced away from its equilibrium position, it is loaded with energy—potential energy (PE) to be exact. Mathematically, the potential energy gained by the system between its equilibrium position ($x = 0$) and its current position ($x = x$) is the integral of the force, F, with respect to displacement:

$$PE = \int_0^x F\,dx \tag{5.15}$$

We know that the force of a spring is $F_{\text{spring}} = kx$, where k is the spring constant. (For simplicity, we have removed the negative sign from Equation 5.9, which just served to indicate direction.) Substituting into Equation 5.15 and solving gives us

$$PE = \int_0^x F\,dx = \int_0^x kx\,dx = \frac{k(x)^2 - k(0)^2}{2} = \frac{1}{2}kx^2 \tag{5.16}$$

Now we know how much energy the mass–spring system has for a given displacement x. This is the first half of the picture.

The other half of the picture is the kinetic energy, KE, which the system has any time it is in motion. We know the spring exerts a force on the mass that makes it move with constant acceleration a. By rearranging Equation 5.7, we can write the acceleration as

$$a = \frac{v^2 - v_0{}^2}{2(x - x_0)}$$

Here, v_0 is the initial velocity, v is the current velocity, x_0 is the initial position, and x is the current position. If the object starts from rest ($v_0 = 0$) at an initial position of $x_0 = 0$, this equation simplifies to

$$a = \frac{v^2}{2x}$$

We know from Equation 5.8 that force, $F = ma$. Therefore,

$$F = ma = m\frac{v^2}{2x}$$

The kinetic energy, KE, is determined by the product of force over a distance, which we obtain by multiplying the above equation by a distance, x:

$$KE = Fx = \frac{1}{2}mv^2 \tag{5.17}$$

Now we have the whole picture—the kinetic energy (KE) and potential energy (PE) of a mass–spring system. To summarize

$$\text{Potential energy, } PE = \frac{1}{2}kx^2$$

$$\text{Kinetic energy, } KE = \frac{1}{2}mv^2$$

Here, x is the displacement of the spring from its equilibrium position, k is the spring constant, m is the mass of the object, and v is the velocity of the object.

In the ideal case where there are no losses due to damping, the total energy, E_{total}, of the system is conserved. That is, $E_{total} = KE + PE = $ constant, or

$$E_{total} = \frac{1}{2}kx^2 + \frac{1}{2}mv^2 = \text{constant} \tag{5.18}$$

What happens, then, is that the energy trades off between potential and kinetic energy during the course of each oscillation. This is shown in Figure 5.5, which represents the total energy of the system as a combination of kinetic and potential energies at various stages of oscillation. The amplitude, A, is the point of maximum displacement, x, from equilibrium ($x = A$ or $-A$). At this point, the mass comes to a complete stop while turning around. With the velocity, $v = 0$, the energy of the system consists entirely of potential energy, or

$$E_{total} = \frac{1}{2} kA^2$$

The mass is then accelerated away from this point, and stored potential energy into kinetic energy. When the mass passes through the equilibrium point ($x = 0$), the conversion is complete and the system now consists entirely of kinetic energy:

$$E_{total} = \frac{1}{2} mv^2$$

The probability that the oscillating mass can be found at a given value of x, or $p(x)$, is inversely proportional to the mass' velocity v. The general equation for this probability is

$$p(x) = \frac{B}{v} \tag{5.19}$$

Here, B is a constant that depends on the particular oscillating system.

It makes sense then that the mass spends the least amount of time at the equilibrium point, where its velocity is always at a maximum. The mass must stop completely for a moment before turning around at A and $-A$, so it spends the most time at the points where the spring is fully compressed or elongated. The probability curve for an undamped harmonic oscillator is shown in Figure 5.6. We will revisit this curve later on in Section 5.3.2.3 when discussing the quantum oscillator.

5.3.1.3 Forced Oscillation

When an external force acts on an oscillating system such as a beam, the system undergoes forced oscillation. This external, or driving, force, F_{drive}, is often repetitive in nature, like the well-timed pushes we give a child on a swing to keep him happily in motion. A driving force can compensate for energy losses due to damping and keep a system oscillating. Examples of damped systems driven by periodically varying external forces are a piston in an engine and a vibrating nanomechanical beam. The equation of motion for a damped system forced to oscillate is

$$F_0 \cos 2\pi f_{drive} t - bv - kx = ma \tag{5.20}$$

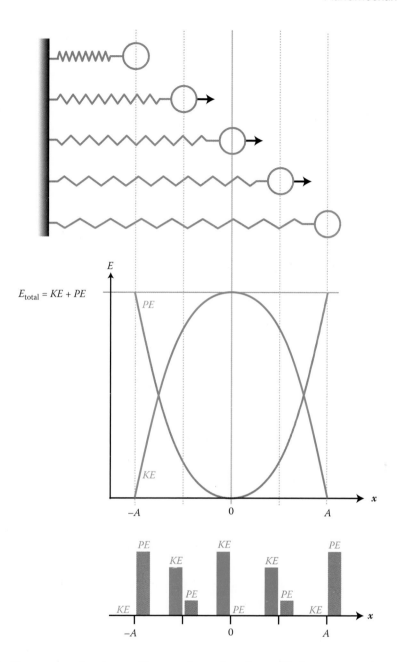

FIGURE 5.5 The energy of undamped harmonic motion. An oscillating mass–spring system (like that of a vibrating beam) is continuously transforming its energy from potential energy, PE, to kinetic energy, KE. Still, at any point during the cycle, the total energy ($E_{total} = PE + KE$) remains constant. The function of the curves is: $KE = 1/2mv^2$, where m is the mass and v is the velocity; $PE = 1/2kx^2$, where k is the spring constant and x is the displacement. The amplitude A (or $-A$) represents the maximum displacement.

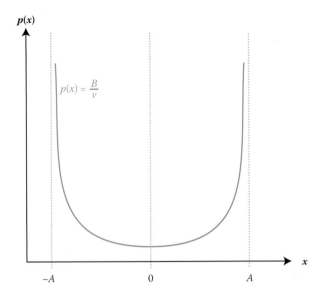

FIGURE 5.6 The probability distribution for an undamped oscillator. The chance of finding the mass at a given point along the x-axis is $p(x)$. It is inversely proportional to the velocity at that point such that $p(x) = B/v$, where B is a constant that varies by system. The mass is always moving quickest at the equilibrium point $(x = 0)$ and spends the majority of its time near the oscillation endpoints $(A$ and $-A)$, where its velocity is small or zero.

This equation tells us the acceleration, a, of an object of mass, m, as a result of the three forces acting on it. These three forces are

1. The spring force, $F_{spring} = -kx$, where k is the spring constant and x the distance from the equilibrium position

2. The damping force, $F_{damp} = -bv$, where b is the damping coefficient and v the object's velocity

3. The driving force, $F_{drive} = F_0 \cos 2\pi f_{drive} t$, where F_0 is the amplitude of the driving force, f_{drive} is the frequency of the driving force measured in Hertz (Hz), and t is time

The amplitude, A, of the oscillations of the system can be calculated using the following equation:

$$A = \frac{F_0}{\sqrt{\left(k - m\left(2\pi f_{drive}\right)^2\right)^2 + b^2 \left(2\pi f_{drive}\right)^2}} \tag{5.21}$$

In the case of forced oscillation, the amplitude does not decay over time. The external force is always working against the damping force to keep the system oscillating with a steady amplitude. The mass oscillates at the frequency of the driving force, f_d.

BACK-OF-ENVELOPE 5.2

What driving frequency, f_{drive}, causes the largest amplitude of oscillation in a given system?

Equation 5.21 gives us the amplitude of a forced oscillator as a function of f_{drive}. To achieve the largest amplitude, we want the denominator of the equation to be as small as possible. The denominator of the equation has a minimum when the drive frequency equals the damped natural frequency, f_d:

$$f_{\text{drive}} = f_d = \frac{1}{2\pi}\sqrt{\frac{\left(k - (b^2/4m)\right)}{m}}$$

You may recognize part of this equation. It is very similar to the equation for the natural frequency of an undamped oscillator, but the frequency shifts down slightly due to the influence of damping on the system. We can see a graph of amplitude versus drive frequency in Figure 5.7. Resonance occurs when the driving force pushes and pulls on the mass at the same frequency the system tends to oscillate at naturally. In the case of beams, we use the effective mass, 0.24 m, throughout the above equation.

As we see in Back-of-the-Envelope 5.2, the amplitude of oscillation for a damped or undamped system will peak when force is applied with a frequency equal to the system's natural frequency (which is lower in the damped case). This condition is known as *resonance*. It makes sense if you think about pushing a child on a swing. The swing goes up highest when you push the child right when he is at his maximum amplitude, at the same

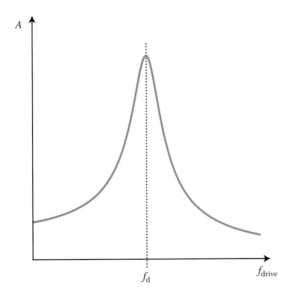

FIGURE 5.7 The resonance curve of a damped oscillator, such as a beam. The amplitude, A, has a maximum when the frequency of the driving force, f_{drive}, is equal to the damped natural frequency, f_d, of the system. This is known as resonance.

frequency that the swing is moving. Still, the peak height and location of this resonance curve can vary, depending on the amount of damping.

A value related to the damping coefficient is the quality (Q) factor:

$$Q = \frac{\sqrt{km}}{b} \tag{5.22}$$

The quality factor tells us the sharpness of the resonance peak and relates oscillation frequency to energy dissipation for a given system, that is, how damped the system is. Figure 5.8 shows the amplitude of an oscillating system as a function of frequency for three different levels of damping: high, low, and zero. Larger values of the damping coefficient, b, lead to smaller, broader resonance peaks at lower frequency because the damping force begins to dominate. If there is no damping, the force drives the oscillator without any loss, building the amplitude without any limit. This lossless situation is not found in real systems—which tend to have at least some damping that robs them of energy. High damping equates to a low quality factor and vice versa.

Nano- and micrometer-scale beams have quality factors of about 100 when oscillating in air; in vacuum, Q can be up to 1000 times higher.

We can also estimate Q for an oscillating system by the ratio of the damped resonance frequency, f_d, to the bandwidth, Δf:

$$Q \approx \frac{f_d}{\Delta f} \tag{5.23}$$

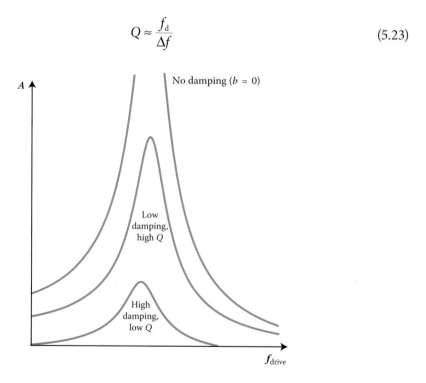

FIGURE 5.8 The amplitude at resonance varies with the damping coefficient, b. Greater values of b equate to lower, broader left-shifted resonance peaks. According to Equation 5.21, an undamped ($b = 0$) system has an amplitude of infinity at the resonance frequency.

On a graph of amplitude versus frequency, the bandwidth, Δf, of the peak is its width at the points where the amplitude equals of $(1/\sqrt{2})$ the maximum amplitude.

BACK-OF-THE-ENVELOPE 5.3

A silicon cantilever beam 500 nm thick, 10 μm wide, and 100 μm long has a quality factor $Q = 60$ in air (see Figure 5.9). A driving force causes the cantilever's tip to resonate at its damped natural frequency ($f_d = 70$ kHz) with an amplitude $A_d = 10$ nm. What is the approximate bandwidth Δf of this system? What frequencies are at either end of the bandwidth? What is the amplitude of the beam's oscillation at these frequencies?

The bandwidth, Δf, is given by

$$\Delta f \approx \frac{f_d}{Q} = \frac{70,000\ \text{Hz}}{60} = 1170\ \text{Hz}$$

Thus, the frequencies on either side of the band are $70 \pm 1.17/2$ kHz, or 70.585 and 69.415 kHz. At these frequencies, the amplitude is

$$\frac{A_d}{\sqrt{2}} = \frac{10\ \text{nm}}{\sqrt{2}} = 7\ \text{nm}$$

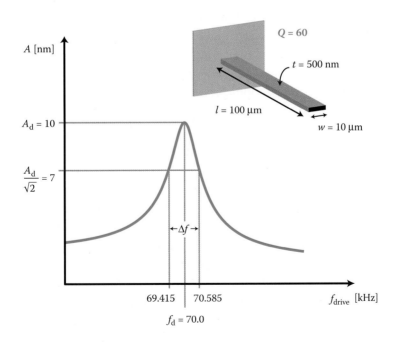

FIGURE 5.9 The quality (Q) factor determines the sharpness of the resonance peak. Q can be estimated for the cantilever beam shown (see Back-of-the-Envelope 5.3) by the ratio of the resonance frequency and the bandwidth: $f_n/\Delta f$.

5.3.2 Atoms

Only at absolute zero do atoms cease to move. Or at least that is what we think: the lowest temperatures ever achieved on Earth are around 1 billionth of a degree above absolute zero, or 1 nanokelvin, and the atoms were still moving around a little so we do not know what happens for sure at 0 K.

Temperature, in a quantum mechanical sense, is a measure of how fast atoms are moving. Atoms in a solid vibrate back and forth; and as the temperature rises, the atoms vibrate with bigger and bigger amplitudes. Sometimes they will vibrate right off the solid lattice into a fluid state, where they are free to bounce around like balls in a bingo spinner. Later on in Chapter 7, the discussion of nanoscale heat transfer will draw on what is presented in this chapter about the motions of atoms.

However, this now-familiar classic mass–spring system does not lose its validity just because we are discussing atoms. The concept of a simple harmonic oscillator can be approached from both classical and quantum viewpoints, and works wonderfully in understanding the vibrations of atoms. We will delve into both viewpoints during our discussion here.

5.3.2.1 The Lennard–Jones Interaction: How an Atomic Bond Is Like a Spring

For now, let us focus our discussion on a pair of atoms bonded together as a molecule. We restrict the motion of these two atoms to a single dimension, along the x-axis. They can move toward each other or away from one another. We will view them as an undamped simple harmonic oscillator. We looked in detail at this situation when discussing the Lennard–Jones potential in Chapter 4, "Nanomaterials." This is also an excellent model for our purposes here—providing us with a helpful description of the interactions between atoms. We use the Lennard–Jones potential to model the potential energy, PE, of a pair of atoms separated by a distance, x. These atoms will be attracted or repelled by each other, depending on x. For the case of two similar atoms, the attractive interaction varies with the inverse-sixth power of the separation distance, while the repulsive interaction tends to vary by the inverse twelfth power of the distance:

$$PE(x) = \frac{C_1}{x^{12}} - \frac{C_2}{x^6} \tag{5.24}$$

The constants C_1 and C_2 represent the strengths of the attractive and repulsive interactions, respectively, and vary by atom. This relationship is shown graphically in Figure 5.10. The Lennard–Jones model applies specifically to van der Waals interactions (and is very accurate for noble gas atoms such as argon, krypton, and xenon), but the shapes of the curves in Figure 5.10 are common to almost every kind of bond type, not just van der Waals bonds.

As we learned in Section 5.3.1.2, energy is a force applied over a distance ($E = Fx$). Thus, if we wish to know the interaction force, we can take the derivative of Equation 5.24. The force F between the atoms is therefore

$$F(x) = -12\frac{C_1}{x^{13}} + 6\frac{C_2}{x^7} \tag{5.25}$$

This relationship is also plotted in Figure 5.10.

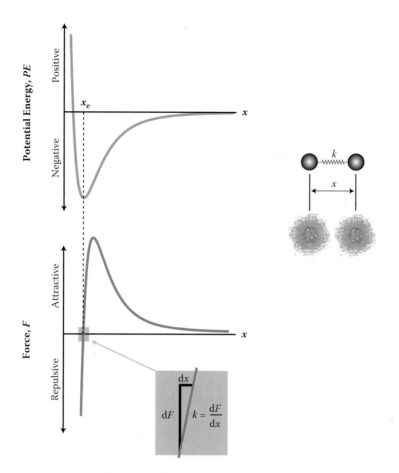

FIGURE 5.10   Potential energy, *PE*, and force, *F*, as a function of separation distance, *x*, for a pair of atoms. The force curve is the derivative of the energy curve. The Lennard–Jones interaction model tells us that at large separations, the atoms do not "feel" one another. In terms of the ball–spring model, this equates to no spring at all between the balls. As the atoms approach one another, attractive forces tug them to the point of least energy at $x = x_e$. The spring constant, k, of the bond between the atoms is determined by the slope of the force curve at the equilibrium point.

At very large separation distances, the atoms do not even "feel" the presence of one another at all. Therefore, both the potential energy and the force between them are zero. In our model, this equates to there being no spring between the balls. As the atoms near one another, they are at first subject to attractive forces (as if there is an elongated spring between the atoms pulling them together). The closer they get, lesser energy is added (the spring is less and less elongated). Eventually, the potential energy curve reaches a minimum at the equilibrium separation distance, $x = x_e$. It is here that the curve of the interaction force also goes to zero. Unless additional energy is added to the system, or the system is in motion exchanging kinetic and potential energy, the atoms remain at this separation distance. In our model, this is just like the equilibrium position of the spring ($x = 0$), where it is neither compressed nor elongated.

The force of a spring is proportional to the spring constant, k, measured in units of force per unit displacement (typically N/m). How then do we determine the spring constant of the bond shared by two atoms? The answer is on the force curve in Figure 5.10. If we zoom in on the region near the equilibrium position, x_e, we see that the force is relatively linear. For values of x less than x_e, the atoms have been forced closer together than they would be naturally—like a compressed spring. Values greater than x_e are the equivalent of an elongated spring. Because we know that the spring constant is directly proportional to the force and inversely proportional to the distance, the spring constant is then the slope of the curve (or dF/dx) in this region.

The spring force, F_{spring}, holding together the atoms in the ball–spring model is

$$F_{spring} = -k(x - x_e) \tag{5.26}$$

Here, x is the separation distance, x_e is the equilibrium separation, and k is the spring constant.

BACK-OF-THE-ENVELOPE 5.4

The two atoms in an argon molecule, Ar_2, have an equilibrium separation $x_e = 0.38$ nm. The Lennard–Jones constants for this particular molecule are $C_1 = 1.51 \times 10^{-134}$ J m^{12} and $C_2 = 1.01 \times 10^{-77}$ J m^6. What is the spring constant, k, for this bond?

The spring constant can be determined by taking the derivative of the force versus separation distance equation, or dF/dx:

$$F(x) = -12\frac{C_1}{x^{13}} + 6\frac{C_2}{x^7}$$

$$\frac{dF(x)}{dx} = 156\frac{C_1}{x^{14}} - 42\frac{C_2}{x^8}$$

Substituting the constants and solving for $x = x_e$ gives

$$\frac{dF(x)}{dx} = 156\frac{(1.51 \times 10^{-134}\,\text{J m}^{12})}{(0.38 \times 10^{-9}\text{m})^{14}} - 42\frac{(1.01 \times 10^{-77}\,\text{J m}^6)}{(0.38 \times 10^{-9}\text{m})^8}$$

$$\frac{dF(x)}{dx} = 0.82\,\text{N/m}$$

The spring constant, k, is therefore 0.82 N/m.

We are very close to a full description of the two-atom model from the classical standpoint of a ball–spring system. A lingering unknown is the mass of the system. We cannot simply add up the masses of the two atoms in this case. When two objects are acted upon by a central force (e.g., the gravitational force between the Earth and the moon), we treat their masses as a single mass, known as the "reduced mass," m_r. In our case, shown in Figure 5.11, there are two atomic masses, m_1 and m_2. The reduced mass is given by

$$m_r = \frac{m_1 m_2}{m_1 + m_2} \tag{5.27}$$

Now that we know the mass, we can write the equation for the kinetic energy, KE, of our atomic mass–spring system:

$$KE = \frac{1}{2} m_r v^2 \tag{5.28}$$

Here, m_r is the reduced mass, and the velocity v is defined as the change in the separation distance, x, during a given change in time, t ($v = \Delta x/\Delta t$).

The potential energy is

$$PE = \frac{1}{2} k \left(x - x_e \right)^2 \tag{5.29}$$

Here, k is the spring constant and x is the separation distance between the atoms. This of course makes the potential energy zero when the atoms are separated by their equilibrium separation, x_e.

Finally, because we now know the reduced mass and the spring constant of our two-atom system, we can calculate the natural frequency, f_n:

$$f_n = \frac{1}{2\pi} \sqrt{\frac{k}{m_r}} \tag{5.30}$$

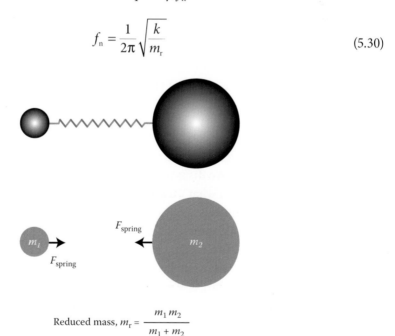

FIGURE 5.11 The mass of two objects acted upon by a central force such as a spring is known as the reduced mass, m_r. Both masses are treated as a single value.

The natural frequency that we just calculated for argon is a little low compared to most of the two-atom systems. This is because it is based on the relatively gentle slope of the van der Waals force curve, as compared to the stronger spring constants—steeper force curves—characteristic of covalent and ionic bonds. Such spring constants are on the order of 10–100 N/m, leading to resonance frequencies closer to 10 terahertz (10^{13} Hz).

BACK-OF-THE-ENVELOPE 5.5

The atomic mass of argon is 39.95 g/mol. What is the natural frequency of two argon atoms bonded together?

First, we determine the mass of one argon atom:

$$m = \frac{0.03995\,\text{kg/mol}}{6.022 \times 10^{23}\,\text{atoms/mol}} = 6.63 \times 10^{-26}\,\text{kg/atom}$$

The reduced mass, m_r, of a pair of identical argon atoms is therefore

$$m_r = \frac{\left(6.63 \times 10^{-26}\,\text{kg}\right)^2}{2\left(6.63 \times 10^{-26}\,\text{kg}\right)} = 3.32 \times 10^{-26}\,\text{kg}$$

We already calculated the spring constant, $k = 0.82$ N/m, for an Ar_2 molecule in "Back-of-the-Envelope 5.4." Using this value we can substitute into Equation 5.30:

$$f_n = \frac{1}{2\pi}\sqrt{\frac{k}{m_r}} = \frac{1}{2\pi}\sqrt{\frac{0.82\,\text{N/m}}{3.32 \times 10^{-26}\,\text{kg}}}$$

$$f_n = 790 \times 10^9 \text{ Hz, or } 790\,\text{gigahertz (GHz)}.$$

5.3.2.2 The Quantum Mechanics of Oscillating Atoms

Can we reconcile what we have discovered about the oscillations of atoms from a classical mechanics point of view with what we know about quantum mechanics? In treating a pair of bonded atoms like two masses attached to a spring, we determined the potential energy to be a continuous function dependent on the separation between the atoms. We relied on the Lennard–Jones potential to model this interaction. The Lennard–Jones curve is shown in Figure 5.12. Note especially that it is continuous and reaches a minimum when the atoms are separated by a distance, $x = x_e$. We used this curve and its derivative to determine the spring constant, k, and later the potential energy, PE, of the two-atom system as a function of separation distance:

$$PE = \frac{1}{2}k\left(x - x_e\right)^2$$

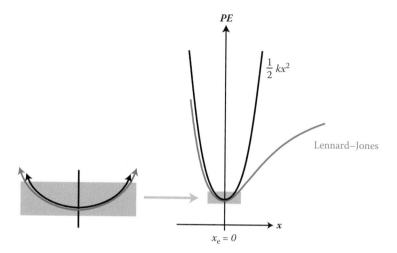

FIGURE 5.12 Approximating the Lennard–Jones potential energy with a parabolic curve. The function $PE = kx^2/2$, where k is the spring constant and x is the displacement from the equilibrium separation, matches the Lennard–Jones curve especially well near the equilibrium separation point, x_e, as shown in the inset.

Figure 5.12 shows this function plotted over the Lennard–Jones curve. It is exactly the same parabolic curve observed when we previously examined the potential energy of a simple harmonic oscillator, such as a beam (see Figure 5.5). In the region near x_e, where vibrations are small, this parabola is a good approximation of the Lennard–Jones potential. If we set up our axes such that x_e (the point of minimum potential energy) occurs at the origin, the graph of potential energy is as shown in Figure 5.13. Since most atomic vibration occurs around this point anyway, the parabolic potential energy curve is an appropriate model. Applicable as well, in fact, is the kinetic energy curve observed in Figure 5.5 (not shown again here).

However, there is an important detail that separates these models from quantum mechanics models—it is that these curves are *continuous*. And since the total energy of an oscillating system is the sum of both potential energy and the kinetic energy, the total energy is also continuous over all values of x. However, we know from quantum mechanics that the energy of atoms can only take on *discrete* values.

So what values of energy are allowed in a "quantum harmonic oscillator?" The answer can be determined using the Uncertainty Principle discussed in Chapter 3, "Introduction to Nanoscale Physics."

We start with the total energy, E, of the system. This is the sum of the kinetic and potential energies:

$$E = KE + PE = \frac{1}{2}m_r v^2 + \frac{1}{2}kx^2 \qquad (5.31)$$

Here, m_r is the reduced mass of the two-atom system; v is the velocity, defined as the change in the separation distance, x, during a given change in time, t ($v = \Delta x/\Delta t$); and k is the

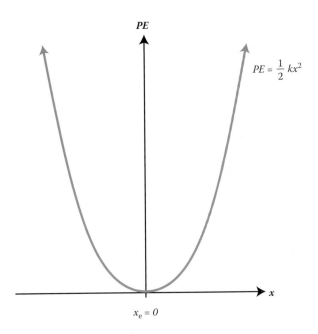

$$PE = \frac{1}{2} kx^2$$

$$x_e = 0$$

FIGURE 5.13 The potential energy PE of a simple harmonic oscillator as a function of separation distance, x. We have assigned the equilibrium separation as $x_e = 0$.

spring constant. (Note that we have simplified the equation for PE by putting x_e at the origin, so $x_e = 0$.)

The Uncertainty Principle states that the uncertainty of an object's position, Δx, and the uncertainty of its momentum, Δp, at any moment in time must be such that

$$\Delta x \Delta p_x \geq \frac{h}{4\pi}$$

Here, h is Planck's constant and $h = 6.626 \times 10^{-34}$ J s. So, according to this principle, the energy of the atoms in the quantum harmonic oscillator model can be no less than a fixed amount. Experimentally, if we knew the position of the atoms enough to say they were located at *exactly* the equilibrium separation, the velocity of the atoms (their change in separation in a given time) would be completely undetermined; if instead, we knew the velocity to be exactly zero, then the atoms' separation could be *any* value. Thus to describe a realistic scenario, we are forced to compromise by saying that the lowest energy allowed is where neither the speed nor the position are exactly zero.

Recall that momentum, $p = mv$, and natural frequency, $\omega_n = (k/m)^{1/2}$, so $k = \omega_n^2 m$. This enables us to substitute into Equation 5.31:

$$E = \frac{1}{2} m_r \left(\frac{p}{m_r} \right)^2 + \frac{1}{2} \omega_n^2 \, m_r x^2$$

$$E = \frac{1}{2} \frac{p^2}{m_r} + \frac{1}{2} \omega_n^2 m_r x^2$$

The minimum energy is determined by the smallest allowable values for p and x—that is, their uncertainties—so we replace x with Δx and p with $\Delta p = h/(4\pi\Delta x)$ so that

$$E = \frac{h^2}{32(\Delta x)^2 \pi^2 m_r} + \frac{1}{2}\omega_n^2 m_r (\Delta x)^2 \tag{5.32}$$

We are looking for the lowest possible energy and this can be found by taking the derivative with respect to position and then setting it equal to zero (corresponding to the bottom of the parabola, where the slope is zero):

$$\frac{dE}{d\Delta x} = -\frac{h^2}{16(\Delta x)^3 \pi^2 m_r} + \omega_n^2 m_r \Delta x = 0$$

Solving for position gives

$$\Delta x = \sqrt{\frac{h}{4\pi m_r \omega_n}}$$

When we plug this uncertainty in position back into the energy equation we find the minimum allowed energy, E_0:

$$E_0 = \frac{h^2}{32(\Delta x)^2 \pi^2 m_r} + \frac{1}{2}\omega_n^2 m_r (\Delta x)^2$$

$$E_0 = \frac{h\omega_n}{8\pi} + \frac{h\omega_n}{8\pi} \tag{5.33}$$

$$E_0 = \frac{h\omega_n}{4\pi}$$

This can also be expressed using f_n, measured in Hertz (Hz):

$$E_0 = \frac{hf_n}{2} \tag{5.34}$$

So that it is, the smallest amount of energy a quantum harmonic oscillator can have. What this suggests of course is that the system cannot have zero energy, even at absolute zero temperature. This is a significant mathematical result. The ground state of a quantum harmonic oscillator is known as the "zero-point energy." The energy values are discrete thereafter, proceeding upward according to the quantum number n, and the allowable quantum energy, hf (see Section 3.3.3), based on Planck's quantum hypothesis:

$$E = \left(n + \frac{1}{2}\right)hf_n \quad n = 0,\ 1,\ 2,\ 3,\dots \tag{5.35}$$

Here, h is Planck's constant and f_n is the natural frequency, expressed in Hertz.

We can see these allowed values of energy on the graph in Figure 5.14. While the classical potential energy curve drops all the way to zero at the equilibrium separation, x_e, the lowest allowed energy for the quantum harmonic oscillator is actually $E_0 = hf_n/2$. From this zero point, the quantized energies increase by the quantum number, as shown.

5.3.2.3 The Schrödinger Equation and the Correspondence Principle

At the dawn of the twentieth century, experiments showed that subatomic particles could have wave-like properties. Electrons, for example, created diffraction patterns when sent through a double slit. So did photons. Scientists scrambled to find an equation that would describe the behavior of atomic particles. This was discussed in Chapter 3, "Introduction to Nanoscale Physics." In that chapter, we learned that a particle of matter can be described by a generalized wave function, denoted by ψ ("psi"). All the mea-

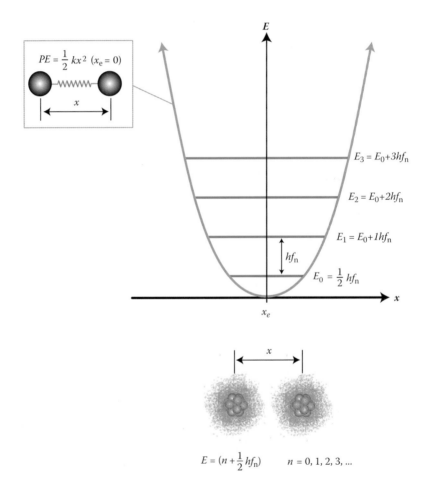

FIGURE 5.14 The quantum harmonic oscillator. The classical potential energy curve drops to zero at the equilibrium separation $x = x_e$. However, it was determined that the lowest allowed energy for the quantum harmonic oscillator is $E_0 = hf_n/2$, where h is Planck's constant and f_n is the natural frequency (in Hertz).

surable quantities of a particle—including its energy and momentum—can be determined using ψ. And the absolute value of the square of the wave function, $|\psi|^2$, is proportional to the probability that the particle occupies a given space at a given time. When we confine a particle inside a "box" of infinite height, the wave function, ψ is exactly the same as for the vertical displacement of a string stretched horizontally between two walls.

The wave function, ψ, satisfies an important equation that describes the behavior of atomic particles. The equation was developed by Erwin Schrödinger (1887–1961), an Austrian theoretical physicist widely regarded as the creator of wave mechanics. The now-famous Schrödinger equation applies to any confined particle and for motion in one dimension along the x-axis. It is written as

$$\frac{d^2\psi}{dx^2} = \frac{-8\pi^2 m}{h^2}(E - PE)\psi \tag{5.36}$$

Here, m is the particle's mass, h is Planck's constant, E is the total energy of the system, and PE is the potential energy of the system. This is known as the "time-independent" Schrödinger equation because it is invariant with time. While solving it can be quite complicated, this equation has proven extremely accurate in explaining the behavior of atomic systems. And it is used to find the allowed energy levels of quantum mechanical systems, including atoms.

The quantum harmonic oscillator system is exactly the kind of system the Schrödinger equation can be used for. We have a pair of atoms, the total mass of which is combined and treated as a single particle with a reduced mass, m_r. The motion of this system is confined to the x-axis, where displacement from the equilibrium position is given by x.

Combining what we know from Equations 5.16 and 5.30, we can find the potential energy of this system to be $PE = 2\pi^2 f_n^2 m_r x^2$, where f_n is the natural frequency of the system (Hz). Substituting this into the Schrödinger equation, we get

$$\frac{d^2\psi}{dx^2} = -\left[\frac{8\pi^2 m_r E}{h^2} - \left(\frac{16\pi^4 f_n^2 m_r^2}{h^2}\right)x^2\right]\psi \tag{5.37}$$

This equation requires a lengthy solution, which we do not show here. In the end, we find a solution that contains wave functions at the corresponding allowed energy levels:

$$E = \left(n + \frac{1}{2}\right)hf_n, \quad n = 0, 1, 2, 3, \ldots$$

The energy levels are the same as presented in Equation 5.35, where we used the Uncertainty Principle as our guide. In addition, the solution of the Schrödinger equation in this case provides the wave functions, ψ, and probability distributions, $|\psi|^2$, for the quantum states of the oscillator. We have graphed solutions in Figure 5.15 for the first three

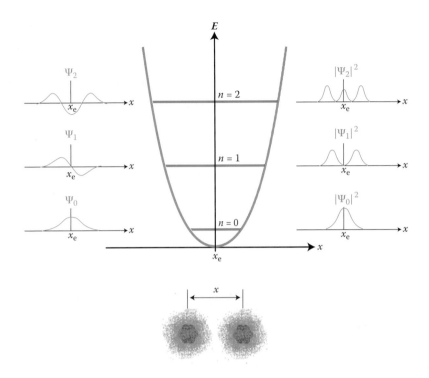

FIGURE 5.15   The wave functions, ψ, and probability distributions, $|\psi|^2$, of the quantum harmonic oscillator. These are determined by solving the Schrödinger equation. The probability of finding the oscillating mass (atom) at a given value of x is given by $|\psi|^2$. Shown here are the lowest three energies, where the quantum number n equals 0, 1, and 2.

quantum states, which correspond to the lowest three energies ($n = 0$, 1, 2). Here we see that the probability of finding the oscillator with a given separation, x, actually drops to zero in certain locations between the oscillator's maximum amplitudes. This of course is completely contrary to classical mechanics, which predicts a continuum of allowed separations where x can be any value between the maximum separation, A, and minimum separation, $-A$.

Where else do the classic and quantum descriptions diverge? Well, recall the classical probability distribution, $p(x)$, which predicted the odds of finding an oscillating mass at a given value of x. This curve is shown again in Figure 5.16. As expected, the probability of finding the mass is smallest near the equilibrium point, where it is always moving the quickest and therefore spending the least time. The probability rises fast near the limits of motion (A and $-A$), where the mass lingers while turning around.

Compare this to the probability density for the lowest energy state of a quantum harmonic oscillator, $|\psi_0|^2$, also shown in Figure 5.16. It is almost the complete opposite of the classical mechanics case. Here, the chance of finding the oscillating atom is highest near the equilibrium point. However, this disagreement becomes less and less striking at higher energies. Figure 5.16 shows $|\psi_2|^2$, with its three humps of probability. Already we see that the probability near the equilibrium point is diminishing while the probability near the

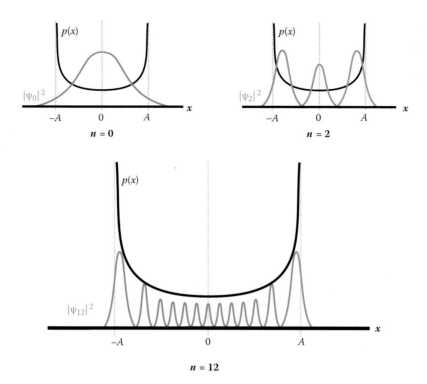

FIGURE 5.16 The probability distribution from quantum mechanics, $|\psi|^2$, compared to that of classical mechanics, $p(x)$. At the zero-point energy ($n = 0$), quantum mechanics predicts that the oscillator is most likely to be near the equilibrium spacing, x_e. Classical mechanics predicts just the opposite: that the oscillator spends more time at the edges of its motion, near A and $-A$. But as n increases, both the quantum and classical probabilities begin to look similar. We can see this progression toward correspondence here, from $n = 0$, to $n = 2$, and then $n = 12$.

edges increases. We can see why this happens if we look again at the time-independent Schrödinger equation:

$$\frac{d^2\psi}{dx^2} = \frac{-8\pi^2 m}{h^2}\left(E - PE\right)\psi$$

From this equation, it is evident why the wave function, ψ, will oscillate less rapidly at the edges than it does near the equilibrium point in the center. The magnitude of $d^2\psi/dx^2$—which governs how fast ψ will oscillate—is proportional to the magnitude of $(E - PE)$. Far from the equilibrium point, the total energy is almost entirely potential energy. This makes $(E - PE)$ very small, or zero, which means that ψ must be large in order for the product of the two, $(E - PE)\psi$, to be large enough for $d^2\psi/dx^2$ to make the curve "bend over." That is why the peaks in the wave are separated more at the edges, and higher as well.

The higher up we go in discrete energy states, the more the probability density resembles the classical curve. Figure 5.16 shows $|\psi_{12}|^2$, corresponding to $n = 12$. If we averaged this function over x, it would have approximately the general shape of the classical probability,

$p(x)$. As we reach higher and higher values of n, the curves of $|\psi n|^2$ and $p(x)$ look more and more similar. In most macroscale systems, energy states are extremely high. (For example, a 1-milligram object traveling a few centimeters per second has an energy state, n, on the order of 10^{23}.) Actually, in many nanoscale systems, the energy states are also very high, typically above 10^5 (see Back-of-the-Envelope 5.6).

Thus, it is crucial that classical and quantum mechanics do not predict *completely* different behavior. The two theories need to jibe. Quantum mechanics is better suited to the nanoscale, where energy comes and goes in packets. However, these packets become more indistinguishable the more atoms there are, until eventually energy is considered continuous. What is important is that there be a regime in which classical and quantum mechanics see eye to eye—where their predictions overlap. This is known as the *Correspondence Principle*.

There is one final discrepancy to clear up. You have probably noticed that the tail ends of the probability curves actually extend beyond the rigid limits of motion predicted by classical physics. When we introduced the concept of the wave function in Chapter 3, we assigned infinite height to the boundaries of the "potential well" in which the particle moved. This meant that the wave function would have to be zero at the edges of the well. The difference here, then, is that the boundaries are not infinitely high. Because they are finite, there is a slim chance that the particle will actually be found outside the confines of the well—or in this case, beyond the classically allowed region of oscillation. We will discuss this phenomenon in greater detail when we cover "tunneling" in Chapter 6, "Nanoelectronics." Here, we will simply note that these tails actually shrink in toward the limits as we go to higher values of n, until the classical and quantum pictures look practically identical.

BACK-OF-THE-ENVELOPE 5.6

The quantum description of an oscillator and the classical description are different at small quantum numbers, n. However, in most cases, we are dealing with high values of n. Consider even a simple spring–mass system consisting of a grain of pollen (mass) at the tip of a small eyelash (beam), as shown in Figure 5.17. The pollen grain weighs 100 ng and the eyelash has an estimated spring constant of 0.1 N/m. The pollen lands on the eyelash, deflects it, and causes it to oscillate. What is the approximate value of n for this system?

The force exerted by the grain of pollen is $F = ma$, where m is the pollen's mass and a is its acceleration. Here, $a = g$, the acceleration due to gravity (9.8 m/s²). The force is therefore

$$F = mg = \left(1 \times 10^{-10} kg\right)\left(9.8 m/s^2\right) = 9.8 \times 10^{-10} N$$

This force, exerted on the beam (spring), causes a deflection, x, from the equilibrium position:

$$x = \frac{F}{k} = \frac{9.8 \times 10^{-10} N}{0.1 N/m} = 9.8 \times 10^{-9} m$$

We know that the energy of a spring, $PE = kx^2/2$. When the pollen initially deflects the beam, this represents the total energy the system has in order to oscillate. The energy of the system is therefore

$$PE = \frac{1}{2}kx^2 = \frac{1}{2}(0.1\text{N/m})(9.8 \times 10^{-9}\text{m})^2 = 4.8 \times 10^{-18}\text{J}$$

The natural frequency of oscillation (assuming zero damping) is f_n:

$$f_n = \frac{1}{2\pi}\sqrt{\frac{k}{m}} = \frac{1}{2\pi}\sqrt{\frac{0.1\text{N/m}}{1 \times 10^{-10}\text{kg}}} = 5000\text{Hz}$$

All we have to do now is solve for n in the equation for the energy of a quantum oscillator:

$$E = \left(n + \frac{1}{2}\right)hf_n$$

$$n = \frac{E}{hf_n} - \frac{1}{2} = \frac{4.8 \times 10^{-18}\text{J}}{(6.626 \times 10^{-34}\text{ Js})(5000\text{Hz})} - \frac{1}{2}$$

$$n \approx 10^{12}$$

Thus, we can see just how "big" a relatively small oscillating system can be in terms of quantum number. At this level, both quantum and classical mechanics correspond very well.

5.3.2.4 Phonons

We know from Chapter 3 that photons are the quanta of electromagnetic radiation. Atoms radiate or absorb energy by absorbing or emitting photons. As we have learned in this chapter, vibrations are quantized too. A nanoscale oscillator such as a pair of bonded atoms can only have discrete amounts of energy. These amounts are equally spaced, with separations of $\Delta E = hf_n$. (Remember: h is Planck's constant and f_n is the natural frequency of the oscillator.) An oscillating system can therefore only gain or lose energy in discrete amounts of energy, equal to hf_n. These quanta of vibrational energy are called phonons.

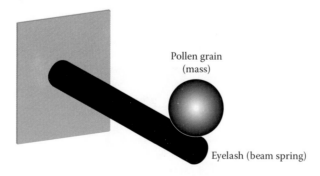

Pollen grain (mass)

Eyelash (beam spring)

FIGURE 5.17   A simple oscillator system made from a grain of pollen at the tip of an eyelash. (See Back-of-the-Envelope 5.6.)

Phonons are key players in numerous phenomena, especially heat and electrical conduction in solids, and conduction of sound—hence the name, *phonon* (after, *phon*, the Greek word for sound). In insulators and semiconductors, phonons are the primary conductors of heat. In metals, heat is conducted mostly by free electrons, but also by phonons. This is discussed in greater detail in Chapter 7, "Nanoscale Heat Transfer."

Whereas photons can travel through vacuum, phonons are only present in solid materials. Atoms in a crystal lattice are held together by bonds similar to springs (see Figure 5.18). Therefore, no atom vibrates independently. Each atom's vibrations are carried throughout the lattice. The way the vibrations propagate through a solid can be visualized by considering a linear chain of atoms like the one shown in Figure 5.19. At time, $t = 0$, vibrational energy is imparted to the bottom atom in the chain, compressing the bond between it and the next atom up. This energy is then transferred from atom to atom, via their bonds, until it leaves the chain at $t = 2$.

We know already that the energy, E, of an oscillator excited to quantum number, n, is

$$E = \left(n + \frac{1}{2} \right) hf_n \quad n = 0, 1, 2, 3, \ldots$$

Note: Lower case n is used as the subscript in f_n to indicate the fundamental oscillator frequency (natural frequency). It is also used to represent the quantum number. These two n's are not the same, but it is common for both natural frequency and quantum number to use this indication. But we can also say that this is the energy of an oscillator that has absorbed n phonons, since each phonon has an energy of hf_n. (Note that at $n = 0$, the oscillator has zero phonons and its energy equals the zero-point energy, $hf_n/2$, as we discussed earlier.)

Phonons propagate among the atoms in a lattice in a periodically varying way—as elastic waves. These waves are constrained between minimum and maximum wavelengths as shown in Figure 5.20. The phonon wavelength, λ_{phonon}, is given by

$$\lambda_{phonon} = \frac{2L}{n} \tag{5.38}$$

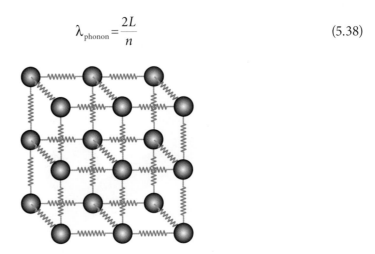

FIGURE 5.18 The bonds between atoms in a crystal lattice can be thought of as springs.

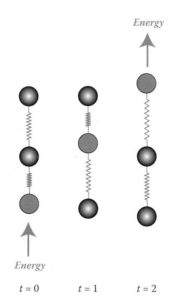

Energy

Energy

$t = 0$ $t = 1$ $t = 2$

FIGURE 5.19 A phonon is the quanta of vibrational energy. Here we see a linear chain of three atoms. At time $t = 0$, vibrational energy is imparted to the bottom-most atom. This energy is passed up the chain as the atomic bonds are displaced from their equilibrium positions, until eventually leaving the chain at $t = 2$.

Here, L is the length of the crystal lattice. The maximum phonon wavelength, $\lambda_{max} = 2L$ at $n = 1$. (At $n = 0$, the lattice has zero phonons.) The minimum phonon wavelength is determined by the spacing, d, between atoms in the lattice. No wavelength smaller than $2d$ is possible, therefore $\lambda_{min} = 2d$. This is relevant to nanoscale devices, which have the potential to limit or filter phonons if structures too small for particular wavelengths are used. (This is discussed in greater detail in Chapter 7, "Nanoscale Heat Transfer.")

If a lattice contains more than one type of atom in its unit cell, as for example in the sodium chloride (NaCl) crystal lattice shown in Figure 5.21, then there will be a heterogeneous distribution of masses, spring constants, and interatomic distances within the lattice—as opposed to the idealized homogenous chain of identical atoms we have been looking at. Such variations within the lattice give rise to additional degrees of vibrational freedom and additional phonon frequencies. Solids like NaCl can therefore support two types of phonons, corresponding to high and low frequency ranges. The lower frequency phonons are called "acoustic" phonons, which carry sound waves through the lattice; the upper frequency phonons are called "optical" phonons, which can be excited by electromagnetic radiation, and occur when positive and negative ions next to one another in the lattice move in opposing directions (see Figure 5.21).

5.3.3 Nanomechanical Oscillator Applications

In the next two sections we will look closer at a pair of developing applications that rely on the unique properties of nanomechanical oscillators—nanomechanical memory elements and nanomechanical mass sensors.

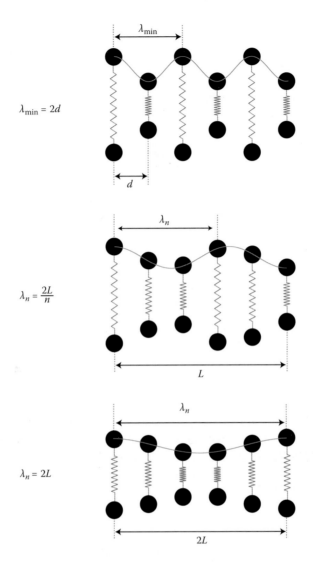

FIGURE 5.20 Phonon wavelength. Phonons propagate among atoms in a crystal lattice as waves, the wavelength of which is $\lambda_{phonon} = 2L/n$, where L is the length of the crystal lattice and n is the quantum number. The maximum phonon wavelength is $2L$ at $n = 1$. (The lattice is occupied by zero phonons at $n = 0$.) No wavelength smaller than $2d$ is possible, d being the spacing between atoms in the lattice.

5.3.3.1 Nanomechanical Memory Elements

The abacus is a counting tool made from a set of bars with sliding beads, invented in China about 21 centuries ago (~300 B.C.). Shopkeepers and street vendors in Asian countries can still be found using the ancient devices for addition, subtraction, division, multiplication, and even to extract square and cubic roots. After calculating an answer, the beads store it as well—remaining in their readout positions until the next calculation. It can probably be called the world's first computer—a *mechanical* computer, capable of storage and manipulation of data.

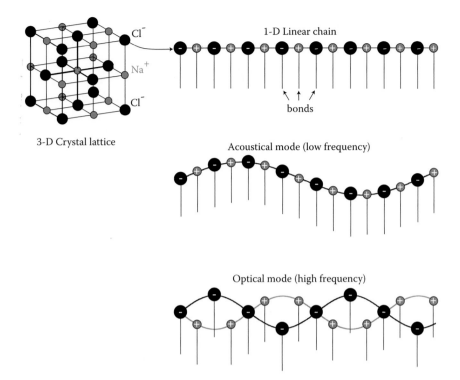

FIGURE 5.21   Acoustic and optical phonons. These two varieties of phonon correspond to low- and high-frequency waves, respectively. Acoustic phonons are present in all solid materials. Optical phonons occur in lattices where the unit cell contains more than one type of atom, as in the sodium chloride lattice shown here, and can be excited by electromagnetic radiation. If we look at a single, one-dimensional chain of alternating positive and negative ions, we can see the difference in these two types of phonons. Acoustic phonons vibrate the atoms in phase, while optical phonons vibrate the ions out of phase, that is, with the ions moving in opposing directions.

The basic unit of computer information is a bit—a 1 or a 0. Today, most computers rely on the presence or absence of electric current (as in a transistor), or positive and negative magnetic poles (as in a hard drive), to represent 1s and 0s. These methods enable high data density and quick manipulation. However, mechanical things can also serve as binary elements. A bead on an abacus can be switched between mechanical states, or store a bit of information.

In fact, mechanical memory elements are once again an attractive option because now we can make them at the nanoscale. Certain devices are being developed with data densities and speeds rivaling electric and magnetic technologies. These mechanical memory elements consist of nanomechanical beams, usually made from silicon, clamped to supports at either end. These beams are typically a few hundred nanometers in thickness and width, and a few micrometers long.

The operation of these elements is based on a nonlinear effect (sometimes called the "jump phenomenon") that arises in oscillating elements when very high driving forces are applied. This effect is shown in Figure 5.22a. What happens is that a region of the resonance curve becomes bistable—meaning that for a given drive frequency f_d, the oscillation amplitude can be one of two values, either large or small, depicted as "state 0" and "state 1" on the graph. One way to observe this bistability is to measure the oscillation amplitude as drive frequency is slowly increased (black curve), and then measure it again, this time slowly decreasing the drive frequency (blue curve). Such a behavior is known as hysteresis. Researchers have shown that by forcing a beam to oscillate at a frequency within the bistable region (f_b on the graph) and then suddenly applying an additional force to the beam (accomplished by running a pulse of current through the beam), they can make it switch between these two states—namely the large-amplitude state and the small-amplitude state. These two states represent 1 and 0 (like a bit). This is depicted in Figure 5.22b. An electromagnetic technique is used to drive the device and detect its deflection. Deflections of less than 0.1 nm can be measured. Amplitudes tend to range from about 0.1 up to 10 nm. With

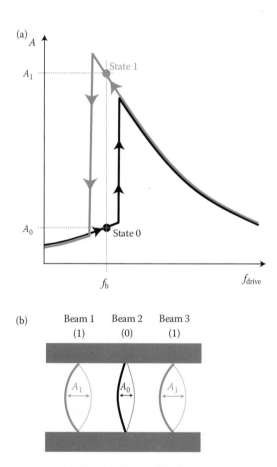

FIGURE 5.22   Nanomechanical memory elements. A nanoscale beam supported at both ends and made to oscillate with a high driving force can exhibit the bistable resonance behavior shown in (a). This creates a region of the resonance curve where the amplitude of oscillation can be either large or small at a given drive frequency fb. Applying a force input to the beam can cause the beam to switch from one state to the other. These two states (state 0 and state 1 on the graph) can represent either a 0 or a 1 and therefore store a single bit of information in a mechanical computer. In (b), three oscillating beams are used to store the sequence 1-0-1.

BACK-OF-THE-ENVELOPE 5.7

Imagine that a silicon cantilever beam (attached at one side to a base) is used as a binary memory element, where an upward deflection represents one state and a downward deflection represents the other state. The beam measures 5 μm long, 250 nm wide, and 150 nm thick. The modulus of elasticity, E_M, for silicon is 110 GPa and the density, $\rho = 2330$ kg/m³. How quickly can the beam switch states?

We know from Equation 5.10 that the spring constant of this beam is

$$k_{beam} = \frac{E_M w t^3}{4L^3} = \frac{\left(110 \times 10^9 \, Pa\right)\left(250 \times 10^{-9} m\right)\left(150 \times 10^{-9} m\right)^3}{4\left(5 \times 10^{-6} m\right)^3} = 0.19 \, N/m$$

The mass of the beam is the product of its density and volume:

$$m = \rho V = \left(2330 \, kg/m^3\right)\left(250 \times 10^{-9} m\right)\left(150 \times 10^{-9} m\right)\left(5 \times 10^{-6} m\right) = 4.4 \times 10^{-16} \, kg$$

The natural frequency, f_n, determines the time the beam takes to oscillate to the opposite state (because the inverse of Hertz is seconds). Natural frequency is given by Equation 5.11, using the effective mass of the beam, $m_{eff} = 0.24 \, m$:

$$f_n = \frac{1}{2\pi}\sqrt{\frac{k}{0.24 \, m}} = \frac{1}{2\pi}\sqrt{\frac{0.19 \, N/m}{0.24\left(4.4 \times 10^{-16} kg\right)}} = 6.8 \, MHz$$

This means that the device can switch from a 1 to a 0 state in about $1/6.8 \times 10^6$ Hz, or 150 ns.

---

some devices, the beams will only remain in a particular state as long as the compressive force is applied—this represents volatile memory, the kind used in computer RAM that is lost when the computer power is turned off. However, other beams have been built with intrinsic stresses that enable them to remain in a particular state even after the small compressive force pulse ends; such beams will not switch states again until a new pulse is applied. This is known as nonvolatile memory.

Because of the geometry of these beams, they have remarkably high natural frequencies. The natural frequency is what determines how fast the beams can oscillate, that is, how quickly they can change from one flexed state to the other.

As we can see in Back-of-the-Envelope 5.7, memory elements made from nanometer-scale beams switch states quickly, on the order of tens of millions of times per second (MHz). Shorter beams could make gigahertz (GHz) speeds possible.

These devices have three key advantages: (1) the beams' miniscule sizes make it possible to store more than 100 gigabytes of information per square inch, a data density over 100 times better than today's memory chips (Note: there are 8 bits to a byte); (2) the state can be switched using just femtowatts of power, orders of magnitude less than electronic memories, which use microwatts; and (3) the beams are more immune to radiation and electromagnetic pulses than electronic and magnetic devices (since they are mechanical devices, after all), making them useful in rugged environments such as those encountered by spacecraft (or laptops in college kids' backpacks).

5.3.3.2 Nanomechanical Mass Sensors: Detecting Low Concentrations

A little girl walks out to the end of a diving board and jounces it. The board bends and starts to oscillate, carrying her with it, up and down. She jumps off the board into the pool. As she leaves it, the board makes a "flong-g-g-g-g!" noise—suddenly oscillating much quicker without the girl's added mass on the end. Why does this happen?

The diving board has a spring constant, k, and a mass, m. Equation 5.11 tells us that the natural frequency, f_n, of an (undamped) oscillating beam is determined solely by these two parameters:

$$f_n = \frac{1}{2\pi}\sqrt{\frac{k}{0.24m}}$$

So when the girl is on the diving board, she adds mass to the system and therefore lowers its natural frequency. If the little girl and her two friends walk out to the end of the board together, the natural frequency would be even lower. Using the equation above, we can determine how much mass the girl added to the system by simply counting the number of oscillations the beam makes in a second and comparing it to the number the beam made without her on it. Of course the sensitivity of this technique is limited. The girl's mass and the board's mass are similar in magnitude. So we probably could not use this diving board sensor to determine whether or not the girl had a coin in her hand.

But what if the diving board were 10 million times smaller?

BACK-OF-THE-ENVELOPE 5.8

A diving board has a characteristic dimension, $D = 1$ m. We reduce it in size by a factor of 10 million. By what factor have we reduced the diving board's mass?

The mass of an object, m, is proportional to its characteristic dimension, D, as

$$m \propto D^3$$

Thus, reducing D by a factor of 10 million corresponds to a reduction in mass by a factor of $10,000,000^3 = 1 \times 10^{21}$.

As we can see, reducing the dimensions of the board drastically reduces its mass. This has important consequences. The less massive we make the board, the more sensitive its natural frequency will be to added mass. Using a diving board we probably would not notice the change in natural frequency due to the incremental mass of a coin. But we certainly could using a popsicle stick.

Nanometer-sized oscillators such as beams have very little mass. Typical lengths of beam-shaped sensors range from a few hundred nanometers up to a few hundred micrometers. Natural frequencies range from hundreds up to billions of hertz. (For example, a silicon beam 100 μm long, 20 μm wide, and 1 μm thick weighs about 5 ng (nanograms) and has a natural frequency of about 140 kHz.) This makes for exquisitely sensitive mass detectors—scales capable of weighing things as small as a single bacterium, or a few thousand atoms. The masses of such things are in the femtogram, attogram, and even zeptogram range. Such sensitivity enables new chemical, physical, and biological mass detection applications where minute quantities of a material must be detected. (See "Lost in Translation" sidebar.)

LOST IN TRANSLATION: USING MASS SENSORS TO MEASURE CONCENTRATION

The U.S. Environmental Protection Agency (EPA) has deemed it unhealthy for a person to breathe air containing 180 ppt (parts per trillion) of lead for a three-month period. But what does that concentration mean? How does a mass sensor figure out how many "parts" of something are in the air?

Parts per million by volume (ppm_v) is a relative measure of the volume occupied by a substance divided by the volume of the carrier (usually a gas or liquid). Lower concentrations can be expressed in parts per billion (ppb_v), parts per trillion (ppt_v), and so on. The Earth's atmosphere is primarily nitrogen (780,800 ppm_v) and oxygen (200,950 ppm_v), with trace amounts of other gases and varying percentages of water vapor. One cubic centimeter of atmospheric air, or roughly the volume of your fingertip, contains about 2.5×10^{19}, or 25 quintillion, molecules. The molecules of N_2, O_2, and other trace gases each measure less than a nanometer in diameter, and tend to be spaced out at a distance of about 10 times the molecular size. This same cubic centimeter of air also holds many particles. Clean air has about 1000 particles per cubic centimeter; polluted air can have 100,000 or more. These particles measure between 1 nm and 50 µm and can be pollen, bacteria, dust, or industrial emissions.

However, mass sensors do not detect concentration—not directly anyway. They detect mass. So, a more appropriate unit for concentration is milligrams per cubic meter (mg/m^3), which measures the amount of mass of a substance in a specific volume (1 m^3) of air. To convert back and forth between a concentration, C, of a substance given in ppm_v and a concentration given in mg/m^3, the following relation can be used:

$$C_{[ppm_v]} = \frac{0.6699 C_{[mg/m^3]}(459.7 + T)}{Mp} \qquad (5.39)$$

Here, M is the molecular mass of the substance (g/mol); p is the absolute pressure of the mixture (substance and carrier), (psia); and T is the mixture temperature (°F).

The EPA's exposure limit for lead can be converted using this equation from 180 ppt_v to 0.0015 mg/m^3, or 1.5 $\mu g/m^3$. Mass sensors have achieved sensitivities better than 1 femtogram (fg). At the EPA's limit, there is about 1 fg of lead in a cubic millimeter of air.

BACK-OF-THE-ENVELOPE 5.9

How much weight do you gain by simply breathing in oxygen?

Oxygen represents 200,950 ppm_v of our air. The tidal volume of adult lungs, or the volume of air inhaled in a regular breath, is about 0.5 L, or 0.0005 m^3. (The maximum capacity of the lungs is about 10 times greater, at 6 L.) Rearranging Equation 5.39 gives

$$C_{[mg/m^3]} = \frac{C_{[ppm_v]} Mp}{0.6699(459.7 + T)}$$

The molar mass of oxygen is 16 g/mol. The pressure of the air is 1 atm = 14.7 psia. We assume the air is at room temperature (70°F). This gives

$$C_{[\text{mg/m}^3]} = \frac{(200,950\,\text{ppm}_v)(16\,\text{g/mol})(14.7\,\text{psia})}{0.6699(459.7 + 70F)} = 133,000\,\text{mg/m}^3$$

Therefore, in a regular breath, the weight gained in oxygen is

$$(133,000\,\text{mg/m}^3)(0.0005\,\text{m}^3) = 67\,\text{mg}$$

That is approximately the mass of a drop of water.

Resonance-based nanometer-scale mass sensors operate mostly on the same principle. The sensor structure, be it a quartz crystal disk or a silicon beam, is made to oscillate at its natural frequency, and this frequency varies according to the amount of mass that accumulates on the structure, as shown in Figure 5.23. The more mass that accumulates, the lower the frequency at which the device naturally oscillates. To "sense" a material is to measure the differences in oscillations, and therefore detect the material's mass. With

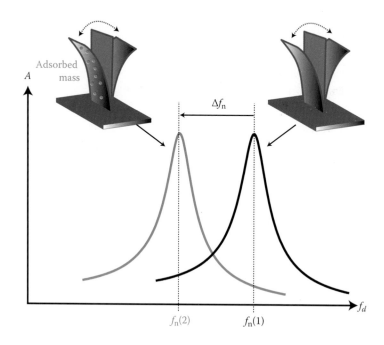

FIGURE 5.23 Mass detection using a nanomechanical oscillator. The resonance curve for a beam (a graph of amplitude A versus drive frequency f_{drive}) peaks at the beam's natural frequency, f_n. Because f_n is inversely proportional to the mass of the beam, the added mass of adsorbed material brings about a detectable shift, Δf_n, from $f_n(1)$ to $f_n(2)$. This technique enables extremely sensitive detection of chemical and biological molecules.

beams, the oscillations can be detected in various ways. One method involves reflecting a laser off the oscillator's surface into a photosensitive detector that counts each time the beam moves up and down in a second. Integrated electrical components can also be used to track oscillation frequency.

For nanomechanical sensor applications, the goal is to make the sensor surface "functionalized" with some kind of coating, such as an adsorptive polymer or a chemical that will bind specifically to whatever molecule the sensor is designed to collect and sense. This would be like putting chocolate ice cream on the end of the diving board to attract certain kids and strawberry ice cream to attract others. Of course, it is possible to be more specific than this. A beam could be functionalized using DNA, for example, such that only a certain genetic sequence of base pairs would bind strongly to the beam.

In general, the smaller the device structure, the more sensitive it is to mass changes because the acquired mass represents a larger percentage of the original mass of the structure and therefore causes a greater relative shift in the resonance frequency. The minimum detectable mass, m_{min}, of a rectangular cantilever beam is given by

$$m_{min} = \frac{\rho^{3/2} L^3 w \Delta f_{min}}{\sqrt{E_M}}$$

(5.40)

Here, ρ is the density of the beam material, L is the length of the beam, w is the width, Δf_{min} is the minimum detectable frequency shift of the system, and E_M is the modulus of elasticity of the beam material. Therefore, it is important to minimize the size of the sensor in order to make it more mass sensitive. However, there are other issues to consider. First, the smaller the sensor, the less likely the target material is to come in contact with it. Also, smaller devices at the nanoscale tend to have lower quality (Q) factors as surface effects begin to dominate damping. This dulls the resonance peak, making it difficult to accurately measure natural frequency shifts. So, Δf_{min} gets worse (larger). For these reasons, we can say that minimum detectable mass scales approximately with the square of the characteristic dimension of the beam, or D^2.

Conflicting design constraints must be reconciled when designing or selecting a mass sensor for a specific application. We want the sensor to be sensitive to very small changes in mass. This means we need to design it to be small. Meanwhile, damping effects must be minimized to achieve a high Q factor, and surface area must be maximized for sample collection. This means we need to design it to be big. The ideal mass sensor is one with the largest possible active area on which the material of interest can be collected and the smallest possible minimum detectable mass. If a scale the size of a diving board were sensitive enough to detect not only the mass of a person, but also the mass of a single hydrogen atom (with a mass of 1.7 yoctograms, or 1.7×10^{-24} g), then this of course would be an ideal scale. In reality, engineers are forced to make trade-offs so as to simultaneously optimize mass sensitivity, minimize damping, and increase the sample collection area. Just as there are different designs of screwdrivers, depending on whether we are changing a watch battery or a car battery, we design different kinds of mass sensors depending on the application.

5.4 FEELING FAINT FORCES

If you lower the tip of a pencil toward the surface of a desk, you will feel the gravitational force from the pencil's weight pulling it downward—but you will not notice the attractive van der Waals forces kick in. Drag the tip of the pencil across the desk and you may notice it hitting little bumps on the desk's surface—features made up of billions and billions of atoms. But with the pencil we do not notice bumps made from single atoms. The forces between atoms are too faint, and their sizes too small, for us to notice without a very small, very sharp pencil.

Currently, we can build such a pencil. New nanoscale tools are sensitive enough to measure the atomic-scale forces exerted by a single layer of atoms as it collects on a surface. These tools can also detect the bump on a surface made by a single atom, or even exert atomic-scale forces with such control that we can pick up individual atoms and move them exactly where we would like. These new tools are discussed in the next sections.

5.4.1 Scanning Probe Microscopes

All microscopes are peepholes into a wonderland where grains of pollen become prickly planets, the jaws of an ant take on man-eating proportions, and a shard of glass turns out to be a landscape—replete with deep chasms, barren deserts, and looming mountains. When we think of microscopes, the image we conjure is most likely that of an optical microscope, where we shine light on a specimen and, using lenses, bend the reflected light so that the specimen appears larger to the human eye. But the workhorse microscopes of nanotechnology do not work this way. They feel around specimens like our fingers would rather than look at them as our eyes would. And such being the case, we can do more than just look at things with them. The next four sections highlight four applications of SPM technology.

5.4.1.1 Pushing Atoms around with the Scanning Tunneling Microscope

The STM was invented in the early 1980s by Gerd Binnig and Heinrich Rohrer after 3 years of collaboration at the IBM Research Laboratory in Zurich, Switzerland. It is the father of all SPMs, and earned its creators the Nobel Prize in 1986.

The STM has a sharp metal stylus, similar to the tip of a pen that scans back and forth across a sample surface. The stylus, however, never touches the surface; instead, it remains a fixed distance away from it. If we apply a voltage bias between the electrically conducting tip and the conducting sample and then bring them extremely close together (within a nanometer), electrons will jump out of the sample onto the tip. This special type of current that can convey electrons across very small gaps is called tunneling—a quantum mechanical phenomenon. (Both tunneling and the STM are discussed in greater depth in Chapter 6, "Nanoelectronics.") Tunneling current varies exponentially with the tip–sample separation distance; therefore, by keeping the tunneling current constant with a feedback loop, the microscope is able to track a surface's topography with picometer resolution. (We will discuss how a feedback loop works in the next section.)

The STM only works on conductive sample surfaces. That is not to say that this microscope is not extremely useful. Very soon after they first used the first STM to "see" surface atoms, Binnig and Rohrer noticed something equally astounding. The tip would often pick up stray atoms, carry them somewhere else on the surface, and set them back down. They knew the atoms were moving because they imaged the surface and, with each raster scan, they saw that some of the bumps (which represent atoms) showed up in different spots. Thus, their microscope did not work perfectly: the tip sometimes nose-dived into the surface and came away with an atom or two on it. But the fact remained: they were moving atoms like Chinese checkers. This was mechanical manipulation at the smallest scale ever. We will now take a look at how it worked.

Not only is there tunneling current between an STM tip and a sample, but there are also other chemical interactions, such as electrostatic forces and van der Waals forces. The topmost layer of surface atoms feels these forces most strongly. At high enough temperatures, the surface atoms are in a state of constant spontaneous diffusion—mingling around from spot to spot on the surface. The voltage bias can influence this diffusion and make these mobile surface atoms clump together near the tip. When the tip is taken away, however, the clump disperses and the surface atoms spread out again. Although this moves atoms around, this type of manipulation is not in any way permanent and is not very useful.

For the manipulated atoms to stay put necessitates temperatures low enough to prevent the spontaneous diffusion of surface atoms. This is typically just a few degrees above absolute zero. To manipulate atoms, we first need to map the surface with the STM. This allows us to understand the surface topography of the object and figure out which atoms we want to move. Then, placing the tip over a region we have mapped, we can vary the tip–sample distance and the voltage bias in order to either pick up and carry, or drag, a surface atom to where we want it (see Figure 5.24a). We can even pluck single atoms out of molecules this way.

As this technique began to mature, scientists around the world took turns demonstrating their prowess with it. They arranged surface atoms into symbols, words, company logos, and pictures. One particularly famous example is shown in Figure 5.24b. This series shows the formation of a "quantum corral." A total of 48 iron atoms were deposited on a copper surface at a temperature of 4.2 K using an STM tip. Note that the STM was used to image the surface as well. Once the ring was formed, it created a type of corral that trapped some electrons on the copper surface. This is evident from the standing wave pattern visible in the image. (The bump in the middle of the ring is not an atom, but rather the middle of the electron standing wave.)

5.4.1.2 Skimming across Atoms with the Atomic Force Microscope

Invented in 1985 by Calvin Quate, Christoph Gerber, and Gerd Binnig, the AFM has become one of the most versatile instruments in nanotechnology. However, it took a few years to catch on. In fact, when the journal article introducing the AFM was initially submitted, it was rejected. Why was it rejected? Because the peer reviewers could not believe the authors' claim that their new instrument could "measure forces on particles as small as single atoms." It turns out, of course, that this instrument can indeed measure such small forces. The article was eventually published in March 1986.

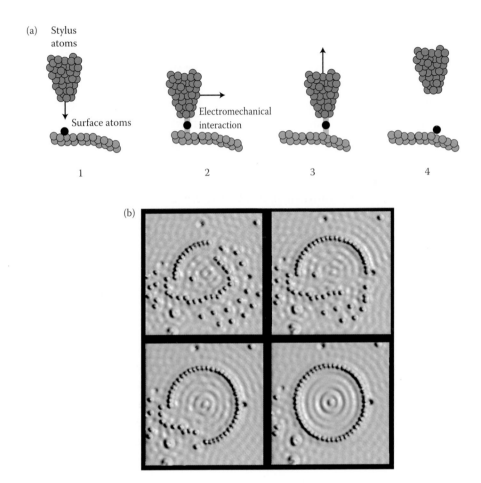

FIGURE 5.24 Arranging atoms with the STM. Placing the tip directly over a surface atom, as shown in (a), increases the interaction between the two. If the electrical and mechanical parameters are just right, the tip can be used to move atoms one at a time to new positions on the surface. The images in (b) were made this way. Here, 48 iron atoms were arranged into a corral that trapped electrons on the surface, as evidenced by the standing wave pattern inside the corral. (Image originally created by IBM Corporation.)

The AFM operates in much the same way a blind person reads a book. However, instead of moving a hypersensitive fingertip over the Braille language, the AFM moves its tiny probing finger over much smaller objects such as DNA molecules, live yeast cells, or the atomic plateaus on a graphite surface. The AFM's "finger" is actually a cantilever beam about a few hundred micrometers long, with a very sharp pointed tip protruding off the bottom, similar to the needle of a record player. This probe is scanned back and forth across a specimen, as shown in Figure 5.25. The first AFM ever built used a piece of gold foil as the cantilever with a crushed piece of diamond mounted at the tip. It also used tunneling as the feedback mechanism instead of beam deflection. Although it achieved a vertical resolution better than 0.1 nm, this first AFM was, like the STM, limited to conductive surfaces only. The AFM has since evolved into a more versatile and sensitive

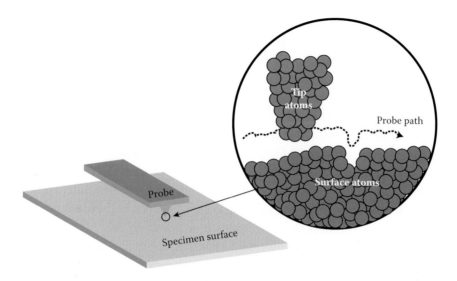

FIGURE 5.25 The AFM cantilever probe. The cantilever has a tip at its free end that is sharp enough to track subnanometer surface details as the probe scans the specimen line by line.

instrument that can be used on just about any kind of surface. The best resolution reported is now on the order of 0.01 nm, measured in vacuum, but the AFM can also be used in air and in liquid. Because hydration is necessary to retain the natural structure and behavior of many materials (especially biological materials), the AFM's ability to operate in liquids makes it unique and valuable to scientific researchers interested in nanometer resolution of "wet" specimens.

Cantilever motion can either be measured optically (by reflecting a laser beam off the cantilever into a photodetector located centimeters away) or by using sensing elements built into the cantilever itself. These sensing elements include piezoresistors (the conductivity of which changes when the beam is stressed), piezoelectric elements (which generate current when the beam is stressed), and capacitive elements (where the beam serves as one of two parallel plates and the capacitance between the parallel plates varies with cantilever deflection).

There are numerous modes in which the cantilever can be used to scan over a surface. Two primary modes are the "contact mode" and the "tapping mode."

In contact mode, the probe's sharp tip stays in touch with the surface of the specimen as it scans. The cantilever beam acts as a spring, so the tip is always pushing very lightly against the sample. As we know, the force of a spring, $F_{spring} = -kx$, where k is the spring constant and x is the displacement from the spring's equilibrium position. The spring constant of AFM cantilevers can vary from 0.001 to 100 N/m, depending on the application; a typical value for contact mode is 0.1 N/m. (The spring constant of a piece of household aluminum foil 4 mm × 1 mm is around 1 N/m.) A typical displacement is 10 nm for an exerted tip–sample force on the order of 1 nN.

As the probe encounters surface features, the microscope adjusts the vertical position of the cantilever's base so that force applied to the sample remains constant. This is done using a feedback loop. The generalized feedback loop is shown in Figure 5.26. To see how it works, consider the case where the probe encounters a bump 12 nm high on the specimen surface. This causes the cantilever to deflect a total of 12 nm, so the "feedback signal" is 12 nm. This signal is subtracted at the beginning of the feedback loop from the set point. Let us say that the set point deflection is 10 nm, a value that ensures that the cantilever tip will remain in contact with the surface without exerting too much force on it. Thus, the differential error signal is $10 - 12 = -2$ nm. That is, the cantilever is deflected 2 nm too far. This error signal is output into the feedback controller, which determines the instructions to be sent to the actuator. The actuator is usually a stack of piezoelectric material attached to the base of the cantilever. The instructions from the feedback controller are in the form of an applied voltage bias, which either elongates or compresses the stack of piezoelectric material, which in turn moves the cantilever up or down. In this example, the actuator is instructed to raise the cantilever by 2 nm. After this, the deflection sensor—in this case a laser/photodetector system—measures the cantilever deflection. The entire loop then repeats. At each point along the scan, the cantilever's vertical position is recorded in order to generate a pixel-by-pixel image of the surface's topography.

In the tapping mode, the cantilever is oscillating at or near its natural frequency, causing it to intermittently "tap" on the sample as it scans. The amplitude of oscillation is typically 20–100 nm. This mode is well-suited to examine soft (biological) samples that are too fragile for the lateral, dragging forces exerted in contact mode. In tapping mode, the feedback loop does not have a set point deflection to maintain; rather, it strives to maintain a set point amplitude.

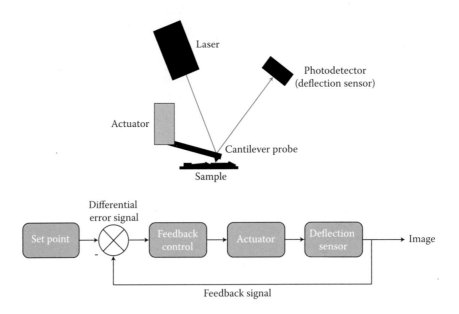

FIGURE 5.26 The AFM feedback loop. In contact mode, this feedback sequence keeps the cantilever's deflection (and hence the tip–sample force) constant by adjusting the vertical position of the cantilever. In tapping mode, the feedback loop keeps the cantilever's oscillation amplitude constant.

Figure 5.27 shows an AFM image of surface atoms on a silicon surface. The image was taken in yet another mode, called the frequency modulation mode. In this mode, the cantilever is made to oscillate at its natural frequency; and when it is brought close to (but not touching) the sample, the long-range forces between the tip and the sample cause the frequency to shift. Thus, the feedback loop in this case works to maintain a set point frequency. This keeps the tip–sample distance constant so that the surface topography can be measured. The image shown here is 256 × 256 pixels; it shows a 10-nm square area and was acquired in 64 s.

Atomic force microscopy promises to evolve and become more versatile. Currently, scan speed is a primary focus of development. Conventional AFMs take a few minutes to acquire a single image, making it impossible to observe many biological or chemical processes as they occur in "real time." So there is much to be gained by a faster scan rate, especially when dealing with dynamic nanoscale systems.

5.4.1.3 Pulling Atoms Apart with the AFM

The forces between molecules govern their interactions. SPMs, especially the AFM, can be used to examine these forces down to the picoNewton range. Of particular interest are the

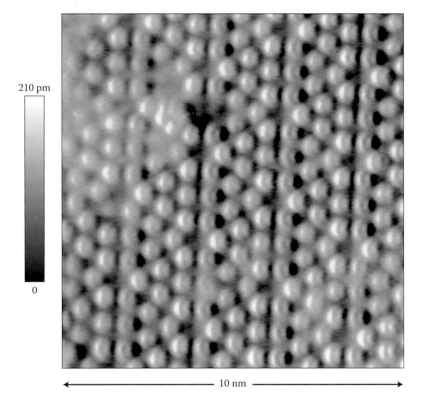

FIGURE 5.27 AFM image of silicon surface atoms. As indicated by the shading key, most of the topographical features shown here are between 0 and 210 pm high. The two crescents seen in each surface atom were probably caused by two atomic orbitals on the endmost atom of the probe tip. These orbitals manifest in the image as a pair of humps when scanning over a surface atom. (Image courtesy of Franz Giessibl, University of Regensburg, Germany.)

highly specific molecular interactions that take place between molecules geometrically complementary to one another—whose unique shapes and surface bonding sites enable them to "recognize" each other and link up. These links are often held together by short-range, noncovalent bonds, such as van der Waals bonds. Although such bonds may be weak on an individual basis, biological molecules tend to form multiple bonds, resulting in a strong connection. Examples of molecular recognition include receptor/ligand pairs such as biotin and avidin, antibody/antigen pairs, and complementary strands of DNA. Without molecular recognition, most of the biochemical processes crucial to life would not take place. (More about this is discussed in Chapter 10, "Nanobiotechnology.")

The AFM enables us to literally get a handle on these molecules and, by very delicately pulling them apart, quantify the binding forces. This gives us a way to better understand how such forces affect the properties of molecules. It is physical chemistry at a fundamental level.

When an AFM cantilever beam bends, it exerts a force. If we know the spring constant of the beam and how far it bends, we can calculate this force. Look at Figure 5.28, where we see a typical "force curve" for an AFM cantilever probe, with "snapshots" of the cantilever at various stages of the measurement. The force exerted by the cantilever on the surface is shown as a function of the probe's vertical position, z.

To obtain a force curve, we first slowly lower a cantilever tip toward a surface (as shown in snapshot 1). When the tip comes within about 1–10 nm of the surface, attractive van der

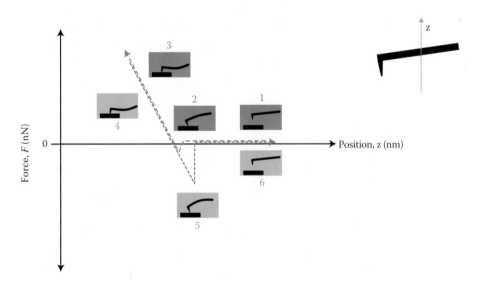

FIGURE 5.28 A typical AFM force curve. The plot shows the force exerted by the cantilever probe as its vertical position varies by z. The blue curve and corresponding "snapshots" of cantilever position relative to the sample are obtained as the cantilever moves toward the surface; the gray curve and snapshots are obtained as it pulls away. On the approach, attractive van der Waals forces snap the tip down into contact with the surface. During retraction, the tip sticks to the surface and stays in contact longer than if there were no adhesion—causing the beam to deflect downward. Eventually, the tip releases and pops off the surface again.

Waals forces between the atoms in the cantilever tip and those on the surface pull the cantilever down, making it snap suddenly into contact (snapshot 2). This causes the small dip to negative force on the approach (blue) section of the curve. As we continue to lower the back end of the beam, the tip end bends up as it pushes against the surface (snapshot 3). The beam exerts a force proportional to how much it is bent. Then the process reverses; the beam is pulled away from the surface (snapshot 4). It tends to retrace the force curve obtained during its approach—with one typical exception. Right at the point where the tip is about to lose contact with the surface, it instead holds on and bends down (snapshot 5). This is due to van der Waals bonds or other, stronger bonds between the surface atoms and tip atoms. These bonds refuse to let go immediately. The more the tip is pushed into the surface, and the "stickier" the surface, the longer the tip will stay attached as the cantilever pulls away. Finally, the force is enough to detach the tip and it snaps back to its equilibrium position (snapshot 6).

As you may have noticed, there are striking similarities in the graphs of (1) the force between an AFM tip and a surface, and (2) the Lennard–Jones force between two atoms. (In Figure 5.28, note that repulsive forces between atoms correspond to a positive AFM force, whereas attractive forces correspond to the negative force that the AFM is exerting.) This is what makes the AFM unique: its tip is sharp enough and its force sensitivity good enough for it to "feel" the forces between atoms in a way no other instrument can.

Now imagine we want to measure not just the forces between a clean probe and a clean surface, but also between two particular molecules. This can be done if we attach one type of molecule to the probe tip and another to the surface. This is known as "functionalizing" the two surfaces. Then we perform measurements to determine the force curve. A region of interest in the curve will be where molecular adhesion prevents the functionalized probe from detaching from the functionalized surface (refer to snapshot 5 in Figure 5.28), as well as the force at the point where the probe eventually does break free. These parts of the curve can be used to measure the rupture force needed to break the bonds of a given molecular pairing.

A good example of this sort of application involves DNA. The famous double-helix structure of DNA, formed by a pair of complementary strands, is fundamental to the storage and implementation of genetic information within living things. The way this molecule works has a lot to do with the forces holding its two strands together. (For an in-depth look at DNA, see Chapter 10.) The strands are complementary because the sequence of base pairs on one strand is chemically matched to the sequence of base pairs on the other—this is the epitome of molecular recognition. Think of DNA as a very selective zipper that will only zip up tight when its two sides match. The zipper is held together by hydrogen bonds between matching base pairs, which are the "teeth" of the zipper.

We can immobilize on the probe a strand of DNA representing one side of the double-helix molecule (say, 30 bases long). On the sample surface we can immobilize the complementary strand (also 30 bases long). This is often done using linker molecules to ensure that only one end of the strand is immobilized. The force curve for such a configuration will have the general shape of the one shown in Figure 5.29. (Only the detraction part of the curve is shown, not the approach.)

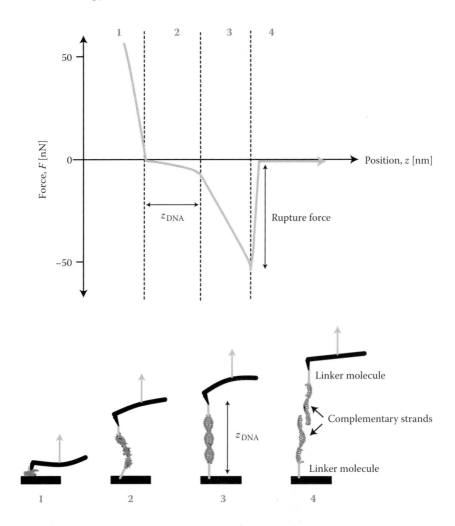

FIGURE 5.29   Pulling a DNA molecule apart. We immobilize complementary strands of DNA on the probe tip and sample surface, using linker molecules as shown, and then measure the force curve. Here, only the detraction portion of the curve is shown. In region 1, the deflected cantilever begins to pull away from the surface. In region 2, the bunched strands exert little force on the cantilever. Raising the cantilever a distance roughly equal to the length of the molecule, zDNA, makes the molecule taut. In region 3, we pull on the molecule until it unbinds. In region 4 the detached strands no longer deflect the beam and the force is zero.

In region 1, the functionalized cantilever is deflected against the functionalized surface. We begin to lift the cantilever's fixed end. In region 2, the complementary strands have bonded and as we pull them in opposing directions, there is very little force because the molecule is bunched up. We pull upward until the molecule is taut, a distance roughly equal to the length of the molecule, z_{DNA}. In region 3, we pull on the taut molecule, exerting more and more force until eventually the bonds linking the strands rupture. In the example shown here, the rupture force is about 50 pN. In region 4, the strands are separated and the force is zero.

The ability to manipulate and separate a single molecule in this manner gives scientists entirely new options for studying nanomechanical phenomena.

5.4.1.4 Rubbing and Mashing Atoms with the AFM

Friction is a measure of how much force it takes to rub one surface across another. Technically, this is *kinetic* friction (as opposed to static friction, which is the friction force between two stationary surfaces that must be overcome to achieve motion). Kinetic friction between various surfaces is expressed as a dimensionless friction coefficient, μ_k, equal to the ratio of the frictional force that opposes motion, $F_{friction}$, and the force pressing the surfaces together, usually called the normal force, F_{normal}:

$$\mu_k = \frac{F_{friction}}{F_{normal}} \tag{5.41}$$

It was previously thought that μ_k was independent of the force between the surfaces. That is, if you pushed them together harder, the frictional force would increase proportionally—so μ_k would not change. Similarly, it was also thought that μ_k was independent of the area of contact. As it turns out, the coefficient we measure *can* vary both with the force between the surfaces as well as the contact area. We know this now because we are able to use very small tools, sensitive to tiny forces.

The AFM probe can even be used to measure friction. When it scans laterally across a surface in contact mode, the frictional force against the tip makes the cantilever twist—and the amount of twisting is proportional to the frictional force (see Figure 5.30). If we have an AFM probe tip made from silicon nitride with a rounded tip about 100 nm in diameter, we would measure a coefficient of friction of about $\mu k = 0.006$ if we drag it across a smooth graphite surface. However, if instead we drag a small silicon nitride ball a few millimeters in diameter over the same surface, we would measure something closer to $\mu_k = 0.1$.

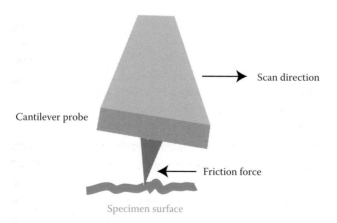

Cantilever probe

Scan direction

Friction force

Specimen surface

FIGURE 5.30 A cantilever probe scanning across a surface twists due to friction. This twisting can be measured with the same techniques used for measuring beam deflection. The amount of twisting is proportional to the frictional force as well as the coefficient of friction.

There are a variety of possible explanations for this large difference in our measurement, including the very light forces that the AFM probes exert, the sometimes increased hardness of surfaces at the nanoscale, the minimal amount of contact area, and the decreased amount of "plowing" (rearrangement of atoms) as the probe drags along. As the AFM tip radius and the force with which the cantilever pushes down are both increased, the coefficient of friction we measure will begin to approach the one we measure using the larger ball. What this tells us is that nano- and microscale devices are capable of sliding against one another very gently, with almost zero friction and therefore practically there is no damage due to wear.

Another interesting aspect of friction measurements with the AFM is the shape of the friction curve during a scan. Because the AFM can measure atomic-scale forces, we can feel the frictional forces of surface atoms individually, giving rise to a typical sawtooth pattern in the graph of frictional force. Look at Figure 5.31. As the probe tip scans across the surface, the periodicity of the atoms creates a unique "stick–slip" behavior. The probe gets stuck, then pops free, again and again.

In addition to all the uses of AFMs that we just mentioned, they can also be used to measure properties such as surface roughness, hardness, and elasticity, as well as adhesion, scratching, wear, indentation, and lubrication. Such studies give us clues about the atomic origins of these mechanisms. Nanoscale variations from bulk behavior have been observed in many of these areas. For example, tests using a probe tip to make dents just a few nanometers deep into silicon surfaces have suggested that nanoscale measurements lead to higher hardness values than typical larger-scale measurements. This may be due to the fact that because the probe deforms such a small volume of material, there is a lower probability that it will encounter defects, which reduce hardness. In general, the coefficients of friction and wear rates have thus far been found to be lower at the nanoscale, and hardness has been found to be larger.

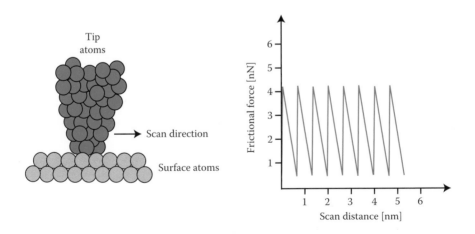

FIGURE 5.31 The "stick–slip" action of an AFM probe rubbing across surface atoms. The force varies due to the periodicity of the surface atoms, giving rise to the sawtooth pattern in the graph.

5.4.2 Mechanical Chemistry: Detecting Molecules with Bending Beams

In 1909, G. Gerald Stoney was depositing thin metal films when he realized that if the object onto which he was depositing the film was thin enough, then the forces distributed within the film could make the object bend. That is, the film added stress to the surface of the object. We know now that the same thing can happen when a chemical accumulates on the top side of a beam. The presence of the chemical can add a distributed force (stress) to the beam's top surface. If the beam is small enough, the added stress will force the beam to bend.

Surface stress (expressed in N/m or in J/m²) is a force distributed within the outermost layer of a solid. Notice that the units of surface stress are the same as those for the spring constant, k. Therefore, for simplification purposes, it may help to visualize the stress as a distribution of springs among the atoms. The stress can stretch or compress the surface. Because nanometer- and micrometer-scale beams tend to have very high surface-to-volume ratios, surface effects are quite pronounced and can lead to noticeable deflections.

Stoney developed a simple relationship between a beam's deflection, z, and the differential surface stress (the difference in stress between the top and bottom surfaces of the beam):

$$\Delta z = \frac{CL^2 (1-\nu)\Delta\sigma}{E_M t^2} \tag{5.42}$$

Here, C is a geometrical constant based on the beam's shape ($C = 3$ for a rectangular beam), L is the length of the beam, t is its thickness, ν is Poisson's ratio (a measure of a material's tendency to get thinner as it elongates), and E_M is the elastic modulus of the beam material (see Figure 5.32).

The radius of curvature, R, is also related to the change in surface stress:

$$R = \frac{E_M t^2}{6(1-\nu)\Delta\sigma} \tag{5.43}$$

These equations can be used to make accurate predictions of beam deformation when the absorption-induced stresses are on a smooth surface or within coatings much thinner than the cantilever thickness, such as a monolayer.

Next, we can use the Shuttleworth equation to relate surface stress to a different parameter known as "surface free energy," represented by γ. This is a measure of the work spent per unit area in forming a surface and holding it together in equilibrium. (Note: work = force × distance.) Put simply, surface free energy can be thought of as the work spent forming the surface, whereas surface stress is the work spent *deforming* it. Like σ, γ is measured in either N/m or in J/m²—again, just like the spring constant, k. It is worth noting that, as opposed to a liquid or gas, it is possible to stretch or compress the surface of a solid without adding or taking away atoms. For small deflections, we assume that essentially the same number of atoms stay on the solid's surface and only the distances separating the atoms change as atomic bonds stretch and compress like springs.

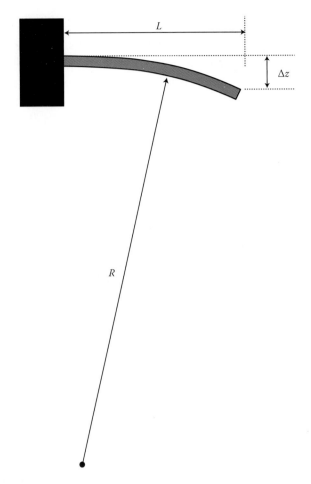

FIGURE 5.32   A cantilever beam undergoing bending. Stoney's equation relates the deflection, Δz, and the radius of curvature R to the change in surface stress of the beam.

The Shuttleworth equation is

$$\sigma = \gamma + \frac{d\gamma}{d\varepsilon} \tag{5.44}$$

This equation states that the total stress in the surface is the existing surface free energy (the first term) plus the additional free energy associated with maintaining the surface in a stretched or compressed state (the second term). We assume that the surface is highly symmetrical and that the stress is isotropic (the same in all directions). The $d\varepsilon$ in the equation is surface strain, the ratio of change in surface area:

$$d\varepsilon = \frac{dA}{A} \tag{5.45}$$

We can see a physical depiction of the Shuttleworth equation in Figure 5.33.

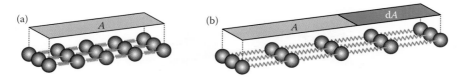

FIGURE 5.33 The Shuttleworth equation relates surface stress σ to surface free energy, γ. The surface free energy is the work done within the area, A, to maintain the surface depicted in (a). The surface stress, in (b), is the sum of the original surface free energy and the additional work needed to stretch the atoms to the slightly larger area, $A + dA$.

The accumulation (adsorption) of molecules on one side of a small beam will make the beam bend. This allows us to perform a kind of mechanical chemistry, where bending beams can be used to detect the presence of particular molecules, or measure molecular properties. As with the mass sensing beams discussed earlier, such beams can be chemically functionalized with chemical coatings that preferentially bind to a target chemical—the molecule(s) we are trying to detect or measure (see Figure 5.34). The amount of bending is proportional to the number of molecules collected, giving us a means of quantifying the concentration of a particular molecule in the cantilever's environment. Bending can also be caused by environmental factors, including temperature changes. Such effects can often be accounted for using a nonfunctionalized reference beam and subtracting its response from that of the functionalized beam.

There are numerous sources of adsorption-induced stresses. The following are sources of surface stress in uncoated beams:

- Electrostatic repulsion among adsorbed molecules (Figure 5.35)

- Dipole repulsion among adsorbed molecules

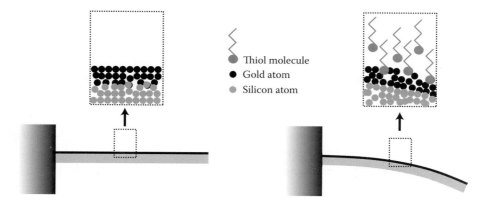

FIGURE 5.34 Nanomechanical chemical detection with a cantilever beam. The beam is coated on its top surface with gold, then exposed to thiols. Thiol molecules contain sulfur, which readily bonds to gold. A spontaneous adsorption process occurs in which the stress state of the surface atoms is altered, causing the beam to bend.

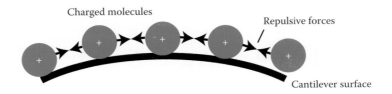

FIGURE 5.35 One source of surface stress is electrostatic repulsion among adsorbed molecules. Here, positively charged molecules are adsorbed onto a cantilever surface but repulsed by the like charges of their neighbors, leading to cantilever bending.

- Change in electron density in the surface layer due to adsorbed molecules
- Charge transfer between adsorbed molecules and surface
- London-type dispersive forces
- Covalent bonding between adsorbed molecules and surface

When the beam is coated with something into which the analytes can absorb, such as a polymer film, the swelling of this coating can also cause stress.

If we assume that the deflection of the cantilever due to the adsorbing of molecules is negligible (although measurable), we can neglect the second term of the Shuttleworth equation (Equation 5.44). This leaves a direct relationship between the surface stress of the cantilever and the surface free energy. Then, if we know how much mass of a certain type of molecule has adsorbed on the microcantilever and the molar mass of this molecule, we can estimate the energy associated with the adsorbed molecule. (Free energy is also expressed in kJ/mol.)

5.5 SUMMARY

Both classical mechanics and quantum mechanics can be used to study the forces that move nanoscale objects. Using a simple ball-spring system and the concepts of simple harmonic motion, we can model the nanoscale oscillatory motion of beams and atoms. We break down the forces acting on an oscillating object into three component forces: (1) the spring force, F_{spring}, (2) the damping force, F_{damp}, and (3) the driving force, F_{drive}. The frequency at which a nanoscale system will oscillate is uniquely high because the masses involved are so small. Higher frequencies open doors for new applications, such as extremely sensitive mass detectors and novel ways to store information.

The forces among atoms and molecules are exclusive to the nanoscale: on any larger scale we measure only the average of many such forces. The technology of nanomechanics enables us to *observe*, *measure*, and *exert* forces on a molecular, or even atomic, level. For example, the accumulation of atoms on one side of a small, chemically functionalized beam will generate forces that make the beam bend, a new type of "mechanical chemistry."

Using classical mechanics, the Lennard–Jones interaction model provides us with a model for describing the energy of a pair of atoms as a function of their separation distance, where we treat the bonds between atoms like springs. Using quantum mechanics, the Schrödinger equation tells us the allowed energy levels of a nanoscale mechanical system. At very small sizes and energy levels, quantum mechanics and classical mechanics each predict very different behavior for an oscillating system—however, the predictions eventually overlap as the size and energy of the system increase.

A phonon is the quanta of vibrational energy, and is transferred from atom to atom through a crystal lattice through the atoms' inherently elastic bonds. We discuss them further in Chapter 7.

SPMs, including the AFM and the STM, can interact with atoms on an individual basis. We use these finger-like tools to create topographical images of surfaces, measure the forces between molecules, move atoms around on a surface one at a time, and measure surface properties such as the coefficient of friction.

HOMEWORK EXERCISES

5.1 Who is generally credited with the founding of classical mechanics?

5.2 True or false? In both classical and quantum mechanics, it is assumed that all motions can be predicted exactly.

5.3 What is an oscillation?

5.4 An electric field in a color television tube exerts a force of 1.7×10^{-13} N on an electron ($m = 9.11 \times 10^{-31}$ kg).
   a. What is the acceleration of the electron?
   b. If the electron starts from rest, how far will it have traveled down the tube after 1 ns?
   c. What will be its velocity at that time?

5.5 The flea, a micromechanical marvel, can hop as high as 70 cm.
   a. What is its initial speed as it jumps from the ground? The acceleration of gravity is -9.8 m/s^2.
   b. How much time does the flea spend in the air?

5.6 How does the natural frequency of a cantilever beam change if the width of the beam is tripled?

5.7 A silicon cantilever beam is 300 μm long, 100 μm wide, and 6 μm thick. Silicon's modulus of elasticity is 110 GPa; its density is 2330 kg/m^3.
   a. Determine the spring constant k of this beam.
   b. Determine the natural frequency in radians per second.
   c. Express your answer from part (b) in Hertz (Hz).

5.8 Name three sources of damping.

5.9 True or false? The damping force varies with the acceleration of the oscillating object.

5.10 How does increasing the amount of damping affect the amplitude of an oscillating system? How does it affect the frequency of oscillation?

5.11 A mass ($m = 350$ μg) coupled to an elastic element with a spring constant $k = 0.4$ N/m is initially displaced 400 nm from its equilibrium position.

    a. How much total energy is stored in this system?

    b. If the mass is then released, what will be its maximum kinetic energy, assuming damping is negligible?

    c. What will be its maximum velocity?

5.12 True or false? The probability that an oscillating mass is located at a given value of x is inversely proportional to the mass' velocity.

5.13 A cantilever beam has length, $l = 100\ \mu m$; width, $w = 40\ \mu m$; thickness, $t = 5\ \mu m$; modulus of elasticity, $E_M = 100$ GPa; density, $\rho = 2000$ kg/m³; and damping coefficient, $b = 10^{-7}$ N s/m.

    a. What is the beam's damped natural frequency, f_d (Hz)? (Do not forget, for a cantilever beam $m_{eff} = 0.24\ m$ in the equation for damped natural frequency.)

    b. What is the quality factor?

    c. A periodically varying driving force with $F_0 = 10$ nN acts on the beam. Plot the amplitude, A, of the beam's oscillation as a function of the drive frequency, f_{drive}. Include f_d on the graph so that the resonance peak is visible.

    d. On the same graph, plot the amplitude if the damping coefficient quadruples. Label this curve "extra damping."

    e. On the same graph, plot the amplitude if, instead, the effective mass of the beam increases by 500 fg. Label this curve "extra mass."

    f. Finally, on the same graph, plot the amplitude if, instead, the spring constant increases by 5 N/m.

5.14 A cantilever beam has a spring constant $k = 80$ N/m and an effective mass $m_{eff} = 7.68 \times 10^{-11}$ kg. The coefficient of damping is $b = 10^{-5}$ N s/m.

    a. Determine the quality (Q) factor of this system.

    b. Plot the amplitude, A, of the beam's oscillation as a function of the drive frequency, f_{drive}, for a driving force $F_0 = 300$ nN. Estimate Q for this system using your graph—labeling the damped natural frequency, f_d, and the bandwidth, Δf, that you measure.

5.15 True or false? The general shape of the Lennard–Jones potential energy curve applies solely to van der Waals interactions.

5.16 The Lennard–Jones potential energy between a pair of atoms is determined to be

$$PE(x) = \frac{2.3 \times 10^{-134}\, \mathrm{J m^{12}}}{x^{12}} - \frac{6.6 \times 10^{-77}\, \mathrm{J m^6}}{x^6}$$

    a. Plot the potential energy curve as a function of separation distance and determine the equilibrium separation, x_e.

    b. Determine the force between these two atoms at x_e.

    c. What is the spring constant k of this bond?

    d. What is the natural frequency, expressed in hertz, of this atomic pair if their masses are 4.12×10^{-26} and 2.78×10^{-26} kg?

5.17 An atom in a lattice has a resonance frequency of 10 THz. According to quantum mechanics, what is the lowest amount of energy this oscillator can have?

5.18 What is the minimum energy of a pendulum that has a period of 2 s?

$$\text{Period}\,\tau = \frac{1}{f_n}, \quad \text{so}\, f_n = 0.5\,\text{Hz}$$

5.19 True or false? Quantum mechanics (using the Schrödinger equation) predicts that an oscillator in its lowest energy state will have a probability distribution, $|\psi_0|^2$, which is almost completely opposite the distribution, $p(x)$, predicted by classical mechanics.

5.20 True or false? The higher the quantum number of an oscillating system, the less the classic mechanical and quantum mechanical predictions of the system's behavior agree.

5.21 A harmonic oscillator with spring constant, k, and mass, m, loses 3 quanta of energy, leading to the emission of a photon.
   a. What is the energy of this photon in terms of k and m?
   b. If the oscillator is a bonded atom with $k = 15$ N/m and $m = 4 \times 10^{-26}$ kg, what is the frequency (Hz) of the emitted photon? (Note: the energy of a photon is $E_{photon} = hf$.)
   c. In which region of the electromagnetic spectrum (x-ray, visible, microwave, etc.) does this photon belong?

5.22 What is a phonon?

5.23 True or false? Phonons are conductors of heat in metals, insulators, and semiconductors.

5.24 What characteristic must an atomic lattice have for it to support optical phonons?

5.25 In a regular breath, how much does the human lung expand due to the nitrogen in the air?

5.26 If you were designing a nanoscale mass sensor to measure minute amounts of a particular chemical in the air, what would be some of your major design considerations?

5.27 A chemistry student accidentally drops a large mercury thermometer and it breaks. The thermometer contained 2 g mercury (Hg), a potent neurotoxin. The Hg leaks out and pools together into a droplet on the floor of the laboratory. The Occupational Safety and Health Administration (OSHA) sets the permissible exposure limit (PEL) for mercury vapor in air at 12 ppb, by volume. At no time should anyone be exposed to mercury vapor above this threshold. The molar mass of Hg is 201 g/mol. The temperature and pressure in the lab are 14.7 psia and 70°F, respectively.
   a. Convert the Hg vapor PEL into mg/m^3.
   b. A professor immediately evacuates the room and seals it off so that no air can flow in or out. The drop begins to evaporate from the drop into the air at a rate of 0.4 µg/min. The laboratory (air) measures 100 m^3. Within 45 min, a cleanup crew arrives. Has the mercury vapor concentration exceeded the PEL by this time? (We assume an even distribution of the evaporated mercury throughout the lab.)
   c. The cleanup crew has a cantilever mass sensor. The cantilever is coated in gold, which binds to mercury vapor. The cantilever is 655 µm long, 10 µm

wide, with a modulus of elasticity $E_M = 100$ GPa and density $\rho = 2330$ kg/m^3 and the minimum detectable frequency change $\Delta f_{min} = 0.1$ Hz. What is the minimum detectable mass of this sensor?

d. What is the minimum number of Hg molecules that must collect on the sensor to equal the minimum detectable mass? (Assume that only Hg collects on the sensor.)

e. The sensor system is coupled to a pump that pulls in a 10-m^3 sample of the contaminated air (after the 45 min of Hg evaporation). What is the minimum percentage of the mass of mercury in this sample that must be collected on the cantilever sensor to detect the mercury?

5.28 The best vertical resolution of an AFM is on the order of _____ nanometers.

5.29 When operating in contact mode, the feedback loop of an AFM works to maintain a constant. In tapping mode, the feedback loop works to maintain a constant.

5.30 True or false? To create an image of a bacterium in its native, aqueous environment, tapping mode would be a preferred imaging mode.

5.31 Name at least three methods of monitoring cantilever deflection in an AFM.

5.32 When measuring a "force curve" with an AFM, what typically causes the cantilever probe to suddenly snap into contact with the surface?

5.33 What two-atom interaction model has a force–distance curve closely resembling the force–distance curve taken with an AFM probe?

5.34 Which microscope came first: the AFM or the STM?

5.35 What does the STM's feedback loop attempt to keep constant?

5.36 True or false? The STM uses only tunneling current to manipulate atoms and move them around on a surface.

5.37 An AFM cantilever probe with a spring constant of 0.3 N/m is lowered onto a surface until it bends 15 nm. The probe is then scanned laterally across the surface, where a coefficient of kinetic friction, $\mu_k = 0.04$, is measured. What was the frictional force on the probe while it scanned?

5.38 What geometrical feature of nano- and microscale cantilever beams makes them so sensitive to surface effects?

5.39 A chemical adsorbs onto the surface of a rectangular silicon beam measuring 200 μm long and 1 μm thick, with an elastic modulus of 100 GPa and a Poisson's ratio of 0.22. This alters the surface stress by 2×10^{-4} N/m.

a. By how much does the beam deflect?

b. Using an AFM with 100 times less vertical resolution than the best demonstrated, would you be able to detect such a deflection?

c. If the deflected beam formed part of the curved surface of a large virtual sphere, what would be the sphere's diameter?

5.40 The equation can be used to estimate the energy associated with molecules adsorbed on the surface of a beam.

RECOMMENDATIONS FOR FURTHER READING

1. M. D. LaHaye et al. 2004. Approaching the quantum limit of a nanomechanical resonator. *Science* 304:74.

2. A. Cleland. 2002. *Foundations of Nanomechanics: From Solid-State Theory to Device Applications.* Springer.

3. S. Rao. 2003. *Mechanical Vibrations.* Prentice Hall.
4. M. Blencowe. 2004. Nanomechanical quantum limits. *Science* 304:56.
5. B. Bhushan. 2004. *Springer Handbook of Nanotechnology.* Springer.
6. F. Giessibl et al. 2000. Subatomic features on the silicon (111)-(7 × 7) surface observed by atomic force microscopy. *Science* 289:422.
7. The Roukes Group: nano.caltech.edu

CHAPTER 6

Nanoelectronics

6.1 BACKGROUND: THE PROBLEM (OPPORTUNITY)

A typical Pentium® chip has 50 million or so transistors occupying about a square inch of space. This seems astounding today, but it will not even seem adequate in a decade when computers will be expected to crank through even more complex tasks in less time. More tasks, more transistors. Less time, less space. The problem (opportunity) is obvious: we need to find a way to cram more transistors into a square inch.

Here is a related problem: heat. Say that a picture taken with a digital camera on a cell phone represents 100 kilobytes (kB) of information. To manage information, a computer uses roughly 160,000 electrons per byte. So, for the computer to manage the entire picture, it needs 16 billion electrons. And as those electrons flow through the labyrinthine conduits and circuits and transistors, they generate heat. A computer processor chip can generate as much heat as the electric range on which you cook your food—100 W. The smaller and faster the transistors, the more unmanageable this heat becomes. Thus, a different approach becomes a matter of necessity.

Electronics is a branch of physics in which circuits are designed and constructed to solve practical problems. Electronics deals in electrons. Or at least it used to. Now it can deal with an *electron*, singular. So how about a computer with components so small they use only one electron per byte?

In our discussion of nanoelectronics, we will dissect the electronic properties of bulk materials to see how we might recreate electronic devices with the functions we have come to expect, except with smaller dimensions and utterly new operating principles.

6.2 ELECTRON ENERGY BANDS

We learned in Chapter 3 that an atom can never have more than two electrons in an orbital. Extra electrons are boosted instantaneously into unfilled orbitals with equal or higher energy. So what happens when one atom is next to another identical atom? That is, what happens in a pure solid material? If the atoms were separated from each other, the electrons in atom 1 would have exactly the same amounts of energy as the electrons in atom 2. But when atom 1 and atom 2 are brought into close proximity and the orbitals begin to overlap, it is as if the electrons occupying each sublevel suddenly come face to face with their clone. How do all the identically energized electrons sort themselves out?

In some ways, what happens is analogous to traffic on a road. Imagine that an electron energy sublevel (such as 1s or 2s or 2p, etc.) is a road on which only one car (electron) is allowed per lane. In the simplest case, the energy sublevel has just one orbital (the 1s orbital), so we will make that a small country road with just two lanes. This road can only accommodate two cars, one going in each direction, just as the 1s orbital can only accommodate two electrons. So what happens when there are more cars? Do we build completely new roads (new energy sublevels)? No. Just because there are more electrons to accommodate does not mean completely new energy sublevels are created, just more "lanes" (and we would widen the road a bit). In other words, the energy sublevels are split.

Let us look at an example. Sodium is an atom with 11 electrons, which means 10 of them are in filled-out sublevels while the highest energy electron is alone in the 3s sublevel. When two sodium atoms are close to each other, these outer electron orbitals start to overlap. And as soon as this happens, the outermost energy sublevel begins to split. Look at Figure 6.1a. It shows the energy, E, of the outermost electrons as a function of the separation distance, r. At large separations, the two electrons have the same energy. As they near one another, the energy sublevel begins to split into two. This happens not only to the 3s energy sublevel, but also to all those below.

As more sodium atoms collect within this separation distance, more splits occur within the energy sublevels. Figure 6.1b shows the splitting of the highest energy sublevel as eight sodium atoms are brought into close proximity. Recall that sodium is a metal, where the atoms are surrounded by a "sea" of electrons (see Chapter 4, Section 4.2.3). With each new sodium atom, a new electron is added to the sea of 3s valence electrons, causing yet another split in the 3s energy sublevel. Notice how the splits are closer in energy to one another when there are eight atoms as opposed to just two. So where does the analogy go from here? Well, eventually we get to a point with our superhighway where we cannot make it any

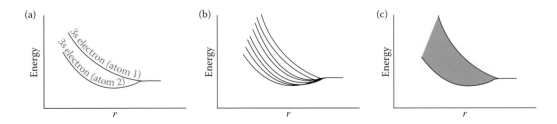

FIGURE 6.1 The energy of the outermost electrons of two sodium atoms as a function of separation distance, r. (a) At large separations, the two electrons have equal energy. As the orbitals overlap, the outermost energy sublevel splits, and there are two different energies for the electrons. (b) The energy sublevel splits eight times as eight atoms are brought together, with different energies for each electron. (c) With numerous atoms, the splits are too close to distinguish and the sublevel can instead be thought of as a continuous energy band.

wider, although we have to keep adding lanes. So the lanes have to get smaller and smaller, until eventually we cannot even tell them apart.

And that is what defines a solid. If we have lots of atoms packed together—for example, 100 sextillion (or 10^{23}) atoms in a cubic centimeter, there will be 100 sextillion splits in the uppermost energy sublevel. There are 100 sextillion splits within each of the lower sublevels as well. Soon the splits are so closely bunched that they cease to be distinct. They are better understood as a continuous energy band, similar to the one in Figure 6.1c.

Remember that these energy bands are broadened versions of the quantized energies of a lone atom, so there is a band for each energy sublevel. Figure 6.2 shows (a) the electron orbital configuration of sodium, and (b) the corresponding energy bands. Between the bands are gaps, which are energy levels forbidden to electrons in the solid. Once the sublevel has split and broadened into a band, it will not get any wider: the width of the bands is independent of the number of atoms in the solid.

Sodium is a case where the four sublevels of the sodium atom (1s, 2s, 2p, and 3s) become four separate bands in a sodium solid. Not all materials are this way. Most notable for this discussion is silicon. Silicon's electron configuration is $1s^2 2s^2 2p^6 3s^2 3p^2$, so all the sublevels up through 3s are full, with two valence electrons left in the 3p sublevel (see Figure 6.3). And this is the configuration when discussing a single silicon atom.

However, when silicon atoms form bonds with other silicon atoms to create a crystal solid, the packing of the atoms is so intricate and the covalent bonds are so tight that the electronic orbitals overlap and share electrons. This is known as orbital hybridization; and in the case of silicon, it causes the 3s orbital (with its two electron states) and the 3p orbital (with its six electron states) to instead behave as a pair of hybrid s–p orbitals, each with four available electron states. Energy wise, this makes for two new hybrid s–p bands. The lower of these two bands ends up accommodating four electrons, leaving the top band

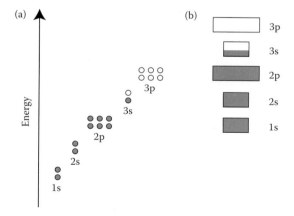

FIGURE 6.2   Energy bands of sodium. In (a), the electronic configuration of sodium as sublevels are filled from low to high energy. When numerous sodium atoms are in close proximity, as in a solid, these can be thought of as energy bands—shown to the right in (b). With sodium, the 3s band is only half full and the 3p band above it is completely empty. The gaps between bands are forbidden energies.

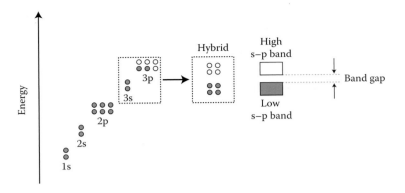

FIGURE 6.3 The hybrid s–p bands of silicon. A silicon atom has the electron configuration $1s^2 2s^2 2p^6 3s^2 3p^2$, so all the sublevels up through 3s are full, with two valence electrons left in the 3p sublevel. In solid crystal form, the atoms' close-packed electron orbitals overlap, causing the 3s and 3p orbitals to hybridize into an s–p orbital. The corresponding energy bands each have four available electron states. In the ground state, the top band is empty; and the bottom band is full. Separating them is a "band gap."

unoccupied, as shown in Figure 6.3. (This also happens with carbon and germanium, two other elements in the same Periodic Table column as silicon.)

6.3 ELECTRONS IN SOLIDS: CONDUCTORS, INSULATORS, AND SEMICONDUCTORS

Electrons fill sublevels in a process that follows strict rules, like long division. And, like long division, there is often a remainder left over. The valence electrons are the remainder. They are the electrons in the outermost occupied energy sublevel. (Note that a noble gas has no "remainder." Its top orbital is completely filled but we still call the electrons in this orbital the valence electrons.)

Electrons in fully occupied energy sublevels are stable and confined to inner orbitals closer to the nucleus and rarely participate in physical processes. It is the valence electrons that give an atom its "personality"—its willingness to enter into relationships with other atoms. And since an atom's chemical properties are largely a reflection of how many valence electrons it has, the columns in the Periodic Table are actually organized by the valence number.

We categorize materials as insulators, semiconductors, and conductors based on their ability to conduct electricity. In Figure 6.4, we generalize the band structure of these materials, the uppermost band with electrons in it at a temperature of 0 K. (At this temperature, the solid is in its lowest energy state—the "ground state"—and so we choose to define the valence band here.) The valence band can be partially or completely full. The unfilled band above the valence band is known as the conduction band. Electrons provided with enough energy are boosted up into the conduction band and move throughout the solid conducting electricity. Separating the valence and conduction bands is a gap in allowed energies called the band gap.

Figure 6.5 shows the band gap of an insulator, a semiconductor, and a conductor.

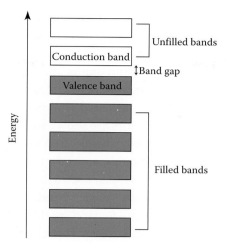

FIGURE 6.4 The band structure of a solid material. Low bands are fully occupied. The uppermost band with electrons in it at the ground state is the valence band, which may be partially or completely full. The unfilled band above the valence band is the conduction band. Separating these two bands is the band gap.

Insulators are materials with exactly enough electrons to keep their energy levels completely occupied. Thus, the valence band of an insulator is completely full—and stable. However, contrary to what one might think, the valence electrons in insulators actually can move within the material. Of course, there is little reason for them to do so since they are fixed in stable chemical bonds within the lattice. Yet if for some reason two electrons occupying filled orbitals happen to swap places within the lattice, it is not a big deal. No change is noticed because the electrons are indistinguishable from each other. Note also that this does not constitute electrical current since there is no net motion of electrons. The reason insulators do not conduct well is because their band gap is so large. It takes a very

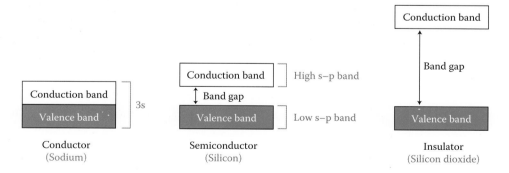

FIGURE 6.5 The band gap of a conductor, a semiconductor, and an insulator. For electrical conduction, electrons need enough energy to move up in the conduction band. In insulators, valence electrons are separated by a wide gap from the conduction band. In semiconductors, the gap is small enough that thermal energy or other types of excitation can be enough to boost electrons across the gap. In conductors such as metals, the valence band and conduction band overlap.

large amount of energy (such as very high heat) to excite the electrons from the valence band to the conduction band. Silicon dioxide, glass, plastic, rubber, and wood are examples of good insulators.

With semiconducting materials, the band gap is smaller. There is no "official" cutoff separating insulating materials from semiconducting materials; conventionally, materials with band gaps less than about 3 eV are considered semiconductors. The band gap energies of a variety of insulators and semiconductors are shown in Table 6.1. The smaller band gap of semiconductors increases the likelihood of an electron's jump across this energy gap due to thermal or other types of excitations. Silicon and germanium are examples of good semiconductors. We use silicon as an exemplary semiconductor in Figure 6.5. The lower hybrid s–p band is full and represents the valence band. The higher hybrid s–p band is empty and represents the conduction band. Separating them is the band gap energy.

Conductors such as metals are a special case where the conduction band and the valence band are not separated by a gap because they occur in the same band. This effectively makes the band gap zero. This is because the Fermi energy, E_F, of conductors occurs in the middle of the highest occupied band. In the ground state, all allowed electron energies below E_F are occupied and all those above it are empty. We use sodium, with its partially filled 3s band, as the exemplary conductor in Figure 6.5. (Other conducting metals, such as

TABLE 6.1   Band Gap Energies of Selected Semiconductors and Insulators at 300 K

Material	Band Gap Energy (eV)
Lithium fluoride (LiF)	12
Calcium fluoride (CaF$_2$)	10
Sapphire (Al$_2$O$_3$)	9.9
Silicon dioxide (SiO$_2$)	8.4
Diamond (C)	5.5
Zinc sulfide (ZnS)	3.6
Copper chloride (CuCl)	3.4
Gallium nitride (GaN)	3.4
Zinc oxide (ZnO)	3.2
Silicon carbide (SiC)	2.8–3.2
Titanium dioxide (TiO$_2$)	2.9
Zinc selenide (ZnSe)	2.58
Cadmium sulfide (CdS)	2.42
Zinc telluride (ZnTe)	2.28
Gallium phosphide (GaP)	2.26
Aluminum arsenide (AlAs)	2.14
Cadmium selenide (CdSe)	1.74
Cadmium telluride (CdTe)	1.45
Gallium arsenide (GaAs)	1.43
Indium phosphide (InP)	1.35
Silicon (Si)	1.14
Germanium (Ge)	0.67
Lead sulfide (PbS)	0.41

aluminum, copper, and gold, each have a single electron in a partially filled valence band.) The bottom half of the band, which contains electrons, represents the valence band. The unoccupied top half is the conduction band. And we know from Chapter 4, Section 4.2.3 that in a metal, the electrons in the outer orbitals are not bound to individual atoms but form a kind of electron "sea." Very little energy is required to turn a valence electron into a conduction electron.

6.4 FERMI ENERGY

All materials have free electrons. Insulators have very few, conductors have many, and semiconductors fall somewhere in between. These free electrons are not entirely free, however; they are beholden to the same laws as other electrons. They must fit into the energy hierarchy—occupy one of the specific, quantized energy states within the solid. The Fermi function, $f(E)$, is a useful statistical formula that gives the probability of finding a free electron in a given energy state, E:

$$f(E) = \frac{1}{e^{((E-E_F)/k_B T)} + 1} \tag{6.1}$$

Here, E_F is the Fermi energy, k_B is Boltzmann's constant (1.38×10^{-23} J/K), and T is the absolute temperature of the solid. The nature of this function dictates that at absolute zero ($T = 0$ K), there is a 100% chance ($f = 1$) of finding a free electron with energy below E_F and no chance ($f = 0$) of finding a free electron with energy greater than E_F. Again, because all materials have at least some free electrons, the Fermi function applies to conductors, semiconductors, and insulators. In conductors, the Fermi energy is in the middle of the highest occupied band. In semiconductors and insulators, the Fermi energy is in the band gap.

Keep in mind that the function, $f(E)$, is merely a mathematical model; and because it is continuous, it indicates that energies inside the band gaps can be occupied by electrons. But we already know that electron energies are actually discrete—separated by the band gaps. In reality, energies inside the band gap are forbidden to electrons. If we were able to zoom in on the *actual* distribution of electron energies, we would see the bands and the band gaps.

The Fermi function is shown in Figure 6.6. Here we see plots of $f(E)$ at absolute zero (0 K) and at a relatively warm temperature (300 K). At any temperature greater than zero, the Fermi function begins to vary at those energies in the vicinity of the Fermi energy. The probability of finding an electron in the valence band begins to drop because higher temperatures provide electrons the extra energy they need to jump into the conduction band. Conductors, which have no band gap for the free electrons to overcome, have more electrons in their conduction band at a given temperature than semiconductors or insulators.

Only those valence electrons very close to the Fermi energy are excited above it. Just *how* close in energy does an electron have to be to the Fermi energy to have a chance of boosting above it and becoming mobile? The answer is that an electron must be within $k_B T$ of the Fermi energy, where k_B is Boltzmann's constant (1.38×10^{-23} J/K), and T is the temperature. An electron at a temperature T has $k_B T$ more energy than it would at 0 K.

(a)

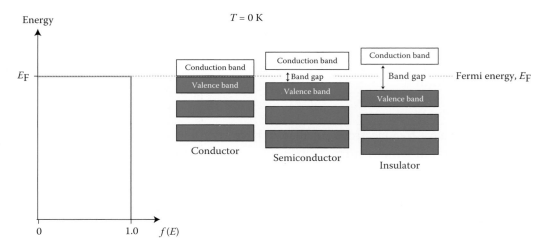

(a)

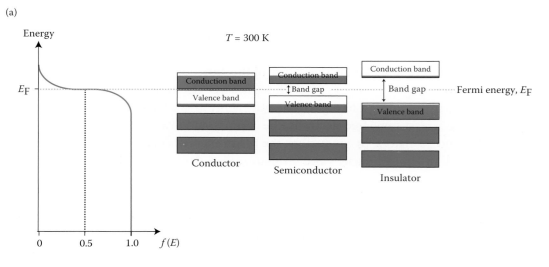

FIGURE 6.6 The Fermi function $f(E)$ and corresponding energy bands in a conductor, a semiconductor, and an insulator. In (a), the material is at 0 K and so $f(E) = 1$. All energy states below the Fermi energy are occupied; all states with energy greater than the Fermi energy are empty. In (b), we can see the effect of adding energy to the material; in this case, the temperature is 300 K. Now some electrons have sufficient energy to jump into the conduction band. While many electrons enter the conduction band in the conductor, only some have enough energy to cross the band gap in the semiconductor, and even fewer in the insulator. At any temperature $T > 0$ K, $f(E_F) = 1/2$.

BACK-OF-THE-ENVELOPE 6.1

At room temperature, roughly what fraction of the electrons in gold have enough energy to become conduction electrons?

Those electrons within k_BT of the Fermi energy are eligible. This range of energies is

$$k_BT = (1.38 \times 10^{-23} \text{ J/K}) (300 \text{ K}) = 4.14 \times 10^{-21} \text{ J}$$

$$k_BT = 0.026 \text{ eV}$$

The Fermi energy, E_F, of gold is 5.53 eV:

$$5.53 - 0.026 \text{ eV} = 5.504 \text{ eV}$$

Thus, any electron with an energy greater than 5.504 eV is close enough to the Fermi energy to become a conduction electron. For simplicity's sake, we assume an even distribution of electrons, energy wise, from zero energy up to E_F at absolute zero (0 K). This way we can calculate that electrons with energies from 5.504 eV up to E_F represent a fraction on the order of 0.026/5.53 = 0.0047 or 0.47% of the total electrons in the gold that can become conduction electrons at 300 K.

6.5 THE DENSITY OF STATES FOR SOLIDS

The minute size of nanomaterials gives them unique electronic properties. One of the major ways in which small-volume materials differ from bulk solids is in the number of available energy states. Realize that any chunk of solid material has a certain amount of energy distributed among its free electrons. The electrons self-organize by incremental energy differences so that there end up being numerous electron energy states within the many energy sublevels. In a bulk solid, the states within each energy sublevel are so close that they blend into a band (as we saw in Figure 6.1).

The total number of electron states, N_S, with energies up to E, can be determined based on quantum mechanics using the following equation:

$$N_S = \left(\frac{8\pi}{3} \right) (2mE)^{3/2} \frac{D^3}{h^3} \tag{6.2}$$

We represent the volume as D^3 (D being the characteristic dimension of the solid), m is the mass of an electron (9.11×10^{-31} kg), and h is Planck's constant. The number of energy states per unit volume, n_s, is then

$$n_s = \frac{N_S}{D^3} = \left(\frac{8\pi}{3} \right) \frac{(2mE)^{3/2}}{h^3} \tag{6.3}$$

If we take the derivative of this equation with respect to energy (dn_s/dE), we arrive at an important relationship known as the density-of-states function, $D_S(E)$. This

---

BACK-OF-THE-ENVELOPE 6.2

How many electron energy states per unit energy are there in a cubic micrometer of gold at an energy of 3 eV?

The density of states is

$$D_S(E) = \frac{8\sqrt{2}\pi m^{3/2}}{h^3}\sqrt{E} = \frac{8\sqrt{2}\pi \left(9.11 \times 10^{-31}\,\text{kg}\right)^{3/2}}{\left(6.626 \times 10^{-34}\,\text{m}^2\,\text{kg/s}\right)^3}\sqrt{4.8 \times 10^{-19}\,\text{J}}$$

$$D_S(E) = 7.36 \times 10^{46}\,\text{J/m}^3$$

We can determine the total number of energy states per unit energy by multiplying by the volume of the solid:

$$\left(7.36 \times 10^{46}\,/\,\text{J/m}^3\right)\left(1 \times 10^{-18}\,\text{m}^3\right) = 7.36 \times 10^{28}\,/\text{J} = 11.8 \times 10^9\,/\text{eV}$$

Thus, at an energy of 3 eV, there are about 11.8 billion electron energy states per 1-eV interval (i.e., 11.8 billion states between 2.5 and 3.5 eV). It is easy to see how little differentiation there is between energy states in even a tiny solid.

---

function gives the number of available electron energy states per unit volume, per unit energy.

$$D_S(E) = \frac{8\sqrt{2}\pi m^{3/2}}{h^3}\sqrt{E} \tag{6.4}$$

The units of the density-of-states function are states "per unit energy per unit volume" (such as /J/ m³). These units can be somewhat puzzling to conceptualize. It might help to think of apartments (states) in a building with a certain number of stories (energies) (see Figure 6.7).

If the volume of the building is 100 m³, we would say that on the first story there are four apartments per story per 100 cubic meter building = 0.04 apts/story m³. On the third story, the density-of-apartments is 0.16 apts/story m³. A deeper or longer building with a corresponding larger volume would contain more apartments per story.

We know that higher energy sublevels have more electrons in them (refer to Figure 3.3) and thus the higher the energy, the higher the density of states. For our purposes, the most important thing to note about the density-of-states equation for a bulk solid is that it is proportional to the square root of the energy:

$$D_S(E) \propto \sqrt{E} \tag{6.5}$$

Various material properties depend on the density of states, such as magnetic susceptibility—the measure of the magnetization per unit volume induced in a solid when a magnetic field is applied. A magnet with a higher density of states at the Fermi level will

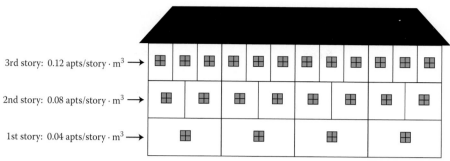

3rd story: 0.12 apts/story · m³ →

2nd story: 0.08 apts/story · m³ →

1st story: 0.04 apts/story · m³ →

Volume of this building = 100 m³

FIGURE 6.7   The density-of-states is analogous to the density of apartments on different stories of a building. In this case, the apartments (states) are in a building (a volume analogous to a particle) with three stories (energies). Each apartment is a place that a person (electron) can occupy. On the first story, there are four apartments per story per 100 cubic meter building = 0.04 apts/story·m³. On the third story the density-of-states is 0.12 apts/story·m³.

have more electrons that enter the conduction band. And since the motion of electrons is what creates magnetic fields, such a magnet will have a higher susceptibility. The intensity of x-ray radiation emitted from a conductor and the specific heat of a conductor at low temperatures both depend on the density of states, as do specific electrical properties of semiconductors and superconductors.

For our purposes, the most relevant application of the density of states is that it provides a means of comparing what is unique, electronically, about nanomaterials. We will see how in the next sections.

6.5.1 Electron Density in a Conductor

Just because an energy state exists does not necessarily mean it is occupied all the time. The Fermi function gives us the probability that a given energy state will be occupied by a free electron. Multiplying the Fermi function by the density-of-states function gives us a more accurate number of free electrons per unit volume per unit energy, based on the statistics of quantum mechanics:

$$f(E)D_S(E) = \frac{1}{e^{((E-E_F)/k_B T)} + 1} \frac{8\sqrt{2}\pi m^{3/2}}{h^3}\sqrt{E} \tag{6.6}$$

Of course, E is the energy, E_F is the Fermi energy, m is the mass of an electron (9.11×10^{-31} kg), and h is Planck's constant. Such being the case, we are able to determine the total number of free electrons per unit volume in a conductor by integrating. To be specific, the number of free electrons per unit volume, n_e, can be found by integrating Equation 6.6 with respect to energy.

$$n_e = \int_0^\infty f(E)D_S(E)dE = \frac{8\sqrt{2}\pi m^{3/2}}{h^3}\int_0^\infty \frac{\sqrt{E}}{e^{((E-E_F)/k_B T)} + 1}dE \tag{6.7}$$

We know that in a conductor at 0 K, the electron distribution goes from zero energy up to the Fermi energy, E_F. So the integral becomes

$$n_e = \frac{8\sqrt{2}\pi m^{3/2}}{h^3} \int_0^{E_F} \frac{\sqrt{E}}{e^{((E-E_F)/k_B)}(0)+1} \, dE \qquad (6.8)$$

Therefore, the number of free electrons per unit volume, or electron density, in a conductor at 0 K is

$$n_e = \frac{8\sqrt{2}\pi m^{3/2}}{h^3} \left(\frac{2}{3} E_F^{3/2} \right) \qquad (6.9)$$

6.6 TURN DOWN THE VOLUME! (HOW TO MAKE A SOLID ACT MORE LIKE AN ATOM)

In terms of the distribution of energy, solids have thick energy bands, whereas atoms have thin, discrete energy states. It seems clear that to make a solid behave electronically more like an atom (and we want to, you will see) we need to make it about the same size as an atom. And indeed this is true. We do not have to get quite that small, however. We can see why mathematically, using what we have concluded so far with the density-of-states function and the Fermi function.

Having solved for the free electron density of a conductor, we can rework Equation 6.9 to solve for the Fermi energy of a conductor:

$$E_F = \frac{h^2}{2m} \left(\frac{3n_e}{8\pi} \right)^{2/3} \qquad (6.10)$$

In this equation, n_e is the only variable—all the other terms are constants. So, the Fermi energy of a conductor just depends on the number of free electrons, N_E, per unit volume, D^3:

$$E_F \propto \frac{N_E}{D^3} \qquad (6.11)$$

Here, the volume is represented by the characteristic dimension of the solid, D, cubed. Values of the Fermi energy for selected metals and the corresponding free electron densities are given in Table 6.2. Since electron density is a property of the material, the Fermi energy does not vary with the material's size. E_F is the same for a particle of copper as it is for a brick of copper.

What this means in terms of nanoelectronics is that for small volumes the energy states span the same range as for large volumes *but with larger spacing between states*. This is the case not only for conductors, but also for semiconductors and insulators too. Let us say that

TABLE 6.2   Electron Density and Fermi Energy of Selected Metals at 300 K

Metal	Free Electron Density, n_e (electrons/m³)	Fermi Energy, E_F (eV)
Potassium (K)	1.40×10^{28}	2.12
Sodium (Na)	2.65×10^{28}	3.23
Silver (Ag)	5.85×10^{28}	5.48
Gold (Au)	5.90×10^{28}	5.53
Copper (Cu)	8.49×10^{28}	7.05

all the states up to E_F are occupied by a total of N_E free electrons. Now we can estimate the average spacing between energy states, ΔE:

$$\Delta E \approx \frac{E_F}{N_E} \tag{6.12}$$

Taking into account Equation 6.11, we see that the spacing between energy states is inversely proportional to the volume of the solid:

$$\Delta E \propto \frac{1}{D^3} \tag{6.13}$$

Figure 6.8 shows an energy sublevel and the spacing between energy states within it, depending on the number of atoms. We have come full circle: at one point we learned that an energy sublevel must be divided as many times as there are atoms in a solid, and that eventually there are too many splits to differentiate so we just refer to each sublevel as a solid energy band. At the other extreme—a lone atom—the sublevel contains one discrete energy state. If we can reduce the volume of a solid or in some way isolate electrons inside a small portion of a larger solid, *the tiny piece of material behaves electronically like an artificial atom* (and we want that, you will see).

6.7 QUANTUM CONFINEMENT

When we reduce the volume of a solid to such an extent that the energy levels inside become discrete, we have achieved quantum confinement. In doing so, we create small "droplets" of isolated electrons. So few electrons, in fact, that we can often *count* them. This is almost always done using a conductor such as gold or a semiconductor such as silicon. The charge and energy of a sufficiently small volume of such materials are quantized just like in an atom. In this way we make fake atoms. And these fake atoms have tunable electrical properties.

It should be noted here that when we confine electrons inside a volume, we are speaking only about the electrons that would ordinarily be free to move about—the excited, conduction electrons.

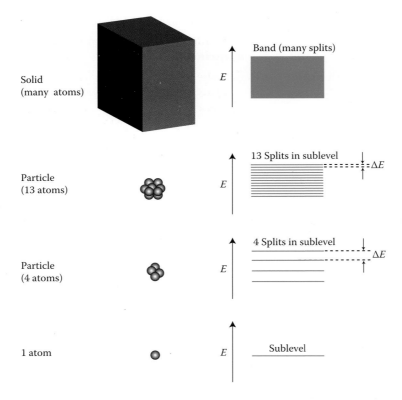

FIGURE 6.8 The spacing between energy states gets larger as the volume gets smaller. At the bottom, a single atom has just one energy state per sublevel. As atoms group together to form particles, there are as many splits per sublevel as there are atoms in the particle. If the particle is small enough, we can discern these splits within the sublevel. As the volume increases to the size of a solid, the spacing between splits gets so tight that the sublevel is best characterized as a band.

BACK-OF-THE-ENVELOPE 6.3

How many electrons are there in a gold particle made from ten atoms? Roughly how many of these electrons might be free conduction electrons?

Gold's electron configuration is $1s^2 2s^2 2p^6 3s^2 3p^6 3d^{10} 4s^2 4p^6 4d^{10} 4f^{14} 5s^2 5p^6 5d^{10} 6s^1$. This means that for each atom, there are $2 + 2 + 6 + 2 + 6 + 10 + 2 + 6 + 10 + 14 + 2 + 6 + 10 + 1 = 79$ electrons. With 10 atoms, that is $(10)(79) = 790$ electrons. The 6s sublevel is the only one left partially unfilled, meaning that there is one free conduction electron per atom. Thus, depending on the temperature and other conditions, there will be about ten or so free conduction electrons in the particle. (At $T = 0$, they would theoretically all be valence and not conduction electrons.)

A single particle can contain hundreds or thousands of electrons, the vast majority of which are tightly bound within inner orbitals. It is the small percentage of electrons that are free to roam that we are interested in confining.

In order to constrain the dimensions of a given volume, we have two fabrication options. In the bottom-up approach, we build low-volume structures atom by atom; in the top-down approach, we remove material from one or more of the three dimensions (length, width, height) of a larger solid. Both of these methods can produce a structure small enough for quantum behavior to manifest. The goal is always to create a conductor or semiconductor island inside an insulator. If the island stands alone, like a particle, the insulator will be the air, or any other insulating gas or liquid that surrounds it. If the island does not stand alone, it may be surrounded by some larger insulating material. Or, electrons can be held captive inside an electric field.

6.7.1 Quantum Structures

Volume is a three-dimensional quantity. To reduce the volume of a box, we can shorten its length, its width, or its height. The same is true of the region occupied by the electrons in a solid. There are three dimensions to confine, and achieving quantum confinement typically necessitates confining at least one of these dimensions to less than 100 nm, or even to just a few nanometers. (In Section 6.9.1, we will discuss exactly how small a volume we need to achieve quantum confinement.)

The more we confine the dimensions, the more the density of states function looks like that of an atom. This progressive discretization gives scientists entirely new ways to understand real atoms and the behavior of electrons. Engineers also can take advantage of quantum confinement in developing novel electronic devices.

When we constrain electrons inside a region of minimal width, we create a "quantum well." Quantum wells can be made from alternating layers of different semiconductors, or by deposition of very thin metal films.

By further constraining the depth of the electron's domain, we create a "quantum wire." Nanotubes and other nanoscale wires can be quantum wires.

Finally, when all three dimensions are minimized, we are left with a "quantum dot." The dot can be a particle located inside a larger structure or on its surface. It can be part of a colloid or entirely isolated. It can also be a place where electrons have been trapped using electric fields.

Looking at Figure 6.9, we can compare the electronic properties of these three quantum-confined structures to those of a bulk material; namely, the plots for the number of free electrons per unit volume at a given energy, $n_e(E)$, and the density of states, $D_S(E)$.

The density-of-states function, while a sufficient mathematical model, is in fact just that—a model. Among the function's inaccuracies is the fact that it is continuous, without gaps separating the various energy bands. If we zoomed in on a plot of the *real* density-of-states function for a bulk material, we would see individual dots instead of a continuous line. Looking at the progression from a bulk material to a quantum dot in Figure 6.9, we see the progression back to discreteness. Because quantum wells and quantum wires each have at least one dimension in which the electrons are free to move, these structures are said to exhibit "partial confinement." Quantum dots exhibit "total confinement."

With quantum dots, the number of electrons per unit volume remains level until the next higher allowed energy state—at which point the number jumps up. The electron

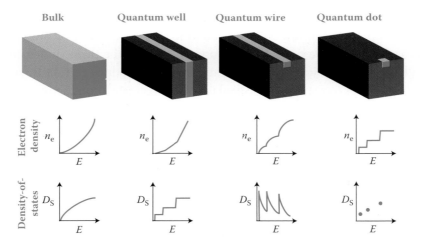

FIGURE 6.9 Three quantum structures and their electronic properties. We can reduce the dimensions of a bulk material in three dimensions to confine electron motion. In constraining width, we create a "quantum well." By also constraining depth, we create a "quantum wire." When all three dimensions are minimized, a "quantum dot" is formed, having electronic properties much like an atom's. As we compare the plots for the number of free electrons per unit volume at a given energy, $n_e(E)$, and the density-of-states, $D_S(E)$, we see that the more the dimensions are confined, the more quantized the electronic characteristics—leading eventually to a quantum dot with a discrete density-of-states function.

density increases with energy because high-energy orbitals usually hold more electrons. (The d and f orbitals hold 10 and 14 electrons, respectively, compared with the s and p orbitals—which hold two and six electrons, respectively—common at lower energies.) We can understand why the electron density must stay the same at each energy if we examine the plot of the density of states. The quantum dot is unique because the density-of-states function ceases to be continuous. There are a specific number of states for the electrons to occupy at a given energy, and there are no more states available until the next energy. Consequently, the number of free electrons in a given volume stays constant between one allowed energy and the next.

6.7.1.1 Uses for Quantum Structures

Quantum structures are unusual little pieces of matter. Given our improving capabilities to define the size and shape of these electronic oddities, we can make mini-laboratories for better understanding atoms and, later, inventing new devices. Think of an atom as a box for electrons, the walls of which are set by the electrons' attraction to the positively charged nucleus. With their quantized energies, atoms have been the sole proving ground for the main ideas of quantum mechanics.

Now we have new tools. Quantum structures have discrete charge states and quantized energy, making them ripe for experimentation and, later, application in real products. Consider this: because we can control the number of electrons in a quantum dot—its atomic number—we are able to recreate the entire Periodic Table of specific, artificial atoms. All in a single device. We can make a fake helium atom that is orders of magnitude larger than the real one and a lot easier to manipulate.

One fascinating application yet to be achieved is the building of an entire lattice of isolated quantum dots—a bulk solid made out of artificial atoms. On a smaller scale, artificial molecules made from pairs of interacting quantum dots have been produced and studied. The small dimensions of quantum dots make them good candidates for high-density data storage. This is just one of the many areas in which quantum dots are being considered for device application. Others include chemical sensing, optics, telecommunications, and computing. We will shed more light on quantum dots' unique optical characteristics in Chapter 8, "Nanophotonics."

As we consider the possibilities and dissect the mechanisms of quantum confinement in the next sections, we will focus much of our attention on dots.

6.7.2 How Small Is Small Enough for Confinement?

Because electrons behave like waves, we trap them according to the rules of waves. In Chapter 3 we showed how this could be done, by mathematically confining an electron in an infinitely deep well. If an electron is trapped inside something, be it an atom or a quantum dot, the key constraint is simple: the electron cannot exist *outside* of the atom or the dot. Such being the case, the wave function that describes the electron must be equal to zero at the boundaries of whatever holds the electron. This, we learned, gives rise to the quantization of wavelengths and energies. When the dimensions of a material are about the same size as the wavelengths of electrons, we rely on quantum mechanics to explain the electronic situation. To confine an electron of wavelength, λ, a material should have a characteristic dimension, D, no smaller than

$$D = \frac{\lambda}{2} \tag{6.14}$$

This rule stems from what we know from Chapter 3 about a particle in a well; that is, the smallest allowed particle wavelength, λ, is

$$\lambda = 2L$$

Here, L is the distance between the opposite walls of the well. Thus, we can say that $D \approx L$ to arrive at Equation 6.14.

BACK-OF-THE-ENVELOPE 6.4

Estimate the minimum diameter of a quantum dot that can effectively confine an electron in copper with a ground state energy equal to the Fermi energy of 7.0 eV (1.12×10^{-18} J).
There is more than one way to determine this. Let us look at two strategies.

Strategy 1: We know from Equation 3.8 that the wavelength, λ, of an electron is Planck's constant, $h = 6.626 \times 10^{-34}$ J s, divided by the electron's momentum, p:

$$\lambda = \frac{h}{p} = \frac{h}{mv}$$

An electron's momentum is equal to its mass, $m = 9.11 \times 10^{-31}$ kg, multiplied by velocity. For copper, the free electrons at the Fermi level move at a speed of 1.56×10^6 m/s.

$$\lambda = \frac{h}{mv} = \frac{\left(6.626 \times 10^{-34} \text{ Js}\right)}{\left(9.11 \times 10^{-31} \text{kg}\right)\left(1.56 \times 10^6 \text{ m/s}\right)} = 4.7 \times 10^{-10} \text{ m}$$

This is 470 pm. Therefore, the characteristic dimension, D (which in this case is the diameter of the quantum dot) must be at least

$$D = \frac{\lambda}{2} = \frac{470 \text{pm}}{2} = 235 \text{pm}$$

Strategy 2: Now we will try the potential well model discussed in Chapter 3, Section 3.3.8, which dictates that the energy of an electron, E, must obey the following relationship:

$$E = \left(\frac{h^2}{8mD^2}\right)n^2, \quad n = 1, 2, 3, \ldots$$

Here, h is Planck's constant, m is the electron mass, and D is the width of the potential well. Therefore, D must be

$$D = \sqrt{\frac{h^2}{8mE}}$$

Solving for the ground-state case of this particular electron (7.0 eV = 1.12×10^{-18} J):

$$D = \sqrt{\frac{\left(6.626 \times 10^{-34} \text{ Js}\right)^2}{8\left(9.11 \times 10^{-31} \text{kg}\right)\left(1.12 \times 10^{-18} \text{ J}\right)}} = 2.32 \times 10^{-10} \text{ m}$$

According to this estimate, the quantum dot should therefore be at least 232 pm in size. So both strategies give practically the same answer. Of course, the quantum dot certainly cannot be much smaller since this is the order of magnitude of single atoms.

In practice, quantum confinement can be obtained in materials having dimensions significantly larger than the smallest-case scenario we estimated in Back-of-the-Envelope 6.4. Confined dimensions are more often the size of a few wavelengths—not just one. In fact, quantum dots typically contain thousands to millions of atoms.

Metal particles with 1–10-nm diameters can behave like quantum dots, while semiconducting particles show quantum confinement up to a few hundred nanometers. For just this reason, applications necessitating quantum confinement more often employ semiconductor materials because the dimensions can be more easily achieved. Next, we will address the specific quantum confinement scenarios of both conductors and semiconductors.

6.7.2.1 Conductors: The Metal-to-Insulator Transition

The essential property of a metal is unquestionably its ability to transport electrons—to be conductive. In conductors, the valence and conduction bands overlap, and so at all plausible temperatures, we expect there to be mobile conducting electrons. As a consequence, it can be difficult to achieve quantum confinement with conductors, especially at normal temperatures.

However, by reducing one or more of the dimensions of the solid to the same order of magnitude as the spacing between atoms in the lattice, we begin to trap the electron waves. Nanomaterials enable us to reach this dimensional threshold of confinement. In doing so, we rob the conductor of its defining characteristic and make it into an entirely different type of material—*an insulator*. In metals, this shift is known as the metal-to-insulator transition.

The average spacing between occupied energy states, ΔE, in a conductor is given by

$$\Delta E = \frac{4E_\mathrm{F}}{3n_\mathrm{v}} \tag{6.15}$$

Here, E_F is the Fermi energy and n_v is the total number of valence electrons. (Remember, in a metal the valence electrons are the free electrons, which at 0 K fill all the energy levels below E_F.) This averaged distribution of electron energies is depicted in Figure 6.10, with electrons arrayed from zero energy up to E_F. We know that smaller dimensions mean fewer atoms and therefore fewer valence electrons. The spacing between energy states, ΔE, grows as we reduce the size of the conductor. Only those electrons within $k_\mathrm{B}T$ of the Fermi energy can become conductors. As soon as $\Delta E > k_\mathrm{B}T$, the conductor no longer conducts. This is shown in Figure 6.10.

Practical application of the metal-to-insulator transition includes development of nanoscale Schottky diodes, which are found in numerous electronic products. Consisting of a metal–semiconductor junction, these diodes could be formed by controlling the size of metal particles in contact with each other such that one of the particles is a good conductor and the other is not.

BACK-OF-THE-ENVELOPE 6.5

At what diameter does a silver particle achieve quantum confinement at 38°C?

Certain electrons, occupying a specific range of energy states, can be excited above the Fermi energy. This range is determined by $k_\mathrm{B}T$. For quantum confinement, the following must be true:

$$\Delta E = \frac{4E_\mathrm{F}}{3n_\mathrm{v}} > k_\mathrm{B}T$$

Rearranging Equation 6.15 gives us the number of valence electrons the particle can have

$$n_\mathrm{v} < \frac{4E_\mathrm{F}}{3k_\mathrm{B}T}$$

$$n_\mathrm{v} < \frac{4\left(8.78 \times 10^{-19}\,\mathrm{J}\right)}{3\left(1.38 \times 10^{-23}\,\mathrm{J/K}\right)\left(311\,\mathrm{K}\right)} = 272$$

The particle must therefore have fewer than 272 valence electrons. A silver atom has the electronic configuration $[Kr]4d^{10}5s^1$, giving it a single valence electron in the 5s sublevel. With a ratio of one valence electron per atom, the particle could therefore have fewer than 272 atoms.

Silver has a density of 10,490 kg/m³ and its atomic mass is 107.9 g/mol. So we are able to calculate the density of atoms in a given volume:

$$\frac{\left(10490 \text{ kg/m}^3\right)\left(6.02 \times 10^{23} \text{ atoms/mol}\right)}{0.1079 \text{ kg/mol}} = 5.85 \times 10^{28} \text{ atoms/m}^3$$

The volume of the particle, V_p, is given by

$$V_p = \frac{272 \text{ atoms}}{5.85 \times 10^{28} \text{ atoms/m}^3} = 4.65 \times 10^{-27} \text{ m}^3$$

We will assume the particle is a perfect sphere, in which case, its diameter, d, is given by

$$d = 2\left(\frac{3V_p}{4\pi}\right)^{1/3} = 2\left(\frac{3\left(4.65 \times 10^{-27} \text{m}^3\right)}{4\pi}\right)^{1/3} = 2.1 \times 10^{-9} \text{ m}$$

By this estimate, the particle behaves like a nonmetal when its diameter is less than 2 nm.

6.7.2.2 Semiconductors: Confining Excitons

Quantum confinement in semiconductors is achievable at much larger dimensions than is necessary for conductors. The electronic configuration of semiconductors, unlike conductors, includes a band gap between the valence and conduction bands. We will need a separate physical model to explain the conditions of quantum confinement.

At normal temperatures, a tiny fraction of the electrons in a semiconductor have sufficient energy to cross the band gap and become mobile conductors. The electrons can also receive energy in forms other than heat, such as from an applied voltage potential or an incoming flux of photons (light).

Regardless of where it gets the energy, once an electron jumps to the conduction band, the empty location it leaves behind is called a "hole." Energy-wise, a hole is a temporary vacancy in the valence band, but it is also a physical location. Because electrons are negatively charged, their absence within the crystal lattice assumes a positive charge and an attraction arises between the two. This electron–hole pair is called an exciton. (Note that there are two types of exciton: (1) the Wannier–Mott, and (2) the Frenkel. Frenkel excitons are tightly bound, with sizes similar to that of crystal's unit cell; the Wannier–Mott excitons are weakly bound and occur far more commonly; it is the latter type that we will discuss here.)

There is much that is common between an exciton and a hydrogen atom. In 1913, Neils Bohr asserted that the atom was a nucleus with orbiting electrons—the simplest case being hydrogen with its one electron and one proton. At the lowest energy state, the distance separating hydrogen's electron from its proton—about 53 pm—has come to be known as the Bohr radius, a_0. As with a hydrogen atom, there is a specific separation between the

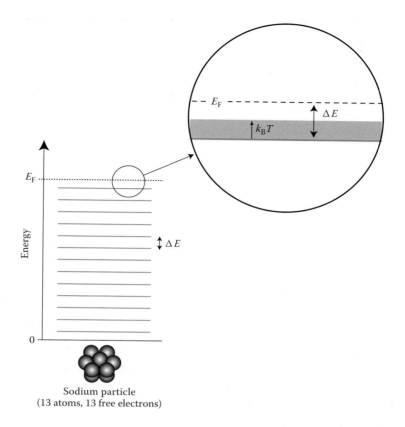

FIGURE 6.10 Quantum confinement makes a metal particle into a nonmetal particle. We can approximate the average spacing, ΔE, between free electron energies using Equation 6.12. Here we show a 13-atom sodium particle, with 13 free electrons and hence 13 energies between 0 and E_F. Random thermal vibrations can boost the energy of a free electron by $k_B T$. Yet unless this added energy boosts at least the topmost electrons above E_F, the metal will not conduct. If ΔE is larger than $k_B T$ (the situation shown here), the electrons remain below E_F and the particle no longer conducts like a metal. So, if the particle is small enough or the temperature is low enough, quantum confinement is achieved.

electron and the hole of an exciton. This distance varies, depending on the material. It is known as the exciton Bohr radius, a_{ex}.

Excitons behave so much like hydrogen atoms that they are even given an effective mass m_{ex}, expressed as the ratio of the exciton mass to that of a free electron. We use this mass and the Bohr radius to calculate the exciton Bohr radius:

$$a_{ex} = a_0 \frac{(\varepsilon / \varepsilon_0)}{m_{ex}} \tag{6.16}$$

Here, a_0 is the Bohr radius ($a_0 = 0.053$ nm), ε is the permittivity of the material, and ε_0 is the permittivity of free space. The effective masses and corresponding exciton Bohr radii of representative semiconductors are listed in Table 6.3.

TABLE 6.3   Dielectric Constant $\varepsilon/\varepsilon_0$, Effective Mass m_{ex}, and Corresponding Bohr Radius a_{ex} of Excitons in Various Semiconductors

Semiconductor	$\varepsilon/\varepsilon_0$	m_{ex}	a_{ex} (nm)
Silicon, Si (longitudinal, transverse)	12.0	0.34, 0.14	2, 5
Germanium, Ge (longitudinal, transverse)	16.0	0.25, 0.06	3, 13
Gallium arsenide, GaAs	13.2	0.06	11
Cadmium sulfide, CdS	8.9	0.16	3
Lead sulfide, PbS	17.3	0.05	18

Note: The exciton Bohr radius makes a convenient yardstick for estimating the degree of confinement in a semiconductor nanomaterial.

Mobile excitons are the means of conduction in bulk semiconductors where the exciton Bohr radius is miniscule in comparison with the dimensions of the solid. But when the solid's dimensions are similar in scale to the exciton Bohr radius, the exciton's motion is constricted. For this reason, the exciton Bohr radius makes a good yardstick for judging the dimensions needed for quantum confinement in a given semiconductor.

The volumetric constriction of an exciton is depicted in Figure 6.11. In the strongly confined regime, the characteristic dimension of the semiconductor is less than the exciton Bohr radius, $D < a_{ex}$. The electron and hole are closer together than they would be in a bulk lattice, and the electrostatic forces between them are higher. In the weakly confined regime, the characteristic dimension is slightly larger than the exciton Bohr radius. The exciton can move about within the lattice, although not very far.

6.7.3 The Band Gap of Nanomaterials

Having very few atoms in a material not only causes the energy states to spread out, as we saw in Figure 6.8, but it also widens the band gap for semiconductors and insulators (see Figure 6.12). When the energy states are discrete, the addition or removal of just a few atoms adjusts the boundaries of the band gap. (As discussed earlier, this is because once two atoms are next to each other, the sublevels split and then begin to broaden into bands as more atoms are added. The broader the bands get, the smaller the band gaps between them get. Eventually, the bands reach a point where they will not get any wider no matter how many atoms are in the solid.) In the case of quantum dots, the smaller the particle, the bigger the band gap. While traditionally considered an innate material property, the band gap is actually tunable to our needs—as long as we make the material small enough.

When an electron is excited across the band gap into the conduction band, it rarely stays there very long. It releases its energy in the form of a photon as it falls back to the valence band. The energy of this photon is equal to the energy the electron loses in its transition: it is the same as the band gap energy. Thus, when we change a material's band gap energy, we change the wavelength of electromagnetic radiation it emits. This has profound effects, especially for optical applications.

(a)

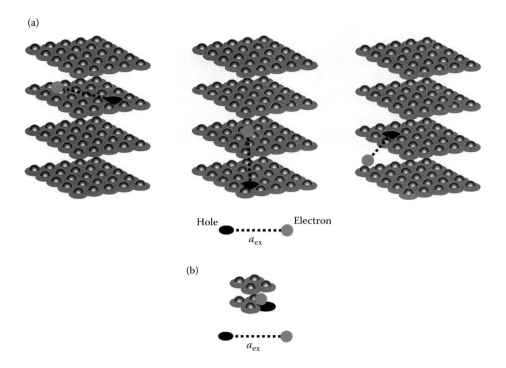

Hole ●┈┈┈┈┈○ Electron
a_{ex}

(b)

FIGURE 6.11 Confinement of an exciton. The exciton is an electron–hole pair created when a valence electron is excited into the conduction band and becomes mobile. In this simplified drawing, electrons occupy planes of a crystal lattice. In (a) the exciton is shown at three separate times as it moves. The hole shifts so as to keep the distance separating it from the free electron no greater than the exciton Bohr radius, a_{ex}. This radius just barely fits within the dimensions of the lattice. This is weak confinement. In (b) the dimensions of the semiconductor are smaller than the exciton Bohr radius. Exciton mobility is severely constricted. This is strong confinement.

6.8 TUNNELING

Quantum structures are what they are because they are isolated. To increasing degrees, a quantum well, quantum wire, and quantum dot have all been electrically severed from bulk material. With quantum dots, there are no physical electrical connections at all to the rest of the world—no wires leading in or out. Quantum dots are islands. Maui is also an island, surrounded by an ocean, a place where people try to disconnect from the hustle and bustle of the mainland. And yet there are thousands of cars on Maui. There is even *traffic* (making it more difficult to feel very disconnected). Cars usually come and go from islands by boat.

Electrons tunnel.

The insulating material that surrounds a quantum dot is an electrical barrier. If electrons were conventional particles, they would hit such a barrier and bounce off of it as if it were a wall. But electrons' wave nature makes it possible for them to hit the barrier and all of a sudden reappear on the opposite side. This phenomenon, called electron tunneling, happens all the time. And it is due to the fact that electrons occupy space in rippling

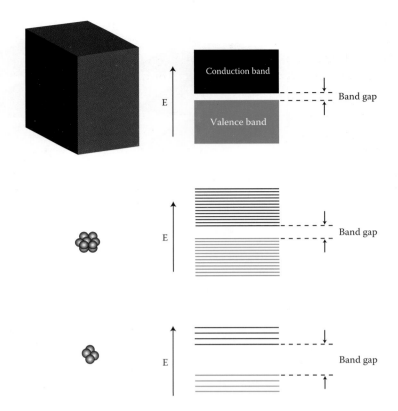

FIGURE 6.12 The band gap gets bigger as the material gets smaller. Once volume is reduced from that of a solid, which has bands for sublevels, to that of a nanomaterial, which has distinct splits in each sublevel, the band gap will widen even if only a few atoms are removed. This allows us to "tune" the electronic and optical characteristics of the nanomaterial.

patterns of probability. (At least that is as clear an explanation as anyone has managed to come up with thus far.)

When a particle such as an electron is confined inside a potential energy well with walls of a given height, the electron's wave function, ψ, extends outside the walls. So too does the probability density, $|\psi|^2$, which is the likelihood of finding an electron at a given location. This means there is a slim chance—but a chance nonetheless—that the electron will be found outside the well.

Before we discuss how exactly this happens, you may find it helpful to review our model of a particle in a potential energy well from Chapter 3, Section 3.3.8. In that model, the well was infinitely high, and the wave function was constrained so as to meet the all-important boundary condition: in all locations beyond the boundaries of the well, both ψ and $|\psi|^2$ had to equal zero. The outside of the well was mathematically off-limits to the electron. This situation is nearly achievable in real devices, and does well to model the energy quantization of atoms and quantum devices. However, there is actually no way of making a potential barrier that is infinitely high. Very, very, high—yes. Infinitely high—no.

The wave functions and probability densities of an electron inside a potential well of finite height are shown in Figure 6.13. The well is L wide, with x being the electron's location; the wave functions shown are those of the two lowest energy states. The potential energy at the walls, while greater than the energy of the electron, is not infinity. For this reason, the prior boundary condition is no longer in effect: *the wave function does not have to equal zero at the walls*. As we can see, this changes both the wave function and the probability density, enabling both of these functions to spill out the sides of the well.

There are three regions of interest here—namely, the two regions outside the well boundaries (regions I and III) and the well itself (region II). The general forms of the wave functions for these three regions are

$$\psi_I = Ae^{Cx} \qquad\qquad \text{for } x \le 0$$

$$\psi_{II} = F\sin(kx) + G\cos(kx) \qquad \text{for } 0 \le x \le L$$

$$\psi_{III} = Be^{-Cx} \qquad\qquad \text{for } x \ge L$$

Here, A, B, C, F, and G are constants. The variable k determines the wavelength λ of the function in region II, such that $k = 2\pi/\lambda$. The wave function is sinusoidal within the well, as expected, and decays exponentially beyond the walls. The probability density $|\psi|^2$ is also

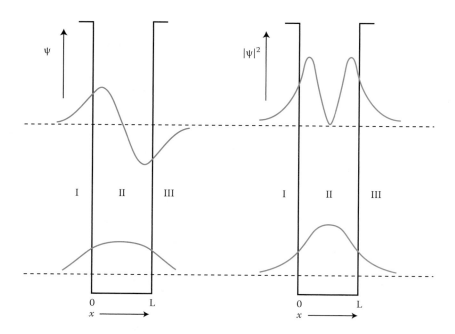

FIGURE 6.13   A particle in a potential energy well of finite depth. Graphed here are the wave function ψ and probability distribution $|\psi|^2$ of the particle. The lowest two energy states are shown. Both ψ and $|\psi|^2$ are nonzero at the well's boundaries and decay exponentially outside the well. The well's width is L.

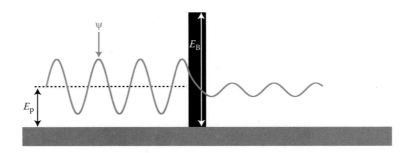

FIGURE 6.14 Tunneling. The wave function ψ of a particle with energy, E_p, arrives at the left side of a barrier with energy, E_B. The function decays exponentially while passing through the barrier but reemerges on the opposite side. In this way, electrons "tunnel" across energy barriers that would otherwise seem to prohibit passage.

nonzero outside the well's boundaries. This does not make sense when viewed through the lens of classical physics, but it does make sense according to quantum mechanics. Most importantly, it is what actually happens with electrons.

When an electron meets with an energy barrier of finite height and width, it can sometimes penetrate it and appear on the opposite side. This is tunneling and it is shown in Figure 6.14. Here we see the wave function of a particle that has an energy, E_p. The potential energy barrier has energy, E_B, which is greater than E_p. (If by chance the particle's energy were instead greater than that of the barrier, the wave could pass right over it.) As it is, the wave function is incident on the barrier, then decays exponentially while propagating through it. If we wanted to measure the motion of the particle inside the barrier, we would be faced with the trade-off posed by the Uncertainty Principle: the surer we were of the particle's position inside the barrier, the less we would know about its velocity. What we do know is that if the barrier is narrow and low enough, the wave reappears on the other side.

In the case of a quantum dot, the barrier is the energy needed to pass across the electrically insulating material that surrounds the dot. The physical distance separating the bulk electrons from the isolated electrons in a dot is often about a nanometer or so—very small. A finite (and occasionally achievable) amount of energy keeps the electrons from moving back and forth from the dot. In Section 6.9.1, we will discuss more about gaining control over electron tunneling.

A particularly innovative and useful application of electron tunneling is the scanning tunneling microscope. To learn more about this high-resolution imaging tool, read "Every *Da* Was One Atom."

"EVERY *DA* WAS ONE ATOM": THE SCANNING TUNNELING MICROSCOPE

Like the Hubble space telescope, the STM offers glimpses of previously unseen worlds. Some of nanotechnology's best pictures have been taken with an STM. With resolution better than its predecessors, the STM has made unique and lasting contributions to science every day since its invention. The first working STM was designed and built by Gerd Binnig and Heinrich Rohrer at the IBM Research Laboratories in Zürich, Switzerland. For their work these two men were winners of the 1986 Nobel Prize in Physics.

Normal light microscopes are limited by the fact that they cannot see anything much smaller than light waves themselves. The smallest features we can see with these scopes depend on the smallest wavelength of visible light we can use and on the lens. It turns out that 200 nm is about the smallest nook or cranny we can "see" (except with a near-field microscope, which uses light in a nanoscale way and can achieve up to 20-nm resolution). (See Chapter 8, "Nanophotonics.")

The STM operates in a completely different way. It has a metal stylus like the tip of a pen that scans back and forth across the surface. It never touches the surface, however, instead remaining a fixed distance from it. This separation distance can be kept constant using the quantum mechanics of tunneling.

As we have learned, an electron can tunnel across an otherwise nonconducting barrier if the gap is narrow enough. If a voltage bias is applied between the conducting surface and the stylus, and if the stylus is kept close enough to the surface, then electrons can tunnel across the empty space, creating current. This current flows from the surface to the stylus.

Tunneling is incredibly sensitive to the barrier distance. (Recall how the probability of finding an electron falls off exponentially outside the potential well.) In fact, this separation distance needs to be less than a nanometer for tunneling to even occur in a typical STM. This is what makes the STM's resolution so good. To keep the distance between stylus and surface constant, the STM monitors the tunneling current. For example, if the scanning stylus comes upon an atom jutting out of the surface, the separation distance shrinks and the current spikes, so it pulls the stylus up a little. All there is to do, then, is keep track of how high the stylus was as it scanned over each feature and we have mapped out an accurate topographical trace of the surface, as shown in Figure 6.15.

Because the tip of the stylus is a single atom, it is sharp enough to get into the nooks and crannies among the surface atoms. The horizontal resolution of the STM is 200 picometers (pm); vertical resolution can be as good as 1 pm, or about 1/100 of an atom's diameter. When the topography lines from the many scans are aligned side by side, amazing 3D images can be rendered. To see just how good the STM's resolution can be, take a look at the atoms in Figure 6.16.

Gerd Bennnig describes the moment in his lab when the first STM began to work:

> I was trying to image a sample that I had tried to image many times before without any success. Suddenly the microscope's like da-da-da-da-da-da-da-da. And I can never forget this noise the plotter made because it really had to work hard to follow the contours of the atoms that the tip was following—da-da-da-da-da-da-da-da. A beautiful noise. At this moment I knew these da's were all atoms, every da was one atom. The plotter drew a picture and you could see it was an atomic structure. And I knew immediately that's it, that's the breakthrough. It was a new world now. [Excerpt from "A Beautiful Noise," Harry Goldstein, *IEEE Spectrum*, May 2004.]

It takes time to convert science to technology. The quantum mechanical phenomenon of tunneling was first discovered in 1920 but the first STM was not built until the 1980s. What quantum devices does nanotechnology still have up its sleeve?

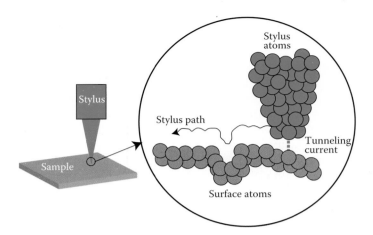

FIGURE 6.15   The STM. A voltage bias between the conducting sample surface and the atomically sharp metal stylus causes electrons to tunnel across the subnanometer gap of empty space. Using a feedback loop to keep this current constant, the STM can very accurately map out the topography of surface atoms.

6.9 SINGLE ELECTRON PHENOMENA

In electronics, the transistor is king. It is arguably the most ubiquitous and versatile little piece of electronics ever created. Transistors are what computers use to compute—tiny switches turning on and off, transferring and amplifying signals, making logic decisions. For a quick primer on how a conventional transistor works, see the sidebar "How the Field Effect Transistor Works Like a Militia."

Today, microchips have over a billion transistors, each one turning on and off a billion times every second. These chips require manufacturing processes with roughly 100-nm resolution. And every year this resolution drops, enabling even smaller transistors, so that even more of them can be squeezed into the same amount of space. So what happens when

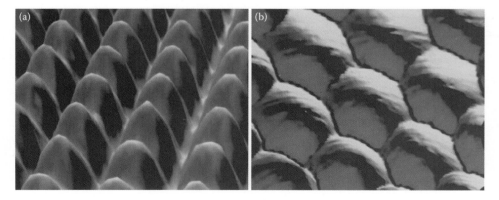

FIGURE 6.16   Surface atoms as seen by the STM. The surfaces are nickel (a) and platinum (b). (Image originally created by IBM Corporation.)

each transistor is reduced to a few atoms, or a single molecule? To be sure, quantum effects will play a role.

In 1970, to switch on a typical silicon transistor required about 10 million electrons. Current transistors require closer to 10,000. You can see where this is headed. Rather than moving torrents of electrons through transistors, it may very well be practical and necessary to move electrons *one at a time*. In fact, we can already build a single electron transistor if we get the dimensions and materials right. We can use transistors to make sensitive amplifiers, electrometers, switches, oscillators, and other digital electronic circuits—all of which operate using single electrons. In this section we will discuss the phenomena that govern such curious, yet promising devices.

6.9.1 Two Rules for Keeping the Quantum in Quantum Dot

Tunneling is the way electrons cross both the physical barriers and the energy barriers separating a quantum dot from the bulk material that surrounds it. Yet the energy needed to tunnel is only part of the story. If any electron on one side of the barrier could just tunnel across it, there would not be any isolation. The dot would not be a quantum dot because it would still essentially be part of the bulk.

So we need to be able to control the addition and subtraction of electrons. We can do this with voltage biases that force the electrons around, as we will discuss later in Section 6.9.2. However, priority number one is ensuring that the electrons stay put when there is no voltage bias at all. There are two rules for preventing electrons from tunneling back and forth from a quantum dot. When followed, these rules help to ensure that the dot remains isolated and quantized.

6.9.1.1 Rule 1: The Coulomb Blockade

Just from the name we can get an idea of what is happening. Coulomb forces are electrostatic. If we have two or more charges near one another, they exert Coulomb forces upon each other. If two charges are the same, the force is repulsive. In the case of a quantum dot, the charges are all negative electrons. Trying to cram even just a few electrons into a tiny piece of real estate creates Coulomb forces. As we would expect, the isolated droplet of electrons does not willingly accept another electron, but repels it. This is the Coulomb blockade and it helps prevent constant tunneling to and from a quantum dot.

HOW THE FIELD-EFFECT TRANSISTOR WORKS LIKE A MILITIA

For their discovery of the "transistor effect" in 1948, William Shockley, John Bardeen, and Walter Brattain earned the 1956 Nobel Prize in Physics. Now found in just about every electronic device imaginable, the transistor is certainly among the most important inventions of the twentieth century. The usefulness and ubiquity of a transistor derive from its simplicity. It is nothing but a switch for turning electricity on and off.

The field-effect transistor (FET) is the most commonly used type of transistor. A schematic of an FET is shown in Figure 6.17. It has three conducting electrodes. One is the "source" and the other is the "drain." Between the source and drain is a semiconductor—a material that can be made to behave like a conductor or an insulator, depending on the voltage applied to it. This voltage is applied via the third electrode, known as the "gate." The gate is separated from the semiconductor by a thin insulating layer.

Whereas a conductor has an army of mobile conduction electrons, semiconductors do not. The electrons they *do* have are sparsely distributed throughout the material, leaving too few between the source and drain to carry current. In this state, the switch is OFF. However, when a positive voltage is applied to the gate, mobile electrons are attracted from all over the semiconductor. It is as if a bugle roused members of a militia from their farms throughout the countryside. The electrons mobilize below the gate, in the thin channel separating the source and the drain. This makeshift army of conduction electrons carries current from source to drain. (Note: the direction of electric current is by convention taken to flow in the direction opposite to the flow of electrons. Here, in the interest of simplifying this discussion, we say that current flows with the electrons; so, the electrode from which the electrons originate is the source and the electrode to which they flow is the drain.) The switch is ON. Turn the gate voltage back to zero and the mobile electrons disperse. The switch is OFF again.

The mode described here, where electrons are attracted to the channel to carry current, is one of two ways an FET can work. In the opposite type, current flows when there is no gate voltage, and when the gate voltage increases, it repels charge carriers and cuts off the current. Either way, these binary ON/OFF states are the digital language of computers (0s and 1s). Transistors can be arranged together to form logic gates—the decision makers of a computer.

The FET uses one electrical signal to control another. Or, since the current from source to drain is proportional to the gate voltage, an FET can also amplify the original signal. In this way, the pattern of a tiny input signal is simultaneously copied and strengthened. This happens in stereos, speakers, microphones, and in boosting telephone signals as they traverse the globe.

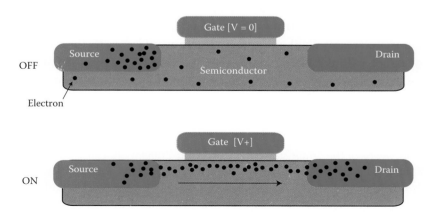

FIGURE 6.17   An FET. Without a voltage bias on the gate electrode, the scattered conduction electrons in the semiconductor do not aid in conduction and the switch is OFF. When a positive voltage on the gate creates an electrical field to attract and mobilize the scattered electrons, they carry electrical current from the source electrode to the drain electrode. This is ON.

We can measure its effect. A quantum dot has a capacitance, C_{dot}, a measure of how much electric charge it can store

$$C_{dot} = G\varepsilon d \qquad (6.17)$$

Here, ε is the permittivity of the material surrounding the dot, d is the diameter of the dot, and G is a geometrical term (if the quantum dot is a disk, $G = 4$; if it is a spherical particle, $G = 2\pi$). It may help to point out that this capacitance is not like the one between two objects, like a pair of parallel plates; an object isolated in space can store charge on its own and therefore can have a capacitance.

The energy needed to add one negatively charged electron to the dot is known as the charging energy, E_C:

$$E_C = \frac{e^2}{2C_{dot}} \qquad (6.18)$$

The charge on an electron, $e = 1.60 \times 10^{-19}$ Coulombs. It makes sense for E_C to be inversely proportional to the dot's capacitance: a large capacitor can quite easily accommodate another electron without too much energy required. However, in the opposite case, with extremely small capacitors (and that is essentially what quantum dots are), the charging energy can be substantial. That is, it can be large enough to "block" tunneling electrons. The question, then, is how much blocking energy must there be in the Coulomb blockade to do that?

The answer is that it needs more energy than a given electron can "spend" trying to tunnel in or out. We have already learned that a free electron in a solid has a certain amount of energy, depending on which band it is in. But it can get extra bursts of energy from the thermal vibrations of the atoms in the lattice—energy that can excite it into a higher band. This extra energy boost is equal to $k_B T$, where k_B is Boltzmann's constant (1.38×10^{-23} J/K). With this extra energy to "spend," an excited electron might be able to tunnel through a small enough barrier. The Coulomb blockade, to be effective, must then "charge a price" the electron cannot afford.

Put plainly, when the charging energy is much higher than the thermal energy of an electron, the Coulomb blockade can prevent unwanted tunneling. The condition of the Coulomb blockade is therefore

$$E_C \gg k_B T \qquad (6.19)$$

or, as a rule of thumb, $E_C > 10\, k_B T$. This criterion can be more easily met the smaller the dot becomes.

BACK-OF-THE-ENVELOPE 6.6

Aluminum gallium arsenide (AlGaAs) is a semiconductor with a lattice structure similar to that of gallium arsenide, but with a bigger band gap. These properties are advantageous when making a GaAs quantum dot surrounded by AlGaAs. At what temperature can electrons from the AlGaAs overcome the Coulomb blockade that exists in a spherical GaAs quantum dot 10 nm in diameter?

First, we determine the dot's capacitance (using 1.16×10^{-19} F/m as the permittivity of AlGaAs):

$$C_{dot} = G\varepsilon d = 2\pi\left(1.16 \times 10^{-10}\,\text{F/m}\right)\left(10 \times 10^{-9}\,\text{m}\right) = 7.3 \times 10^{-18}\,\text{F}$$

This gives the quantum dot a charging energy of

$$E_C = \frac{e^2}{2C_{dot}} = \frac{\left(1.6 \times 10^{-19}\,\text{C}\right)^2}{2\left(7.3 \times 10^{-18}\,\text{F}\right)} = 1.75 \times 10^{-21}\,\text{J}$$

By satisfying the rule of thumb and setting $E_C = 10k_BT$, we can solve for the unknown temperature:

$$T = \frac{E_C}{10k_B} = \frac{1.75 \times 10^{-21}\,\text{J}}{10\left(1.38 \times 10^{-23}\,\text{J/K}\right)} = 13\,\text{K}$$

Remember that this is a rule of thumb. Even if we only meet the condition that $E_C > k_BT$, giving us a temperature of 126 K, it may still prove impossible to add or remove a single electron to or from the quantum dot without using a voltage bias to force an electron across. Also, with different materials and different dimensions than those used in this example, it is possible to achieve a Coulomb blockade in a quantum dot at room temperature.

6.9.1.2 Rule 2: Overcoming Uncertainty

For the second condition that must be met in order to keep quantum dots electronically isolated, we look to the Uncertainty Principle. The central tenet of the Uncertainty Principle is: the more precisely we know something's position, the less precisely we can know its momentum. A corollary to this rule has to do with energy. It states that the uncertainty in the energy of a system is inversely proportional to how much time we have to measure it. Specifically, the energy uncertainty, ΔE, adheres to this relationship:

$$\Delta E \approx \frac{h}{\Delta t} \tag{6.20}$$

Here, h is Planck's constant and Δt is the measurement time.

Since it is a tiny capacitor, the time we use for Δt is the capacitor's time constant (the characteristic time a capacitor takes to acquire most of its charge). The time constant of a capacitor is RC, where R is the resistance and C is the capacitance. In our case, the resistance is the tunneling resistance, R_t, and the capacitance is C_{dot}. This gives us

$$\Delta t = R_t C_{dot} \tag{6.21}$$

The uncertainty of the system we are interested in—the quantum dot—must, like all other systems in the known universe, adhere to the constraints of the Uncertainty Principle. That being said, remember the goal: to keep electrons from tunneling freely back and forth to and from the dot. To ensure this, *the uncertainty of the charging energy must be less than the charging energy itself.* After all, if what we determine to be the amount of energy needed to add an electron to the dot can be off by as much as, or more than, the actual amount, random tunneling can occur without our knowledge. (Imagine if an amusement park determined that people needed be at least 50 in. tall to ride a particular roller coaster safely, but the tool used for measuring height had a precision of plus-or-minus 75 in. The park cannot be certain who ends up getting on the ride, or if they will be safe.)

We must require that $\Delta E_C < E_C$. Knowing this, we combine Equations 6.18, 6.20, and 6.21 to arrive at the following relationship:

$$\frac{h}{R_t C_{dot}} < \frac{e^2}{2C_{dot}} \tag{6.22}$$

Boiling this equation down gives the second rule for maintaining electron isolation in a quantum dot:

$$R_t \gg \frac{h}{e^2} \tag{6.23}$$

Here, h is Planck's constant and the charge on an electron is $e = 1.60 \times 10^{-19}$ C. Meeting this criterion is often as simple as making sure the insulating material surrounding the dot is thick enough. (As it turns out, this particular resistance, $h/e^2 = 25.813$ kΩ, is the "resistance quantum"—a value fundamental to quantum mechanics.)

Now that we know the two rules, let's play the game ... and build a single-electron transistor (SET).

6.9.2 Single-Electron Transistor

The SET is built like a conventional FET. The difference is that instead of a semiconductor channel between the source and drain electrodes, there is a quantum dot. This dot can take on numerous geometries. It can be a particle on an insulating surface, a disk sandwiched between insulators, or even just a section of semiconducting material where electric fields effectively isolate electrons. A generalized schematic of such a device and its operation is shown in Figure 6.18.

The purpose of the SET is to individually control the tunneling of electrons into and out of the quantum dot. To do this, we must first prohibit random tunneling (as discussed at length in previous sections) by selecting the right circuit geometry and materials. If an electron comes or goes from the dot, it will be on purpose.

To control tunneling, we apply a voltage bias to the gate electrode. There is also a voltage difference between the source and the drain that dictates the direction of the current. (As mentioned previously, the direction of electric current is by convention taken to flow in the

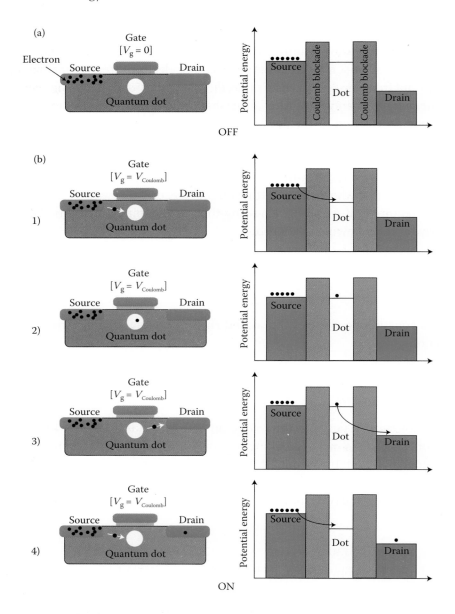

FIGURE 6.18 A single-electron transistor (SET). As opposed to the semiconductor channel in a field-effect transistor, the SET has an electrically isolated quantum dot located between the source and drain. (a) The SET in "off" mode. The corresponding potential energy diagram shows that it is not energetically favorable for electrons in the source to tunnel to the dot. (b) The SET in "on" mode. At the lowest setting, electrons tunnel one at a time, via the dot, from source to drain. This is made possible by first applying the proper gate voltage, $V_{gate} = e/2C_{dot}$, so that the potential energy of the dot is made low enough to encourage an electron to tunnel through the Coulomb blockade energy barrier to the quantum dot. Once the electron is on it, the dot's potential energy rises. The electron then tunnels through the Coulomb blockade on the other side to reach the lower potential energy at the drain. With the dot empty and the potential lower again, the process repeats.

direction opposite to the flow of electrons. Here we say that current and electrons flow in the same direction and we will consider the electrode from which the electrons originate, the source, and the electrode to which they flow, the drain, for simplicity's sake.) This is analogous to the operation of an FET, where the gate voltage creates an electric field that alters the conductivity of the semiconducting channel below it, enabling current to flow in from source to drain.

Applying a voltage to the gate in an SET creates an electric field and changes the potential energy of the dot with respect to the source and drain. This gate voltage-controlled potential difference can make electrons in the source attracted to the dot and, simultaneously, electrons in the dot attracted to the drain. For current to flow, this potential difference must be at least large enough to overcome the energy of the Coulomb blockade.

The energy, E, needed to move a charge, Q, across a potential energy difference, V, is given by $E = VQ$. In our case, the charge is an electron ($Q = e$). So we set the energy needed equal to the energy of the Coulomb blockade ($E = E_C$), and determine the voltage that will move an electron onto or off the dot:

$$V = \frac{E_C}{e} = \frac{\left(e^2 / 2C_{dot}\right)}{e} = \frac{e}{2C_{dot}} \tag{6.24}$$

With this voltage applied to it, an electron can tunnel through the Coulomb blockade of the quantum dot.

BACK-OF-THE-ENVELOPE 6.7

A gallium arsenide quantum dot has a capacitance of 7.3×10^{-18} F. If used in a SET, what voltage potential will cause an electron to tunnel between the source and the dot?

The voltage necessary is

$$V = \frac{e}{2C_{dot}} = \frac{1.6 \times 10^{-19} \, C}{2\left(7.3 \times 10^{-18} \, F\right)} = 11 \, mV$$

We can use the gate voltage to tune the number of electrons on the dot at any one time. At the lowest gate voltage for tunneling to occur, this current corresponds to one electron. Increasing the current is as simple as increasing the gate voltage. The voltage on a capacitor (like a quantum dot) is the charge divided by the capacitance. Because the smallest unit of charge is a single electron, the smallest change we can make to the voltage is

$$\Delta V = \Delta Q/C = e/C_{dot}$$

This is a key characteristic of the SET. The voltages at which current will flow are discrete and increase only by this increment. We control this voltage using the gate electrode, so the gate voltages at which conduction (tunneling) occurs are quantized also.

When the gate voltage, V_g, is zero, no current flows. The first gate voltage large enough to move an electron through the Coulomb blockade is called $V_{Coulomb}$. For single-electron tunneling, we require that $V_g = V_{Coulomb}$. At this setting, an electron can tunnel from the source electrode to the quantum dot. When it does, it reestablishes the Coulomb blockade and prevents additional electrons from tunneling in. The electron *can* tunnel *out* of the dot to the drain; and when it does, another electron leaves the source and tunnels to the newly vacated dot. This repeated tunneling pattern represents the current through the transistor.

The next highest gate voltage (which causes there to be two electrons on the dot at a time) equals $V_{Coulomb} + e/(2C_{dot})$. The next (three electrons on the dot at a time) equals $V_{Coulomb} + e/(2C_{dot}) + e/(2C_{dot}) = V_{Coulomb} + 2e/(2C_{dot})$. And so on, with each allowable gate voltage thereafter always $e/(2C_{dot})$ higher than the previous one.

The electrical behavior of an ideal SET is shown in Figure 6.19. The number of electrons in the quantum dot is controlled using the gate voltage. Figure 6.19 also shows the spikes in current through the transistor (from source to drain), which occur at discrete gate voltages. These ON and OFF states can be utilized to make an effective switch out of a SET.

6.10 MOLECULAR ELECTRONICS

Traveling to the ultimate limit of electronic miniaturization, we arrive at the molecule. Molecular electronics is the use of individual molecules as active components in electronic devices such as wires, switches, logic gates, transistors, and memory storage elements. This is not to be confused with bulk molecular systems, like films, or any other device that

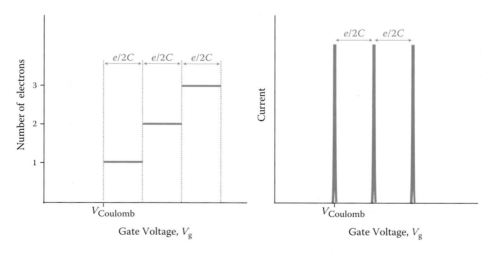

FIGURE 6.19 The electrical behavior of an ideal SET as a function of gate voltage. The number of electrons on the quantum dot is zero until the Coulomb blockade is overcome at $V_{Coulomb}$, at which point electrons tunnel one at a time from the source to the drain via the quantum dot. Because the energy on the dot is quantized, only discrete gate voltages enable the tunneling of electrons and consequent increases in the number of electrons in the dot. The separation between these gate voltages is $e/2C$. This is the voltage necessary to increase the number of electrons on the dot by one. The current-versus-voltage chart at the bottom shows the corresponding spikes in the source-to-drain current through the transistor at discrete gate voltages. Between the spikes, the number of electrons on the dot remains fixed. Typical gate voltages for such a device are a few millivolts; typical source-to-drain currents are in the picoampere range.

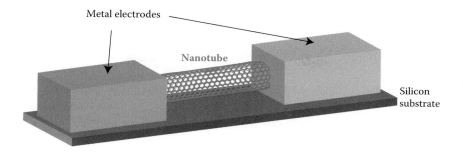

FIGURE 6.20 A nanotube wire. Depending on its chirality, a nanotube can be either semiconducting or metallic. The latter variety can have conductivities 50 times higher than that of copper—making for efficient nanometer-scale wires.

moves electrons through trillions of molecules. We are interested here in the devices that use molecules on a more individual basis. This can mean anywhere from one to a few thousand molecules per device. In this way, the devices *are* molecules, and vice versa.

The molecules worth considering initially are the ones with physical properties we can customize by tweaking their chemical structure—adding a methylthiophene unit here, a hexafluorocyclopentene bridge there—turning an otherwise ordinary molecule into an electrical switch. That is the true beauty of this approach, and also the opportunity to try out such novel device architectures motivates enthusiastic researchers. There is a chance that the guts of tomorrow's electronic devices will look very different from the right-angled matrix of wires and transistors we see today. Things may look more organic. Tailor-made molecules, created in a lab and cooked up in a factory, will ideally be capable of self-assembly. They will template themselves onto surfaces in complex processor patterns, and we will add electrodes to interact with them. This is "bottom-up" engineering.

While many factors favor development of molecular electronics, extensive research remains. What may likely emerge in the meantime are hybrid technologies that use both traditional silicon and molecular electronics components. Nanotubes with the correct chirality are conductive and can be used as wires between metal electrodes on silicon substrates, as shown in Figure 6.20. Research in molecular electronics may also foster tangential breakthroughs in areas such as sensors, nanometer-scale machines, or novel ways to interface with biological systems.

One practical advantage of molecules is their uniformity. Electronics made using super-high-resolution lithography or incorporating quantum dots have one thing in common—they can differ very slightly in the total number of atoms from device to device. We have seen how just a single atom can make a difference in the electronic characteristics of a nanoscale device. If the number and arrangement of atoms in a nanoelectronic device cannot be controlled, the device's characteristics and performance cannot be predicted and it becomes difficult to mass produce identical products. Meanwhile, molecules of a given compound have always been, and will always be, completely identical. Electronic functions, quantum energy spacing, conductivity—take a trillion rotaxane molecules and every single one will have the same electrical characteristics.

With typical molecules measuring 1–100 nm, size is an obvious advantage because smaller size tends to improve the cost, efficiency, and power dissipation in electronics. However, a disadvantage of many molecules is their instability at high temperatures.

Most of the work we do today with molecules happens in a beaker, not on the surface of chips. Liquid-phase synthesis and characterization steps with which we are already familiar will need to be translated to solid-state devices. To even make basic electrical connections is a monumental challenge with atomic-scale dimensions.

BACK-OF-THE-ENVELOPE 6.8

We will discuss shortly how a molecule can be made into a switch (like a transistor). What we must first appreciate, however, is just how small and densely packed molecular switches can be. One potential switch molecule is azobenzene, the electronic properties of which can be changed using light. Its chemical formula is $C_{12}H_{10}N_2$. How many switches are there in 1 pound (0.45 kg) of azobenzene?

The molecular mass is: 12(12 g/mol) + 10(1 g/mol) + 2(14 g/mol) = 182 g/mol.

Next, we determine how many moles are in 1 pound:

$$\frac{450\,g}{182\,g/mol} = 2.47\,mol$$

Since each mole contains 6.02×10^{23} molecules, that is about 1.5×10^{24} switches—more than the total number of all the transistors ever made in the history of the world. In a 1-pound bucket! Granted, it is truly unlikely that we would be able to hook up a power supply and other necessary connections to every one of these molecules. Still, we can see just how small and densely packed molecular switches can be, and why they are so promising as a means of miniaturizing computer chips.

If the electrode–molecule bond is a weak van der Waals bond, we may be able to treat the molecule like a quantum dot, with electrons tunneling through a Coulomb blockade. (In fact, if a gate electrode is positioned nearby, the molecule can behave like a SET.) But much of the time, stronger covalent bonds are used (such as those between thiol groups and gold electrodes) because they last longer. And a molecule's electronic configuration can change in complex ways as soon as it is "hooked up" and its orbitals overlap and interact with the orbitals of the conducting electrodes—in which case the bonds must be taken into account to understand how the device will work, or will not work.

6.10.1 Molecular Switches and Memory Storage

Certain types of molecules have dual personalities. They can be stable in two different states, each with differing physical properties. If one of these alternating properties is conductivity, then the molecule is a great candidate for an electronic switch. By changing the physical state back and forth, we can store a bit (a 1 or a 0) of information in the molecule to be read back later (perhaps using some kind of scanning probe). Or, we could use the molecule in a transistor, where current is switched on and off based on the molecule's dual physical states.

Figure 6.21 shows five types of hypothetical bistable molecules, classified by the physical change each can controllably undergo. Often, more than one type of change happens at a time—such as a molecule that changes shape when ionized. The stimuli that trigger each

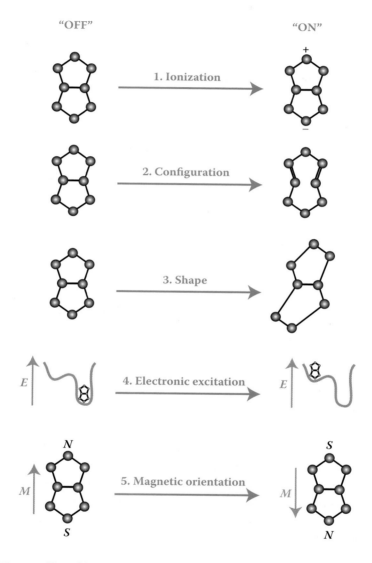

FIGURE 6.21 Types of bistable molecules. Some molecules have two physical states, each with a different conductivity. Such molecules make good switches or memory storage elements because they can be turned "on" or "off." The five hypothetical examples shown here are (1) molecules that can be either neutral or ionized; (2) molecules whose atoms can have two different bonding configurations; (3) molecules (like proteins) that can twist and fold upon themselves into two different shapes; (4) molecules that can be stable in either a ground or an excited electronic energy state; and (5) molecules that have dual magnetic orientations.

of these physical changes include light energy (photons), electrical energy (electrons), and chemical energy (from protons, ions, other molecules, etc.). Light and electrical stimuli tend to work faster than chemical stimulus.

6.11 SUMMARY

The spacing between the allowed electron energies of a solid increases as the size of the material decreases. If we can reduce the volume of a solid or in some way isolate electrons inside a small portion of a larger solid, the tiny piece of material behaves electronically like an artificial atom.

There are two "rules" that must be followed to ensure that an artificial atom, often called a quantum dot, maintains its quantized electronic properties:

- *Rule* 1: The energy needed to add one electron to the dot, or charging energy, E_C, must be significantly higher than the thermal energy of an electron:

$$E_C \gg k_B T$$

  Here, k_B is Boltzmann's constant (1.38×10^{-23} J/K) and T is the temperature.
- *Rule* 2: The uncertainty of the charging energy must be less than the charging energy itself. This can be accomplished if

$$R_t \gg \frac{h}{e^2}$$

  Here, R_t is the tunneling resistance, h is Planck's constant, and the charge on an electron is $e = 1.60 \times 10^{-19}$ C.

By following these rules, it becomes possible to manipulate electrons one at a time, such as in devices like the SET.

Finally, individual molecules may be useful as wires, switches, logic gates, transistors, and memory storage elements.

HOMEWORK EXERCISES

6.1 Your DNA represents about 10 billion bytes of genetic information. Roughly how many electrons would a typical computer use to process that much information?

6.2 True or false? When numerous atoms are brought into close proximity, only the highest occupied energy sublevel forms a band.

6.3 True or false? Once an energy band has formed, its width is proportional to the number of atoms in the solid.

6.4 What do carbon, silicon, and germanium have in common?

6.5 At which temperature do we determine the "ground state" of an atom? What is the chance of finding a free electron in every energy state below the Fermi level at this temperature?

6.6 For an atom in its ground state, what is the lowest, completely unfilled energy band?

6.7 True or false? Conductors have no band gap.

6.8 The Boltzmann constant (k_B) appears in many important equations governing electronics. To which two physical quantities does this constant relate?

6.9 What does the Fermi function fail to account for?

6.10 If a block of sodium is put in a pot of boiling water at sea level, will the temperature be high enough to make one in every hundred sodium electron mobile?

6.11 Calculate how much energy (in eV) an electron in a copper penny on the sidewalk loses when the sun sets and the temperature drops from 90°F to 68°F.

6.12 How many electron energy states are there below 5 eV in a 1-in. cube of gold? How does this number compare with the number of states below 5 eV in the same volume of copper?

6.13 What is the largest diameter a spherical silver particle can have for there to be no more than 1 billion electron energy states over a 1-eV interval at 2 eV?

6.14 We can look at the band structure of an element to get an idea of how many electrons per atom participate in conduction. We can also determine this number based on the free electron density, the atomic mass, and the mass density of the material.
   a. Calculate the free electron density of lithium. The Fermi energy is 4.72 eV.
   b. Derive a formula for the number of atoms per unit volume.
   c. Calculate the number of atoms per unit volume of lithium (535 kg/m³, 6.9 g/mol) using the formula that you derived in part (b).
   d. Determine the number of free electrons per lithium atom.
   e. What is the electron configuration for Li? Explain whether your answer to (d) makes sense according to this configuration. The free electron(s) of Li is (are) in which sublevel?

6.15 The average spacing between energy states in a solid depends on
   a. The number of free electrons
   b. The Fermi energy
   c. The volume
   d. Only a and b
   e. All of the above

6.16 True or false? Quantum confinement relates only to those electrons that would otherwise be mobile—the conduction electrons.

6.17 List three ways to confine electrons.

6.18 Name the structures in which partial confinement is achieved.

6.19 What is unique about the density-of-states function of a quantum dot compared to quantum wells, quantum wires, and bulk materials?

6.20 True or false? A quantum dot containing thousands of atoms can be used to trap a single free electron—in effect creating a gigantic, artificial hydrogen atom.

6.21 A quantum dot contains a confined electron. A photon enters the dot and imparts additional energy to the confined electron, causing it to jump from the ground state to the $n = 3$ state. What would the diameter, D, of the dot have to be in order to keep the excited electron confined?

6.22 Determine the temperature at which a cube of gold 3 nm on a side becomes quantum confined (via the metal-to-insulator transition). Gold has a density of 19,300 kg/m³ and its atomic mass is 197 g/mol.

6.23 Why is a physical model different from the one used for conductors needed to determine the conditions of quantum confinement for semiconductors?

6.24 What name is given to the distance separating an electron–hole pair in a semiconductor?

6.25 True or false? The band gap of a quantum dot is directly proportional to its size.

6.26 How must the boundary conditions of a potential energy well change to allow for tunneling?

6.27 The wave function, $\psi(x)$, of an electron in a potential well of finite depth and width, L, is given by

$$\psi(x) = Ae^{Cx} \qquad\qquad \text{for } x \le 0$$

$$\psi(x) = F\sin(kx) + G\cos(kx) \qquad\qquad \text{for } 0 \le x \le L$$

$$\psi(x) = Be^{-Cx} \qquad\qquad \text{for } x \ge L$$

a. Given $F = G = 5$, $k = 1$, and $C = 2$, determine A.
b. What is $\psi(0)$?
c. Given $\psi(L) = \psi(0)$, determine the values of L and B whether this wave function corresponds to the ground energy state. See Figure 6.22.
d. Given $\psi(L) = \psi(0)$, determine the values of L and B if this wave function corresponds to the third lowest energy state.
e. Use a spreadsheet program to graph the wave function from part (d) over the range $-2 \le x \le (L+2)$. Indicate the location and value of L along the x-axis.

6.28 About what size are the smallest features we can see with a light microscope?

6.29 What variable does the STM attempt to keep constant as it scans?

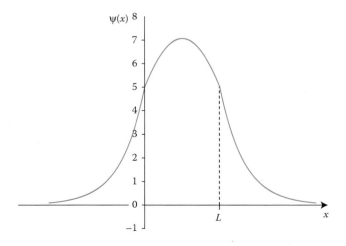

FIGURE 6.22   Homework Exercise 6.27c.

6.30 True or false? In a field-effect transistor, the current between the source and the drain is proportional to the gate voltage.

6.31 A disk-shaped quantum dot 8 nm in diameter is separated from the nearest electrode by glass with a permittivity of 4.43×10^{-11} F/m.

    a. Determine the energy needed to increase the charge on the dot by one electron.

    b. Using the rule of thumb for the Coulomb blockade, determine the temperature below which quantum confinement is possible.

6.32 A spherical quantum dot embedded in silicon dioxide (which has a dielectric constant of 4.5) is found to have a charging energy of 96 meV. What is its diameter?

6.33 At room temperature, what values must the charging energy and the tunneling resistance be significantly greater than to ensure no unwanted electron tunneling occurs to or from a quantum dot?

6.34 True or false? The purpose of a single electron transistor is to prevent all tunneling of electrons into or out of a quantum dot.

6.35 A quantum dot has a capacitance of 2 aF.

    a. What is the smallest incremental change that can be made to the voltage bias between the dot and the source electrode?

    b. When this quantum dot is incorporated into a single electron transistor, it is found that the gate voltage at which single electron tunneling occurs is 110 mV. Over what range of gate voltages can we expect there to be three electrons on the dot?

6.36 True or false? An electrically conductive polymer film is an example of molecular electronics.

6.37 Explain how molecules may be better suited than quantum dots for mass-produced electronics.

6.38 Why must the chemical bonds between molecules and electrodes be considered when evaluating the electronic properties of molecular-scale electronics?

6.39 Name five ways in which molecules can be bistable. What important electronic property of these molecules can change when the physical state changes?

RECOMMENDATIONS FOR FURTHER READING

1. P. Yu and M. Cardona. 2010. *Fundamentals of Semiconductors: Physics and Materials Properties.* Springer-Verlag Telos.
2. C. Kittel. 1995. *Introduction to Solid State Physics.* Wiley.
3. Lieber Research Group: http://cmliris.harvard.edu
4. J. Heath and M. Ratner. 2003. Molecular electronics. *Physics Today*: 43.
5. J. M. Tour. 2003. *Molecular Electronics: Commercial Insights, Chemistry, Devices, Architecture, and Programming.* World Scientific Publishing Co. Inc.
6. R. Waser. 2003. *Nanoelectronics and Information Technology: Advanced Electronic Materials and Novel Devices.* Wiley-VCH.
7. L. Jacak, P. Hawrylak, and A. Wójs. 1998. *Quantum Dots.* Springer.
8. M. A. Kastner. 2000. The single electron transistor and artificial atoms. *Annals of Physics (Leipzig)*, 9:885.
9. D. Ferry, S. Goodnick, and J. Bird. 2009. *Transport in Nanostructures.* Cambridge University Press.

CHAPTER 7

Nanoscale Heat Transfer

7.1 BACKGROUND: HOT TOPIC

Do more with less—this is the mantra of nanotech. We have come to expect smaller devices that handle more tasks more efficiently than their larger predecessors. Often, this requires taking advantage of special physical behaviors that arise only at small scales. The gate length in field-effect transistors, for example, is becoming comparable to the mean free path and wavelength of electrons. Devices this small offer new opportunities and unforeseen problems in the areas of heat transfer and energy conversion. Computers and communication systems use electricity and light to relay information, but the motion of electrons and photons often generates excess heat. So we must find ways to move increasing amounts of heat through smaller areas. We must convert energy more efficiently. Tomorrow's computers will be miraculously fast, but only if they do not melt when we turn them on.

Here we will discuss the underlying principles of heat transfer. Although heat is most often discussed from a macroscale perspective, we are going to zoom in all the way to the nanoscale to see the stuff that heat is actually made of. We will also discuss how the physics of heat and energy vary at the nanoscale. To do so, we can apply what we already know about electrons, phonons, photons, and atoms.

7.2 ALL HEAT IS NANOSCALE HEAT

Heat is traveling energy. This energy travels within objects and also between them. Because the universe behaves in an inherently benevolent and fair-minded way, heat "shares the wealth" by redistributing energy whenever possible. Hot things give away enough energy to cold things so that everything will eventually have the same "hotness." We use temperature as a gauge for this "hotness," that is, how hot or cold something is. More specifically, temperature is a measure of an object's tendency to spontaneously release energy to its environment. If an object is hot compared to its surroundings, it will let go of some energy. And if a particular region of an object is hotter than the rest of the object, the energy will automatically redistribute so as to equalize the temperature. Thermal equilibrium is always the goal.

Heat can be transferred in three different ways: (1) conduction, (2) convection, and (3) radiation. Conduction is the transfer of heat through atoms that are in contact with each other, convection is the transfer of heat due to the bulk motion of a fluid, and radiation is the transfer of heat via electromagnetic waves between atoms not in contact with each other.

Most often, heat is thought of in a generalized, amorphous way—a gush of energy on its way to somewhere else. But if we zoom in, we see that the energy is individualized. Just as water flowing down a pipe is, at the fundamental level, actually a vast number of H_2O molecules tumbling in the same direction, heat has to take on some particular form. In fact, it takes on four: atoms, electrons, photons, and phonons. Heat is always carried by at least one of these four nanoscale things. For this reason, all heat is nanoscale heat. In this chapter we will highlight the role of the four heat carriers in the three categories of heat transfer.

First, let us examine an important constant—Boltzmann's constant—to see more specifically how temperature and energy are related at the nanoscale.

7.2.1 Boltzmann's Constant

Austrian physicist Ludwig Boltzmann is widely regarded as the man who successfully married mechanics and statistics. And, like Planck's constant relates energy to frequency, Boltzmann's constant relates thermal energy to temperature. The Boltzmann constant is $k_B = 1.38 \times 10^{-23}$ J/K. For a system at a temperature, T, the Boltzmann constant can tell us roughly the amount of thermal energy, E, of a single heat carrier in the system:

$$E \approx k_B T \tag{7.1}$$

Thus, the average air molecule at room temperature has about k_B (300 K), or 4.14×10^{-21} J of energy. The ideal gas law is well known: $PV = nRT$, where P is the pressure of the gas, V its volume, n is the number of moles of gas, R is the universal gas constant (8.31 J/K/mol), and T is the absolute temperature. Because $n = N/N_A$, where N is the total number of gas molecules and N_A is Avogadro's number (6.02×10^{23}), we can rewrite the ideal gas law using Boltzmann's constant, transforming it into an equation about the energy of each molecule in a gas:

$$k_B T = \frac{PV}{N} \tag{7.2}$$

We can see then that Boltzmann's constant, $k_B = R/N_A$. The quantity $k_B T$ in the equation equals the average energy per molecule (since $V \times P$ has units of $m^3 \times N/m^2 = N\ m = J$). We will find these relations important when discussing how our four carriers transfer heat. Next, we will discuss each of the three categories of heat transfer: conduction, convection, and radiation.

7.3 CONDUCTION

Heat conduction occurs when two atoms with different temperatures are in contact with one another. In a solid, the atoms are stacked together, while in fluids they come in contact occasionally. Either way, for conduction to occur there must be some type of matter for the heat to move through and some temperature gradient to endow the heat with direction. Heat, of course, naturally flows from hot areas toward cold.

The rate of heat conduction, $Q_{conduct}$, in one dimension, x, is given by

$$Q_{conduct} = -k_c A \frac{\Delta T}{\Delta x}$$ (7.3)

Here, k_c is the thermal conductivity of the medium (measured in W/m/K), A is the cross-sectional area of the heat flow (m²), and ΔT is the temperature difference (K) across the medium, Δx (m). $Q_{conduct}$ is measured in watts. Note that 1 W = 1 J (the unit of energy) per second. The negative sign in the equation reminds us that if temperature increases from left to right in the solid, then the heat will flow from right to left. This equation, known as Fourier's law, forms the basis of classical (continuum) heat transfer. The thermal conductivity, k_c, is a material property, dependent on temperature, that varies from the order of 0.01 W/m/K for gases up through the order of 100,000 W/m/K for very cold solids. The tight covalent bonds of carbon structures make diamonds and nanotubes the best thermal conductors we know of.

An example of heat transfer by conduction is a gas trapped between two walls, as depicted in Figure 7.1. At time zero, the wall on the left side is instantaneously heated and the cold wall instantaneously cooled. This means that the atoms in the hot wall vibrate more rigorously (with bigger amplitudes) than the atoms in the cold wall. The gas molecules between the walls are cold at $t = 0$. But as time passes ($t = 1$), some of the molecules collide with the hot wall. Kinetic energy is transferred during these collisions. When the molecules come away from the collisions, they have larger velocities than they did earlier.

BACK-OF-THE-ENVELOPE 7.1

A geologist drills a temperature probe deep into the earth and finds that the temperature rises by 20°C per kilometer of depth. The average thermal conductivity of the rock is 2.5 W/m/K. If we assume the heat conduction where he drills is typical of the Earth's surface in general, at approximately what rate is the planet losing heat from conduction?

The heat conduction per square meter in this location is determined by rearranging Fourier's law:

$$\frac{Q_{conduct}}{A} = -k_c \frac{\Delta T}{\Delta x} = \left(2.5 \frac{W}{mK}\right)\left(\frac{20K}{1000m}\right) = 0.05 \text{ W/m}^2$$

The Earth's radius is 6400 km, giving it a surface area of $4\pi(6{,}400{,}000 \text{ m})^2 = 5.15 \times 10^{14} \text{ m}^2$. Thus, the estimated heat conduction from the entire planet is

$$(Q_{conduct})(A) = (0.05 \text{ W/m}^2)(5.15 \times 10^{14} \text{ m}^2) = 2.6 \times 10^{13} \text{ W}$$

Some of this heat warms Earth's surface and the atmosphere and some is radiated into space.

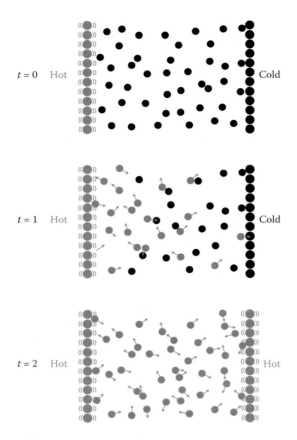

FIGURE 7.1 Heat conduction through a gas. At time $t = 0$, the left wall is instantaneously heated and the right wall instantaneously cooled. Atoms in the hot wall vibrate with large amplitude. This energy is soon imparted to gas molecules that come in contact with the wall, making them move around faster ($t = 1$). The random motion of the hot molecules (blue) brings them into contact with slower, cooler ones (black), and kinetic energy is shared yet again. This process carries heat, one collision at a time, through the gas to the cold wall. If heat continues to be added to the left wall to keep it a constant temperature, then both walls and the gas will eventually reach thermal equilibrium ($t = 3$).

These hot gas molecules zip about randomly, colliding with more sluggish, colder molecules, and passing along energy. Eventually, this energy is handed off so many times throughout the gas that it reaches the other wall. Gas molecules with high kinetic energy collide with the cold wall, passing heat to the atoms of the solid lattice. In this way, energy in the hot wall is conducted via the gas to the cold wall. If heat continued to be added to the hot wall to keep it at a constant temperature, then eventually both walls and the gas would reach the same temperature, as we see at $t = 3$. Remember: the molecules do not necessarily travel from the hot to the cold wall. Their trajectories are random. But the energy that the molecules carry *does* travel from one wall to the other via the collisions.

Thus, in a gas or a liquid, the principal carriers of heat are the atoms (and molecules) themselves.

Solids are different. Atoms in a solid do not float around randomly but generally stay fixed at one location within the crystal lattice. The bonds between the atoms can be approximated as springs, and energy undulates through the lattice via these springs. Such waves of vibrational energy are known as phonons. (For more on phonons, see Chapter 5, "Nanomechanics.") The amplitude of an atom's vibration is proportional to temperature. If one part of the lattice is hotter than another part, the large amplitudes of the hot atoms' vibrations will be carried by phonons to the colder parts. This is one way heat can be carried through a solid. Figure 7.2 shows a lattice of atoms, one of which is heated up with energy at time zero. Its large oscillations are then transmitted throughout the lattice by phonons.

In insulating solids, phonons are the primary carriers of heat.

Phonons carry some heat in conductors also. However, they are not the primary carriers; electrons are. In a conductor, a sea of free electrons ebbs and flows throughout the lattice. These electrons behave in much the same way as the gas molecules that we just discussed.

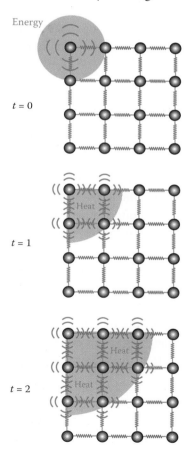

FIGURE 7.2   How phonons conduct heat in a solid. The atom in the upper left corner of the lattice is given energy at time $t = 0$. This energy takes the form of large oscillations, which are then transferred to nearby atoms through the springy bonds that hold the lattice together. In this way, heat is transferred in waves of vibrational energy (phonons) throughout the solid.

Their motions are random but energized electrons pass along their excess energy to nearby electrons—thereby carrying energy from hot areas to colder areas in the conductor. An electron in a conductor moves about a thousand times faster than a phonon.

Semiconductors also have free electrons, although not enough to handle much heat. A typical metal has about 10^{17} free electrons per cubic meter; semiconductors have about 100,000 times less than that. So phonons are the primary heat carrier. However, semiconductors heavily doped with impurity atoms have a much larger population of free electrons and thus begin to handle heat more like a conductor.

7.3.1 The Thermal Conductivity of Nanoscale Structures

Thermal conductivity is like a ranking—a means of comparing materials like we compare sports teams. The higher the value of k_c, the better the material is at conducting heat. And we know that heat is conducted by atoms, phonons, and electrons. These are the players on the team. So, in order to get a better feel for the ranking of the team at the nanoscale, we have to see how each of the players handles being confined to a very tiny space. Imagine if we shrunk a basketball court to the size of a ping pong table. The whole game would change. Players who had mastered the original version might not fare too well in the smaller format. Team rankings would change.

7.3.1.1 The Mean Free Path and Scattering of Heat Carriers

We can calculate the thermal conductivity of a heat carrier as

$$k_{c,\text{carrier}} = \frac{\rho c v \Lambda}{3} \tag{7.4}$$

This equation applies to molecules, phonons, and electrons—the three carriers used in conduction. Here, v is the average random velocity of the carrier (m/s), Λ is the mean free path of the carrier (m), ρ is the density of the carrier, and c is the specific heat of the carrier (which is the amount of heat per unit mass needed to raise the carrier's temperature by 1°K). For molecules in a gas, we express ρ in units of kg/m³ and c in J/kg/K. For phonons and electrons, we typically replace these two terms, ρ and c, in the equation with a single term—the volumetric specific heat, $C = \rho c$, which is the amount of heat per unit volume needed to raise the carrier's temperature by 1°K, expressed in J m⁻³K⁻¹.

The mean free path is the average distance a carrier travels before losing its extra energy due to scattering (see Figure 7.3). It is given by

$$\Lambda = v\tau \tag{7.5}$$

Here, v is the average velocity of the carrier and τ is the average time it travels before being scattered, also known as the relaxation time. (We discuss scattering briefly.) For a gas, the mean free path can also be calculated using this equation:

$$\Lambda = \frac{k_B T}{\pi \sqrt{2} P d^2} \tag{7.6}$$

FIGURE 7.3   The "mean" free path.

Here, k_B is the Boltzmann constant (1.38×10^{-23} J/K), T is the temperature (K), P is the pressure (Pa), and d is the effective diameter of an atom (or molecule) of the gas. Typical mean free paths and velocities (by order of magnitude) of heat carriers are provided in Table 7.1. Note that the vast range of values for the mean free path of photons is due to the fact that photons can travel through solid materials, where the mean free path is very short, and also through a vacuum, where it is very long. We can see a similarly large range for air molecules: the mean free path for air molecules at sea level (where there are many molecules and therefore less free space for the molecules to move without colliding with each other) is much smaller than the mean free path much higher in the atmosphere, where there are very few air molecules.

TABLE 7.1   Approximate Mean Free Paths and Order-of-Magnitude Velocities of the Four Heat Carriers

Carrier	Mean Free Path, Δ (nm)	Velocity, v (m/s)
Air (sea level)	60	~100
Air (100 km altitude)	100,000,000	~100
Selected gases (0°C, 1 atm)		
Carbon dioxide	39	~100
Argon	63	~100
Benzene	148	~100
Oxygen	63	~100
Nitrogen	59	~100
Hydrogen	110	~100
Electrons	10–100	~1,000,000
Phonons	10–100	~1000
Photons	10–1,000,000,000,000	~100,000,000

BACK-OF-THE-ENVELOPE 7.2

Free electrons in copper at room temperature travel with a speed of 1.6×10^6 m/s and collide with one another every 25 fs, on average. Molecules in the air at room temperature travel with an average speed of 524 m/s and collide about every 267 ps. Which of these two heat carriers has a longer mean free path?
    For the electrons:

$$\Lambda = v\tau$$
$$\Lambda = \left(1.6 \times 10^6 \, \text{m/s}\right)\left(25 \times 10^{-15}\,\text{s}\right) = 40 \, \text{nm}$$

For the gas molecules:

$$\Lambda = \left(524 \, \text{m/s}\right)\left(267 \times 10^{-12}\,\text{s}\right) = 140 \, \text{nm}$$

So the gas molecules, as we may have expected, travel farther between collisions.

Scattering is a general description of what happens when a carrier encounters something that makes it deviate from its straight trajectory. This usually causes the carrier to give up excess energy. A gas molecule is scattered when it collides with another gas molecule; an electron can be scattered by another electron. A phonon wave can interfere with other phonon waves. But there is crossover also. For example, phonons have a tendency to scatter electrons, hindering electrical conduction. This is why the electrical resistance of a solid tends to increase at higher temperatures. Solid surfaces can cause scattering, as can defects, impurities, and grain boundaries in a crystal lattice.

The mean free path is *not* the distance between carriers. The value of Λ for a dilute gas is much greater than the average distance between molecules. A molecule can fly by many of its neighbors before colliding with one. As we just observed in Back-of-the-Envelope 7.2, electrons in copper at room temperature have a mean free path of 40 nm. Copper atoms in a solid are spaced out by 0.2 nm, meaning the free electrons fly past about 200 atoms between collisions.

We can see from Equation 7.4 that the rate of heat conduction is limited by how far a carrier can travel before it collides with something, including another carrier. Imagine how much more efficient the heat conduction depicted in Figure 7.1 would be if the hot molecules could fly straight across to the cold wall, unimpeded, to deliver their energy instead of passing the energy along in countless random collisions, with molecules flying every which way.

Suffice it to say that the more scattering that occurs within a conductor, the lower the thermal conductivity will be. This makes the mean free path one of the limiting factors in conductivity, at least until you get to the nanoscale. In nanoscale devices, the structures can have features about the same size as the mean free path of the carriers. Sometimes the features are actually smaller than Λ. This means that the carriers not only bounce off each other, but they also smack against the constraining surfaces of the solid more often—leading to even more scattering.

This affects numerous nanoscale technologies enabled by ultrathin solids. Consider the silicon device layer in modern transistors, as thin as 10–100 nm. The thermal conductivity of this layer affects device reliability, thermal aging, and performance. Microelectromechanical systems (MEMS) such as microcantilever sensors and actuators also rely on thin silicon features. Silicon is a semiconductor, so the dominant heat carriers are phonons. And the thermal conductivity of silicon layers measuring 10 to hundreds of nanometers thick can be an order of magnitude lower than silicon's bulk conductivity (148 W/m/K) due to the extra phonon-boundary scattering.

The thermal conductivity of thin metal layers and wires is also adversely affected by the added scattering of electrons at the surfaces of the layer. Copper, for example, shows diminished thermal conductivity below about 500 nm thicknesses.

7.3.1.2 Thermoelectrics: Better Energy Conversion with Nanostructures

Thermoelectric devices either use power to cool things or convert heat into power. One is shown schematically in Figure 7.4. Here the thermoelectric material is a semiconductor, sandwiched between metal conductors. A voltage bias is applied, causing electric current to flow through the loop as shown. Electrons travel through the metal until reaching the left side of the semiconductor sandwich. In order to enter the semiconductor, the electrons must overcome an energy barrier. This is because semiconductors have a band gap and metals do not. The barrier is therefore the difference in energy between the metal's Fermi energy, E_F, and the bottom of the semiconductor's conduction band. (For a thorough discussion of electron energy bands, see Chapter 6, "Nanoelectronics.") Only the highest-energy electrons are likely to jump into the semiconductor. These are the "hottest" electrons. The result is that heat leaves the left side of the sandwich and it cools off. The hot electrons flow across the semiconductor. When they reach the metal on the opposite side, they are carrying more energy than they need in order to be conduction electrons in the metal and thus they release their extra energy as heat. So, the right side of the sandwich gets hot.

Heat in a thermoelectric device is carried by electric current. To refrigerate something, a voltage bias forces electrons to make the journey described above. To generate power, a heat source on the hot side of the sandwich excites the electrons, causing them to "escape" into the semiconductor and travel to the opposite side where they can release their energy—thus creating an electric current to power things.

A good thermoelectric material needs a high electrical conductivity in order to carry as much heat-bearing current as possible and prevent Joule heating (heat created when current encounters resistance.) It also needs a low thermal conductivity in order to keep heat from leaking backward from the hot to the cold side. The figure-of-merit for a thermoelectric material is Z:

$$Z = \frac{S^2 \sigma}{k} \tag{7.7}$$

Here, S is the material's Seebeck coefficient (expressed in V/K or V/°C), a measure of the voltage induced by a temperature gradient along the material, σ is the material's electrical conductivity, and k is its thermal conductivity. For silicon, the figure-of-merit, $Z = 440$ V/°C

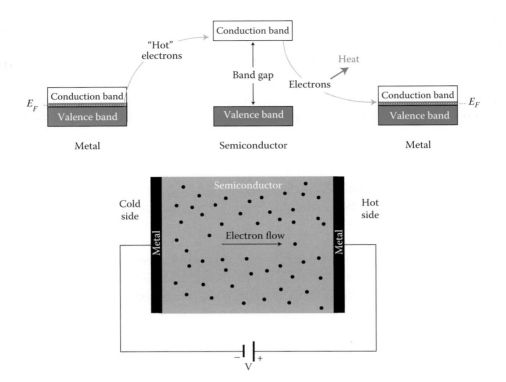

FIGURE 7.4 How a thermoelectric cooler works. A voltage bias is applied to a semiconductor sandwiched between metal conductors. "Hot" electrons with enough energy to surmount the energy barrier separating the metal's Fermi energy, E_F, and the bottom of the semiconductor's conductance band are carried away—making the left side cooler. Upon reentering metal on the other side, electrons dump their excess energy as heat—making the right side hotter. The thermoelectric material needs a high electrical conductivity to efficiently transport electric current and a low thermal conductivity to prevent heat from leaking back from the hot side to the cold side.

at 0°C, whereas gold, copper, and silver all have $Z = 6.5$ V/°C at 0°C. Metals are further disadvantaged for this application because electrons are their dominant carrier of both heat *and* electricity, so k is proportional to σ. Insulators have poor electrical conductivity and thus are not a viable option. In semiconductors, the dominant heat carriers are phonons. This is good. Phonon thermal conductivity in semiconductors can be reduced without too much reduction in electrical conductivity by alloying (adding metal). The resulting heterogeneous mass distribution in the lattice disrupts the flow of phonons.

We also know that nanostructures have diminished thermal conductivities due to the added scattering that occurs at their surfaces. It has further been demonstrated that quantum structures (dots, wells, wires) tend to have large Seebeck coefficients due to the unique way that their electron energies are distributed (their density of states). Experimental and theoretical studies of nanostructured devices have revealed unprecedented values for Z, higher than any bulk material. Often, these devices are made by layering thin films (quantum wells) or creating quantum dot superlattices—complex arrangements that are often difficult to build. However, their performance justifies continued development.

7.3.1.3 The Quantum of Thermal Conduction

As we have seen, heat flows through a solid by means of collective, wave-like vibrations of the atoms known as phonons. Phonons are found in all three types of material: conductors, semiconductors, and insulators. In most cases, these waves are multitudinous, like countless ripples upon the surface of a sea. We cannot perceive of any quantization of the heat conduction—unless we make a structure so incredibly small that only a very limited number of phonons can be active within it.

In fact, experimental results suggest there exists a strict fundamental limit to the amount of heat that can be conducted through structures of atomic dimensions—a "quantum of thermal conductance." Once a structure gets small enough, only this certain amount of heat will ever be able to flow through it at a time—no more. It does not matter what material is used. The only way to make more heat flow is to make the structure larger. It is believed that this amount is universal, applicable not only to phonons, but also any type of heat flow (including electrons).

The quantum of thermal conductance, g_0, (with units of W/K) is a value based solely on fundamental quantum mechanical constants and absolute temperature:

$$g_0 = \frac{\pi^2 k_b^2 T}{3h} \tag{7.8}$$

Here, k_b is Boltzmann's constant (1.38×10^{-23} J/K), T is the ambient temperature (K), and h is Planck's constant (6.626×10^{-34} m^2 kg/s). When a structure is small enough for g_0 to manifest, this equation can tell us the maximum power (in watts) that it will conduct for a given temperature gradient across the structure.

BACK-OF-THE-ENVELOPE 7.3

A nanoscale wire bridging a gap between a heat reservoir and a bulk solid is thin enough to observe the quantum of thermal conductance. If the ambient temperature is 10 K and a 1-picowatt power source is applied to the reservoir, how much will the reservoir's temperature rise?

The quantum of thermal conductance, g_0, gives the maximum power the wire conducts for a given temperature gradient at an ambient temperature T:

$$g_0 = \frac{\pi^2 k_b^2 T}{3h} = \frac{(\pi)^2 \left(1.38 \times 10^{-23} \text{ m}^2 \text{ kg/s}^2/\text{K}^2\right)^2 (10\,\text{K})}{3\left(6.626 \times 10^{-34} \text{ m}^2 \text{ kg/s}\right)}$$

$$g_0 = 9.46 \times 10^{-12} \text{ W/K}$$

So, for an input power of 1 picowatt, the temperature gradient between the reservoir and the bulk solid (i.e., the rise in temperature of the reservoir) would be

$$\Delta T = \frac{1.0 \times 10^{-12} \text{ W}}{9.46 \times 10^{-12} \text{ W/K}} = 0.1\,\text{K}$$

The quantum nature of heat conductance has been observed in experiments performed at temperatures near absolute zero using silicon nitride beams about 100 nm (300 atoms) wide. The beams bridged a gap separating a tiny heat reservoir from a bulk solid. For heat to escape the reservoir and reach the solid, it could only flow via the narrow beams. However, heat could not escape as fast as it was added to the reservoir, so the reservoir heated up. By measuring how quickly the reservoir's temperature rose, the limited thermal conductance of the beams could be determined. This experimental device was powered at about 1 femtowatt during experiments—about the same amount of power your eye receives from a 100-W light bulb 60 miles away.

The exact dimensions that a structure must have for g_0 to manifest have not been clearly defined, and research in this area is still in its infancy. Still, g_0 has important implications for the way electronic devices will be made in the future—transistors especially, which could soon be so small that power dissipation is limited by g_0. Just turning the transistor on could make it too hot to work if it is unable to dissipate its heat fast enough.

7.4 CONVECTION

In heat conduction, the energy is handed off from carrier to carrier during random collisions. In convection, these random motions still occur but the overall average motion of the carriers is in a particular direction. If we zoom in on one H_2O molecule in a pipe and watch it move, it will zip and bounce every which way; but if we zoom back out a little and look at lots of water molecules, we would see they are flowing down the pipe. Convection occurs when a fluid flow overlaps with a temperature gradient. Conduction within the fluid still occurs but the heat is additionally carried along by the bulk motion of the fluid.

In "free" convection, fluid motion arises solely due to temperature differences within the fluid. We all have heard the adage that "heat rises." But why? What does that mean with respect to the nanoscale heat carriers? Well, the heat carriers of convection are fluid molecules. As these molecules acquire energy, they zip around faster, bouncing off each other more often. This extra motion requires extra space. A given mass of hot molecules will fill a larger volume than the same mass of cold molecules, for a given pressure. Viewed another way, a given volume of hot air will be less dense than the same volume of cold air (again, for a given pressure). As long as there is gravity, there will be buoyancy differences that drive the hot air to the top of the denser cold air. This creates free convection.

Forced convection, on the other hand, is driven artificially—usually by a pump or a fan. Blowing on hot soup to cool it is an example of forced convection.

Newton's law of cooling describes the convection heat transfer rate, Q, expressed in watts, between two points in a fluid, the temperatures of which are T_1 and T_2:

$$Q = h_t A (T_1 - T_2) \qquad (7.9)$$

Here, A is the cross-sectional area of the heat flow and h_t is the heat transfer coefficient (with units of W/m/K). In the case of convection from a solid surface to a fluid, we can say that T_1 is the temperature of the surface (equal to the temperature of the fluid molecules in contact with the surface), T_2 is the temperature of the fluid, and A is the area of the solid–fluid interface.

Unlike thermal conductivity, h_t is not a material property. It depends on many things: the properties of the fluid flow, the properties of the fluid itself, and the geometry of the solid object over which the fluid is moving. Is the fluid gushing around a sphere, or gently meandering down a tube, or swirling above a flat surface? Such considerations matter a great deal in determining h_t. There are a few generic geometries and flows for which h_t can be calculated using dimensionless parameters such as the Reynolds number, the Prandtl number, and the Nusselt number. But at the nanoscale, such calculations are often inaccurate.

To see how, visit Earth's outer atmosphere. Here, the air is so thin and the pressure so low that molecules are few and far between. This is known as a rarefied gas, in which the mean free path of the molecules is very long. Whereas a ball thrown through the air on Earth is much larger than the mean free path of the molecules it encounters, spacecraft in the upper atmosphere are comparable to, or smaller than the mean free path of the molecules. A nanoscale structure is actually the same way—often having a characteristic dimension, D, equal to, or smaller than, the mean free path of the gas molecules around it, even at atmospheric pressure. Rarefied gas flow can be characterized by yet another dimensionless parameter called the Knudsen number, Kn, which is the ratio of a gas' mean free path, Λ, to the characteristic dimension of the structure with which the gas interacts:

$$Kn = \frac{\Lambda}{D} \tag{7.10}$$

If $Kn \geq 10$, scattering among the molecules can be entirely neglected; this is the rarefied gas regime in which only the interaction between the gas molecules and the structure's surface are accounted for in heat transfer calculations. If $Kn \leq 0.01$, the gas is not rarefied but treated as a continuum. Values of Kn between 0.01 and 10 are transitional regimes between these two extremes.

System-specific details can make h_t difficult to pin down at any size scale. It is often established based on experiment. But we see that the complexities of convection are in large part due to the complexities of the fluid flow.

7.5 RADIATION

When two objects are separated by empty space—like the Sun and Earth, or the inner and outer shells of a thermos—heat is transferred by electromagnetic radiation (photons). As opposed to conduction and convection, radiation requires no medium; it can carry heat across a vacuum. But it can also carry heat through fluids and solids. The photons that carry heat are the same as those that carry TV and radio signals.

Photons are generated whenever a charged particle accelerates. Radio stations create photons by quickly alternating the voltage applied to an antenna, making the electrons in the antenna oscillate. The atoms of a crystal lattice also have electrons, and when the atoms get warmer, they vibrate with greater amplitude. These accelerations create photons—thereby converting heat energy into electromagnetic energy. Radiation generated by a heat source (as opposed to an antenna, for example) is known as thermal radiation. The hotter an object gets, the higher the frequencies of thermal radiation it will emit. Room-temperature objects and people tend to emit photons within the infrared region of the electromagnetic spectrum. Heat them up and they will glow in the visible region.

7.5.1 Increased Radiation Heat Transfer: Mind the Gap!

In vacuum, radiation heat transfer is straightforward—literally. The photons propagate unimpeded. But in a solid there are many things with which the photon can interact. A photon has oscillating electric and magnetic fields—waves. As these waves encounter a solid, polar objects like atoms and electrons within the material oscillate in response.

Of course, the atoms and electrons in a solid are already oscillating on their own, in waves we call phonons. Phonons have much in common with photons: just as the photon is the smallest quantum of electromagnetic energy and equals hf, where h is Planck's constant and f is the frequency (Hz) of the electromagnetic wave, the phonon is the smallest quantum of vibrational energy and also equals hf, where f is the frequency of the lattice vibrational wave.

In a conductor, the free electrons also oscillate collectively, as a kind of plasma. The natural frequency of their oscillation is often called the plasma frequency, f_p. (See Chapter 8, Section 8.2.4.1 for details.)

It should then come as little surprise to learn that all of these crisscrossing, fluctuating patterns of energy can intermingle. Indeed, there is a great deal of coupling that occurs between the various oscillators in a solid material. (We already know, for example, that photons can push atoms around, creating high-frequency "optical phonons.") When the electrons or phonons oscillate at frequencies similar to those of passing photons, resonant modes arise. Photon–phonon couplings are often called "polaritons" and photon–electron couplings are called "plasmons." Polaritons tend to be at infrared frequencies, while plasmons are in the visible and ultraviolet domain. Both are examples of electromagnetic energy acting in tandem with mechanical energy. The details of these pairings are more complex than we need to cover here. What we will say is that this hybrid energy is typically most dense near the surface of an object, decaying exponentially away from the surface.

This matters for radiation heat transfer at the nanoscale. To see how, let us look at a pair of parallel metal plates. The plates are separated by a gap of a few micrometers. There is no air in the gap—it is a vacuum. Photons radiating from one plate carry heat across the gap and are absorbed by the other plate. If one plate happens to be hotter than the other, the net heat transferred will be from the hot plate to the cooler one. The heat transfer rate remains relatively constant if we move the plates a little closer together or farther apart. After all, in the vacuum gap, there is nothing to impede or enhance the photons on their journey.

But as the gap begins to close to the point where the plates are nearly touching, the heat transfer rate can shoot up by several orders of magnitude. This is because the polaritons and plasmons, like auras hovering just above the surface of the plates, are close enough to play a dominant role as heat carriers. These auras are sometimes called "evanescent waves."

BACK-OF-THE-ENVELOPE 7.4

At approximately what separation distance will heat transfer between two objects at room temperature (293 K) be dominated by photon tunneling?

Using Equation 7.11, we determine λ_T:

$$\lambda_T = \frac{0.2898 \times 10^{-2}\,mK}{293K}$$

$$\lambda_T = 10 \times 10^{-6} m$$

So at room temperature, photon tunneling will become the primary means of heat transfer at separation distances less than roughly 10 μm.

At such small separations, they can couple with the surface of the opposing plate, transferring energy (see Figure 7.5). This phenomenon is called photon tunneling.

As a rule of thumb, heat transfer between two bodies separated by a very small gap will be dominated by photon tunneling when the gap distance is shorter than the dominant photon wavelength, λ_T. The Wein displacement law gives λ_T for an object at temperature, T, emitting thermal radiation:

$$\lambda_T = \frac{0.2898 \times 10^{-2}\,m\,K}{T} \tag{7.11}$$

The boost provided by these surface-based energy carriers might prove important in applications requiring rapid and highly confined heat transfer. While our example involved metal plates, this phenomenon also has been observed in semiconductors and insulators, as well as in much smaller, nonplanar objects separated by nanoscale gaps. It might be a good way to rapidly heat a tiny area with pinpoint accuracy or, alternatively, to detect temperature with high spatial resolution.

It could even improve the efficiency of thermophotovoltaic generators. These energy conversion systems consist of a thermal emitter separated by a small gap from a photovoltaic device. At temperatures of hundreds to thousands of degrees Celsius, the emitter spontaneously radiates photons that are then absorbed and converted to electrons (electric current) by the photovoltaic device. Minimizing the gap between the hot photon emitter and the photovoltaic absorber could make thermophotovoltaic generators more efficient.

7.6 SUMMARY

Tiny new devices must move more heat through smaller channels than ever before. They must use and convert energy with better efficiency. Engineers need a better understanding of heat, especially the physics of heat transfer at the nanoscale.

Heat is traveling energy. It travels in three ways: by (1) conduction, (2) convection, or (3) radiation. In conduction, the heat is carried primarily by atoms (molecules) in fluids, by phonons in insulating and semiconducting solids, and by electrons in conductors.

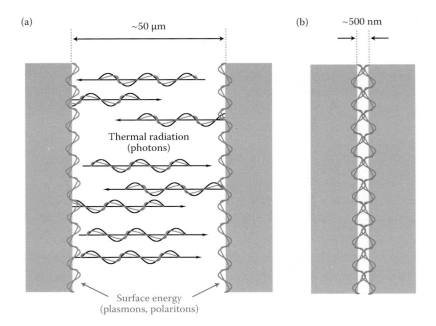

FIGURE 7.5   Radiation heat transfer across a small gap. Two parallel metal plates are separated by a vacuum gap. In (a), the gap is roughly 50 μm. Photons travel across the gap but the surface-concentrated polaritons and plasmons—both of which arise from the coupling of electromagnetic and mechanical energy—play no role in heat transfer. In (b), the plates are separated by about 500 nm. The surface energies overlap across the gap and the heat transfer rate increases. This is known as photon tunneling.

In nanoscale devices, the size of the conductor can be comparable to the mean free path of these carriers—leading to increased scattering and lower thermal conductivity. Research has shown that there is a quantum of thermal conductance—a limit on the amount of heat a very small structure can conduct. Reduced thermal conductivity can enable higher performance in thermoelectric devices, which require materials with high electrical conductivity but low thermal conductivity.

In convection, a fluid flow overlaps with a temperature gradient. Because fluids are involved, the primary heat carriers are atoms and molecules. Convective heat transfer is complicated by flow characteristics and the geometry of the system under investigation.

Heat is transferred by radiation when two objects are separated by empty space. The size of the gap separating the objects typically does not affect the heat transfer rate; however, at the micrometer and nanometer size scales, the rate can increase by several orders of magnitude due to the phenomenon of photon tunneling.

HOMEWORK EXERCISES

7.1   Heat is traveling _____.

7.2   Define the concept of temperature.

7.3   What are the three modes by which heat is transferred?

7.4 A free electron in a block of gold has an energy of 3.8×10^{-21} J. What is the approximate temperature (°C) of the block?

7.5 What is the volume of a canister containing 1 kg of oxygen molecules (O_2) at 1 atm and 75°C? (O_2 has a molecular mass of 32 g/mol.)

7.6 True or false? Heat can be transferred by conduction through a vacuum.

7.7 When you heat a solid material, do the atoms within it vibrate with greater amplitude or greater frequency?

7.8 Name the principal carrier(s) of heat in (a) a solid material, (b) a gas, (c) a vacuum.

7.9 Which are faster carriers of heat through a metal: electrons or phonons? Approximately how much faster?

7.10 A copper wire 3 μm in diameter and 300 μm long has a temperature at one end of 35°C and 32°C at the other. Calculate the rate of heat flow though the wire, in watts. (For copper, $k_c = 400$ W/m/K.)

7.11 The thermal conductivity of air at room temperature is 0.028 W/m/K and its density is 1.16 kg/m³. The average "air" molecule travels at 524 m/s and collides about every 267 ps with another air molecule. How much heat (in Joules) is needed to raise the temperature of a cubic meter of air by 10°C?

7.12 Very thin films are usually deposited under vacuum conditions to prevent contamination and ensure that atoms can fly directly from the source to the depositing surface without being scattered along the way.
   a. To get an idea of how few and far between the air molecules are in a thin-film deposition chamber, determine the mean free path of a generic "air" molecule with an effective diameter of 0.25 nm at a pressure of 1.5×10^{-6} Pa and temperature of 300 K.
   b. If the chamber is spherical with a diameter of 10 cm, estimate how many times a given molecule will collide with the chamber before colliding with another air molecule.
   c. How many air molecules are in the chamber (treating "air" as an ideal gas)?

7.13 Give a brief explanation for why the electrical resistance of a solid tends to increase as it gets hotter.

7.14 Provide a brief explanation of why a 10-nm thin layer of silicon can have a different thermal conductivity than a 1000-nm layer.

7.15 What are the dominant carriers of heat in a thermoelectric device?

7.16 True or false? Nanostructures show promise for thermoelectric applications due in part to their diminished thermal conductivity.

7.17 A nanoscale beam is thin enough to support only the quantum of thermal conductance. At an ambient temperature of 300 K, what is the heat transfer rate (J/s) from one end of the beam to the other if the temperature difference between the two ends is 10 K?

7.18 True or false? Using a fan to keep a computer processor cool is an application of forced convection.

7.19 True or false? A spacecraft in the vacuum of outer space is characterized by a Knudsen number similar in magnitude to that of a nanoscale structure at atmospheric pressure because the mean free path of the molecules in these two scenarios is similar in magnitude.

7.20 If "air" molecules with an effective diameter of 0.25 nm are at room temperature and pressure (293 K and 101,325 N/m², respectively), what is the Knudsen number of a particle 10 nm in diameter surrounded by air? Is the Knudsen number of this system similar to that of a rarefied gas?

7.21 The polariton and the plasmon each represent a combination of which two types of energy?

7.22 True or false? The hotter an object gets, the shorter the wavelengths of photons it emits.

7.23 True or false? Photons emitted by cold objects can tunnel across wider gaps than those emitted by hotter objects.

RECOMMENDATIONS FOR FURTHER READING

1. L. Kouwenhoven and L. Venema. 2000. Heat flow through nanobridges. *Nature*, 404:943.
2. K. Schwab et al. 2000. Measurement of the quantum of thermal conductance. *Nature*, 404:974.
3. NanoEngineering Group at MIT: http://web.mit.edu/nanoengineering.
4. G. Chen. 2005. *Nanoscale Energy Transport and Conversion: A Parallel Treatment of Electrons, Molecules, Phonons, and Photons.* Oxford University Press, USA.

CHAPTER 8

Nanophotonics

8.1 BACKGROUND: THE LYCURGUS CUP AND THE BIRTH OF THE PHOTON

Let us go back 1700 years or so, to the Roman Empire, inside a workshop. An artisan is molding a cup out of glass near an open fire. The cup depicts King Lycurgus being dragged into the underworld by Ambrosia. The glass mixture he is molding contains traces of metal. The artisan sets the cooling glass on his workbench and admires his handiwork. The glass is green, like jade. He lifts a twig from the fire and puts the burning end into the cup, so that the light from the flame shines through the glass. The glass turns from green to red. He grins approvingly.

The Lycurgus Cup is one of only a handful of such cups that remain from Roman times. (Today you can see this particular cup at the British Museum.) The decorative pigment used in these cups is actually a suspension of gold and silver particles roughly 70 nm in diameter. The particles make the glass look green in reflected light, but red in transmitted light. Later, in the seventeenth century, another colloid made of tin dioxide and gold particles, known as Purple of Cassius, was used to color glass.

A case of technology far outpacing science, artisans were using metal particles as colorants for more than a century before scientists tried to figure out why it worked. Why was gold not always golden? The scientist who figured that out was Michael Faraday. In 1857, he took a systematic look at the unusual colors of gold particles, spawning research that has since shaped our understanding of how light interacts with small bits of matter. The optical properties do not depend entirely on the material, as was long taken for granted—they also depend on the size of the material. If we take a gleaming, yellow brick of gold and make it very thin, it turns blue. Or, if we take nanometer-scale particles from the same block of gold, they can be orange, or purple, or red.

This is an example of nanophotonics. We define nanophotonics as the nanoscale interaction of photons and materials. It is a field still in its infancy.

Long before Romans made magical glasses, Aristotle pondered the nature of the stuff that came out of the Sun. Newton also tried to make sense of it. Finally, Max Planck's groundbreaking work in 1900 suggested that energy radiated from objects in discrete quantities. It was an observation that defied the laws of thermodynamics and physics, and received very little attention until 1905, when Albert Einstein published a paper providing a physical explanation for the "bundles" of light Planck had observed.

And thus the photon was born. These enigmatic and counterintuitive little packets of energy are the best way we have to explain how EM radiation (including visible light) works. Yet there is no single word that can properly describe what a photon is. This is probably because we do not understand them well enough yet. Photons act like waves *and* they act like particles, depending on what experiment we are running. They lack mass; however, they carry momentum. They travel as a wave. In empty space, the speed of a photon is about 300 million meters per second (m/s).

To reconcile the wave and particle attributes of photons (and electrons) requires abstract thinking and faith in experimental explanations. The data tell the story—photons are what they are, and the best we can do is remake the laws of physics in order to fit what the data tell us is true. The laws of physics are like the laws of a nation: they must be updated and rewritten regularly to reflect new realities.

Nanophotonics is helping to rewrite the rules governing what can be done with photons. For example, near-field optical microscopes achieve resolution an order of magnitude lower than normal light microscopes by "cheating" the diffraction limit—using light to see things smaller than the wavelength of light.

In this chapter we will shed light on this and other nanophotonic feats, focusing on physical explanations and applications.

8.2 PHOTONIC PROPERTIES OF NANOMATERIALS

The photonic properties of materials are a reflection, so to speak, of their electronic properties. The arrangement, energy-wise, of a material's electrons sets the rules for how that material is allowed to interact with photons. Conductors, semiconductors, and insulators each have unique electron energy band arrangements (as discussed in Chapter 6, "Nanoelectronics"). Our treatment of the photonic properties of nanoscale materials will take into account the basic electronic differences among these three types of material.

8.2.1 Photon Absorption

With insulators and semiconductors, a band gap of forbidden energies separates a completely full energy band from a completely empty one. Because the valence band has no unoccupied energy states, a valence electron can typically absorb only those photons having enough energy to induce a transition to an unoccupied state in a higher band. There is really nowhere within its own energy band for the electron to go.

Photon absorption by insulators and semiconductors is likely to occur only if the photon's energy exceeds that of the band gap. In Figure 8.1 we see a photon of short wavelength (and hence high energy) exciting an electron across the band gap separating the unfilled conduction band of silicon from the filled valence band. This is known as "interband" absorption. It is the reason insulators and semiconductors tend to be transparent,

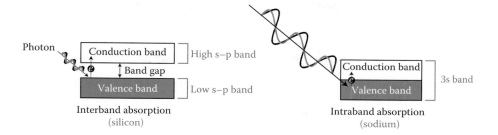

FIGURE 8.1   Interband and intraband absorption. In insulators and semiconductors (like the semiconductor, silicon), a valence electron can typically absorb only those photons having enough energy to induce a transition to an unoccupied state in the conduction band. Here, the electron is boosted from silicon's filled, low s–p band to its unfilled, high s–p band. This is known as interband absorption. In conductors (like the metal sodium), there are numerous available energy states within the partially filled valence band, and valence electrons can also absorb lower-energy (longer-wavelength) photons. This is known as intraband absorption.

and weakly reflecting for photon energies less than the band gap: photons in the visible wavelength range can pass right through, unabsorbed. Because insulators have such a large band gap, interband transitions are rare.

Interband absorption occurs in conductors as well, where electrons in filled or partially filled bands jump into unfilled, high-energy bands. However, in conductors, the conduction and valence bands overlap, in effect making the band gap zero (as with the 3s band of sodium, for example). The electrons in outer orbitals form an electron "sea." Because of all the available energy states in a conductor's partially filled valence band, the valence electrons are also able to absorb low-energy photons. Instead of jumping across an energy gap into an unfilled band when it absorbs a photon, the electron instead makes a mini-jump into one of the unoccupied states within its own band. This is known as "intraband" absorption. It is what makes metals highly absorbing and reflecting at visible and infrared wavelengths. In Figure 8.1, the photon with the longer wavelength (hence lower energy) excites an electron from the valence band to the conduction band, both of which are within the 3s band of sodium.

It should be noted that intraband transitions are possible in semiconductors, although unlikely and only under special conditions. If the semiconductor has been doped with impurities, or if it is at high temperature and the available thermal energy is greater than the band gap, then there will be some free electrons in a partially filled conduction band. The semiconductor will behave more like a conductor than an insulator and intraband transitions (such as within the conduction band) may occur.

8.2.2 Photon Emission

Once excited, electrons eventually release their extra energy and drop back down to their ground state. This usually happens immediately, with the electron emitting a photon about 100 ns after absorbing one. That is called *spontaneous* emission. (Reflected light can be thought of as spontaneously emitted light.) Or, a photon may encounter an already excited electron and, by perturbing it, cause the electron to emit a photon with the same phase and in the same direction as the photon that caused the emission. This is called *stimulated* (or induced) emission. Stimulated emission is used in lasers, as we will discuss later. Both types of emission are depicted in Figure 8.2.

An excited electron does not always emit a photon, however. Sometimes it contributes its energy to the countless atomic vibrations and collisions going on around it instead. No photon is emitted in this case.

8.2.3 Photon Scattering

So, we have covered absorption and emission. A third interaction between photons and matter that we should understand is *scattering*. Scattering occurs when a photon changes direction after it strikes a bit of matter. This is usually a type of scattering called "elastic scattering," where no energy is exchanged between the photon and the matter. Fewer than one in a million of the scattered photons will instead undergo "inelastic scattering." As opposed to absorption, where the photon is entirely annihilated and gives up the whole of its energy, inelastic scattering involves either the transfer of a fraction of the photon's energy to the matter—often in the form of heat—or the transfer of some of the matter's energy to the photon. If the photon loses energy, it carries on with a lower frequency and

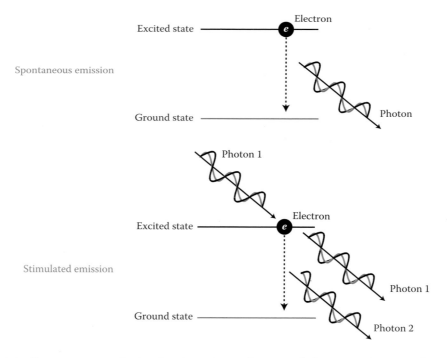

FIGURE 8.2 Spontaneous and stimulated emission. An excited electron can emit a photon on its own or when prompted by an incoming photon. Either way, the electron drops back to its ground state. In stimulated emission, the incoming photon 1 is not absorbed by the electron but continues on, in phase and in the same direction as the newly emitted photon 2.

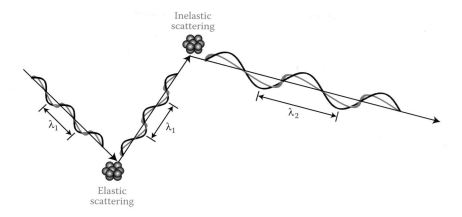

FIGURE 8.3   Elastic and inelastic scattering. A photon strikes the leftmost particle and is scattered without any loss of energy toward the second particle, where it is again scattered. However, the second particle acquires a portion of the photon's energy, so the redirected photon ends up with a somewhat lower frequency (and therefore a longer wavelength, $\lambda_2 > \lambda_1$).

in a new direction. If it gains energy, it carries on with a higher frequency. Both elastic and inelastic scattering are shown in Figure 8.3.

8.2.4 Metals

We use the metric of *conductivity* to measure how easily charge (electric current) can flow through a material. Photons also move through materials. In this case, we use *permittivity* to measure the degree to which a material enables the propagation of photons. As its name implies, permittivity is a measure of how strong of an electric field a material will "permit" inside of it. The permittivity of free space (a vacuum) is $\varepsilon_o = 8.854 \times 10^{-12}$ Farads per meter (F/m). The permittivity of other materials is usually given in terms of relative permittivity, ε_r (also called the dielectric constant). A material's actual permittivity is determined by multiplying the two:

$$\varepsilon = \varepsilon_r \varepsilon_o \tag{8.1}$$

The relative permittivities of selected materials are listed in Table 8.1. These values correspond to static electric fields. In a capacitor, a high permittivity material is better at storing charge. (Note that no conductors appear in the table. A metal will not work like the material inside a capacitor because it will short out the capacitor.)

However, when the field is varying in time, as is the case with photons, a material's permittivity depends on how polarizable it is. A photon comprises oscillating electric and magnetic fields—waves. As these waves encounter a material, polarities within the material oscillate in response. The more polarizable the material, the stronger the field that can be established inside it. So in this case, permittivity has both real and imaginary parts, both of which are functions of the frequency, f, of the field:

$$\varepsilon(f) = \varepsilon'(f) - i\varepsilon''(f) \tag{8.2}$$

TABLE 8.1   Relative Permittivity of Selected Materials

Material	Relative Permittivity, ε_r (Dielectric Constant)
Vacuum	1
Air	1.0006
Polyethylene	2.5
Quartz	3.7–4.5
Glass	3.8–14.5
Rubber	2–7
Water	80
Barium titanate	100–1250

The imaginary part, ε'', expresses the energy lost in the material as the waves move through. The real part, ε', expresses the relationship between the forward speed of the wave and the material's capacitance. If the real part is positive, the material is a "dielectric." A good dielectric material will have a high permittivity (dielectric constant). If instead the real part of the permittivity is negative, the material is considered a metal. Knowing this, next we will examine how the free electron "sea" in a metal affects its permittivity and how the metal–photon interaction varies at the nanoscale.

8.2.4.1 Permittivity and the Free Electron Plasma
The free electrons in a metal determine its permittivity. The equation we use to calculate the permittivity of these free electrons is

$$\varepsilon = 1 - \frac{f_p^2}{f^2 + i\gamma f} \tag{8.3}$$

This equation can also be expressed by real and imaginary parts:

$$\varepsilon' = 1 - \frac{f_p^2}{f^2 + \gamma^2} \quad [\text{real part}] \tag{8.4}$$

$$\varepsilon'' = \frac{f_p^2 \gamma}{f(f^2 + \gamma^2)} \quad [\text{imaginary part}] \tag{8.5}$$

Here, f is the frequency of the electric field (the photon) and γ is the damping frequency (determined by the inverse of the time between collisions for the free electrons). Just as with the analogy of the electron "sea," we treat the free electrons collectively—in this case as a kind of plasma. By definition, a plasma is a highly ionized gas, composed not only of electrons but also ions and neutral particles, and found in lightning bolts and candle flames. Here, however, our "plasma" contains only free electrons. This plasma oscillates at a particular frequency, f_p, which can be estimated by

$$f_p = \sqrt{\frac{n_e e^2}{m\varepsilon_o}}$$

(8.6)

Here, n_e is the density of free electrons, e is the elemental charge, and m is the effective mass of an electron. (Note: While this equation provides quite accurate results for many metals, including aluminum, lithium, sodium, potassium, rubidium, magnesium, and lead, it does not take into account the bound electrons' effect on the plasma. Electrons bound within the crystal lattice can in certain cases inhibit free electrons' motion. Gold and silver are such cases. In gold, this bound-electron effect lowers f_p from ultraviolet to visible wavelengths, and in silver it lowers f_p from far ultraviolet to near-ultraviolet wavelengths.)

The plasma frequency greatly influences the photonic properties of metals. And it can be explained using classical physics. Photons of frequency less than f_p are reflected because the rapidly oscillating, high-energy electrons serve to screen the more slowly oscillating, low-energy electric field; photons of frequency greater than f_p are transmitted because the electrons do not respond quickly enough to "steal" the energy from the electric field. We can see from Equations 8.4 and 8.5 that when $f \gg f_p$, the real part of the permittivity is positive, so the material behaves like a dielectric; meanwhile, the imaginary part is very small, indicating that minimal energy is lost as the EM waves move through the material. On the other hand, when $f \ll f_p$, the real part is negative, and the material behaves as a metal; the imaginary part is large so the EM field cannot penetrate very far into the material; photons are either reflected back out by free electrons near the surface or give up their energy to the electron plasma deeper in.

With most metals, f_p is in the ultraviolet range, hence their characteristic metallic sheen in visible light. However, certain metals, including copper and gold, have f_p in the visible range, leading to reflectance of red light more than blue light and endowing these metals with a more golden luster.

BACK-OF-THE-ENVELOPE 8.1

Aluminum (Al) has a free electron density, $n_e = 4.08 \times 10^{27}$ electrons/m³. Assume the effective mass is $m = 9.11 \times 10^{-31}$ kg. Determine f_p.

$$f_p = \sqrt{\frac{n_e e^2}{m\varepsilon_o}} = \sqrt{\frac{\left(4.08 \times 10^{27} \text{ electrons/m}^3\right)\left(1.60 \times 10^{-19} \text{ Coulombs}\right)^2}{\left(9.11 \times 10^{-31} \text{kg}\right)\left(8.854 \times 10^{-12} \text{ F/m}\right)}} = 3.6 \times 10^{15} \text{ Hz}$$

This can also be expressed as 3.6 petahertz (PHz). This frequency corresponds to ultraviolet light in the EM spectrum.

8.2.4.2 The Extinction Coefficient of Metal Particles

We are most interested in metal objects small enough to gain unique photonic properties. As a simple example, let us examine a spherical metal particle. The collective oscillation of the electron plasma is generally considered a bulk phenomenon, unbounded and unaffected by boundaries. This assumption holds true above the nanoscale, where objects are so gigantic compared to the electrons that surface interactions do not play much of a role. However, things change at the nanoscale. As we have learned, the smaller the object, the higher its surface-to-volume ratio. So, with a very small particle, we have to adjust our thinking and now assume that the electrons "sense" their boundaries, which are the surfaces of the object. And we can guess that this will somehow modify their collective behavior.

Electrons are always colliding with one another. The average distance they travel between collisions is called the mean free path, Λ. (We discussed the mean free path in terms of heat conduction in Chapter 7, Section 7.3.1.1.) However, a particle can be made smaller than the mean free path of electrons. Within this tinier particle volume, the electrons will have to collide more often with the particle surface than they would in a larger particle. In Equation 8.3, the damping frequency, γ, is determined by the inverse of the time between collisions for the free electrons. But in a nanoscale particle, this frequency has to be adjusted to account for the added collisions. The damping frequency for the free electrons of a particle becomes

$$\gamma_{\text{particle}} = \gamma + \frac{v_{\text{F}}}{\Lambda_{\text{adjusted}}} \tag{8.7}$$

Here, γ is the damping frequency of the bulk metal, v_{F} is the electron velocity at the Fermi energy, and $\Lambda_{\text{adjusted}}$ is the adjusted mean free path. (The Fermi energy is the energy level between occupied and unoccupied electron energies at a temperature of 0 K; for more information see Chapter 6, "Nanoelectronics.") For particles with diameters smaller than the mean free path of the bulk solid, $\Lambda_{\text{adjusted}}$ generally varies with particle radius, r_{p}, such that

$$\Lambda_{\text{adjusted}} = \frac{4r_{\text{p}}}{3} \tag{8.8}$$

Therefore, as the radius decreases, the damping frequency of the particle increases.

We are now armed with the equations needed to examine the effect of particle size on the photonic properties of a metal—all the equations, that is, but one.

Suppose some particles are placed in the way of a beam of photons, as shown in Figure 8.4. The intensity of the EM energy (number of photons) incident on the particles is E_1. The intensity of this energy downstream from the particles is E_2. Three things happen to the photons.

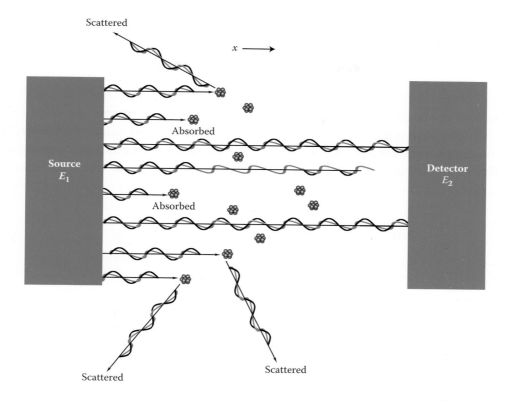

FIGURE 8.4   The extinction coefficient, C_{ext}. A beam of photons with intensity E_1 is directed from a photon source toward a group of particles. Because some of the photon energy is absorbed by the particles and some is scattered, the intensity measured at the detector is E_2, such that $E_2 < E_1$. The fraction of photons lost per unit meter (in the x direction) is the extinction coefficient. In this depiction, the particles are smaller than the wavelength of incoming photons. Such would be the case for particles a few hundred nanometers in diameter interacting with photons of visible light, the wavelengths of which span from about 400–750 nm.

BACK-OF-THE-ENVELOPE 8.2

A free electron inside a copper wire collides with another electron every 25 femtoseconds (fs), on average. The average speed of a free electron is 1.6×10^6 m/s. How small must a copper particle be before it begins to limit the mean free path of the electrons?

The mean free path is equal to the free electron speed multiplied by the time between collisions:

$$\left(1.6 \times 10^6 \, \text{m/s}\right)\left(25 \times 10^{-15} \, \text{s}\right) = 4 \times 10^{-8} \, \text{m}$$

This is 40 nm—thus, a copper particle with a diameter less than that is constricting the mean free path of the electrons. Note that we worked out this same result in Back-of-the-Envelope 7.2. Because copper atoms in the wire occur about every 0.2 nm, the electrons fly past about 200 atoms between collisions.

1. Some of them pass unobstructed through the particles.

2. Some are absorbed and excite electrons within the particles.

3. Some are scattered.

Because the particles absorb or scatter some of the photons, we can say that $E_2 < E_1$. The particles have brought about an "extinction" of the incident beam of photons. The fraction of photons lost per unit distance (in the x direction) is known as the extinction coefficient (sometimes called the extinction cross section), C_{ext}. For metal particles, we can express C_{ext} as

$$C_{ext} = \frac{24\pi^2 r_p^3 \varepsilon_m^{3/2}}{\lambda} \frac{\varepsilon''}{\left(\varepsilon' + \left(2 + (12/5)x^2\right)\varepsilon_m\right)^2 + \left(\varepsilon''\right)^2} \tag{8.9}$$

Here, r_p is the particle radius, ε_m is the relative permittivity of the medium surrounding the particles, λ is the photon wavelength, ε' and ε'' are, respectively, the real and imaginary parts of the permittivity of the metal, and $x = 2\pi N r_p/\lambda$, where N is the refractive index of the surrounding medium.

Let us take a closer look at a real example—spherical particles of aluminum metal. Accounting for the increased surface effects using $\gamma_{particle}$ we can plot the extinction coefficient (in air) as a function of photon wavelength for four particles between 2 and 20 nm in radius. Specifically, we have plotted the particle volume-normalized C_{ext} (i.e., $C_{ext}/V_{particle}$). This plot is shown in Figure 8.5.

We can learn a great deal from this plot. First, let us examine the 20-nm particle. The peak extinction wavelength of this particle is approximately 175 nm. As the particle size decreases, two things happen. First, the location of maximum extinction shifts leftward, toward smaller (higher energy) wavelengths. Smaller metal particles absorb smaller wavelengths. Second, as the particle shrinks, the peak height shrinks along with it. Also, the peaks broaden. Smaller metal particles have increased damping due to the increased interactions of the electron plasma with the particle surface. This is the effect of Equation 8.7 manifesting in the plot. Both of these effects—the shift to smaller wavelengths and the increased damping—have been verified in experiments with metal particles.

Nonspherical metallic particles, shaped more like rods, have both a longitudinal and a transverse plasma frequency (along and perpendicular to the long axis of the rod), giving the extinction coefficient curve two peaks instead of one. Typically, the transverse plasma frequency is similar to that of a particle, while the longitudinal frequency (corresponding to the longer axis) is lower.

Particle "shells," consisting of a dielectric core like silica surrounded by a metal film, also have photonic characteristics similar to purely metallic particles. However, the plasma resonance wavelength of these shells (as well as the peak in the extinction coefficient curve) is often higher than those of solid metal particles. As the thickness of the metal shell increases, leaving the size of the dielectric core constant, the resonance wavelength decreases.

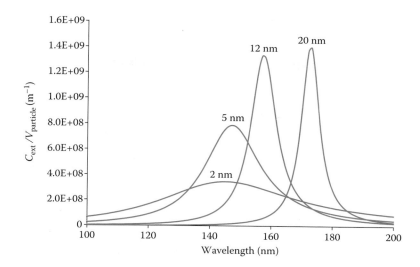

FIGURE 8.5   The effect of particle size on the extinction coefficient, C_{ext}, of aluminum particles. The particle volume-normalized extinction coefficient (or $C_{ext}/V_{particle}$) is plotted here versus photon wavelength for four particles between 2 and 20 nm in radius. The particles are in air, having a relative permittivity $\varepsilon_m = 1.006$ and a refractive index $N = 1.0008$. Other values used in calculating the real, ε', and imaginary, ε'', parts of the permittivity are: free electron plasma frequency, $f_p = 3.6 \times 10^{15}$ Hz, damping frequency, $\gamma = 1.94 \times 10^{13}$ Hz, and Fermi velocity, $v_F = 2 \times 10^6$ m/s.

8.2.4.3 Colors and Uses of Gold and Silver Particles

As we saw in the previous section, varying the size of an aluminum particle changes the wavelengths of light that it will absorb and scatter. The wavelengths in Figure 8.5 were lower than those of visible light (which are about 400–750 nm). However, if the extinction curves were instead to peak in the visible range, we could vary the particle size so as to tailor which colors of light the particles absorbed. As it turns out, gold and silver can turn color in just that way. When we confine the free electron plasma of these metals into tinier and tinier particles—effectively limiting the mean free path of the electrons due to a higher number of surface interactions—we change the photonic behavior. We change the particles' colors.

Medieval artisans made red stained glass windows by mixing gold chloride into molten glass. The gold particles, roughly 6–20 nm in diameter, absorbed the shorter wavelengths of blue and yellow light and not the longer-wavelength reds. These red wavelengths passed through the window, making the window appear red. Above diameters of about 20–25 nm, the absorption peak broadens and shifts to longer wavelengths—just as we saw with larger and larger aluminum particles. The gold color changes from ruby red to purple, then violet, and finally pale blue for particles about 160 nm in diameter. Above this size, the particles become gold in color again.

Size effects in silver particles are similar to those in gold. Silver particles 10 nm in diameter stain glass a bright yellow color. As the size increases up to about 130 nm, the color transitions through reds, purples, violets, dark and light blues, and finally to gray-green, before turning back to the color silver. Silver particles have long been incorporated into photosensitive materials used in photography. (Photographic film contains billions of silver halide crystals suspended in gelatin. Photons hit these crystals and break them up, leaving behind silver atoms that aggregate into tiny particles too small to see. During the developing process, these particles are amplified into the visible grains of a photograph.)

Colored metal particles have applications in biological sensing. In one application, we can use them to stain biological tissues so as to identify or watch specific biological processes in action. Metal particles can be coated with a specific type of molecule that is known to bind to another specific target molecule. Then, when these "functionalized" particles are introduced to an environment where the target molecule may be present, we can monitor color changes or just keep track of where the colored particles end up in order to detect, identify, and track all sorts of molecules. For example, it has been shown that when numerous gold particles are brought into close proximity (as in a colloid where the particles are separated by distances less than the average particle diameter), the plasma frequencies of the electrons in the particles tend to couple with one another and shift to higher energy (turning the colloid from red to blue). If we chemically coat gold particles with a snippet of DNA, and then add the particles to a solution containing an unknown strand of DNA, the solution will turn from red to blue if the sequences match and the dots aggregate. Otherwise, the solution remains red. Detecting specific DNA sequences is critical to diagnosing genetic and pathogenic diseases.

8.2.5 Semiconductors

Unlike the photonic interactions of metals, which can be explained accurately using classical physics laws, the interaction of semiconductors and photons is inherently quantum mechanical. When a photon with the right amount of energy encounters a semiconductor, it can boost an electron from an occupied state in the valence band to an unoccupied state in the conduction band. This transition requires a specific amount of energy, represented by the band gap. In the next section we will discuss how adjusting the band gap gives us control of the type of photons with which the material can interact.

8.2.5.1 Tuning the Band Gap of Nanoscale Semiconductors

When a material is reduced to a low number of atoms, the energy bands spread out and break into discrete energy levels. That is, the band gap widens. (For a review of this phenomenon, see Chapter 6, Section 6.7.3.) Recall that semiconductor particles at the size scale where this is possible are known as quantum dots. The smaller the quantum dot gets, the bigger the band gap. This enables us to adjust the boundaries of the band gap simply by adding or removing material.

When an excited electron drops from the conduction band back down to the valence band, the energy of the photon that is emitted equals the energy the electron loses in its transition—that is, it is equal to the band gap energy. So, when we change a material's band gap energy by changing its size, we change the energy of the photon it emits (the wavelength of the EM radiation). This means that we also change the energy of the photon it absorbs.

Recall from Chapter 3 the relationship between a photon's energy, E, and its wavelength, λ:

$$\lambda = \frac{hc}{E}$$

Here, c is the speed of light and h is Planck's constant.

This relationship clarifies the reason for the shift in the optical absorption and emission of quantum dots toward shorter wavelengths ("blue shift") as the dots get smaller. This is shown in Figure 8.6.

BACK-OF-THE-ENVELOPE 8.3

A conduction electron in a quantum dot emits a photon with a frequency of 600 THz as it drops to the valence band. Determine the band gap.

$$\lambda = \frac{c}{f} = \frac{299{,}792{,}458 \text{ m/s}}{600 \times 10^{12}\, s^{-1}} = 500 \text{ nm}$$

$$E = \frac{hc}{\lambda} = \frac{\left(6.626 \times 10^{-34}\, m^2\, kg/s\right)\left(299{,}792{,}458\, m/s\right)}{500 \times 10^{-9}\, m} = 4 \times 10^{-19}\, J$$

A useful unit of energy is the electron-volt (eV), which is the amount of energy an electron acquires moving through a potential difference of 1 V. A single electron-volt equates to 1.602×10^{-19} J. So our answer converts to 2.5 eV.

8.2.5.2 The Colors and Uses of Quantum Dots

By varying the size and composition of quantum dots, the emission wavelength can be just about any color, from blue up through the infrared. For example, cadmium sulfide (CdS)

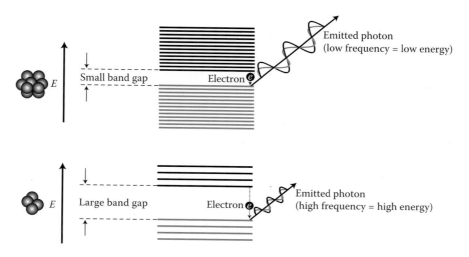

FIGURE 8.6 The energy of photons emitted (or absorbed) by very small semiconductors varies with band gap. Because the band gap of nanoscale semiconductors varies with their physical size, we can "tune" the boundaries of the band gap by adding or removing material. In doing so, we control the energy of photons emitted when excited electrons release their energy and drop back across the band gap. If the band gap is small, the emitted photon will have less energy (longer wavelength). This direct proportionality of band gap size and photon energy also holds true for absorption: to be absorbed, a photon must have at least the band gap energy.

and zinc selenide (ZnSe) dots can be sized to emit blue to near-UV light; cadmium selenide (CdSe) dots can emit light across the entire visible spectrum; and indium phosphide (InP) and indium arsenide (InAs) dots can emit in the far-red and near-infrared.

Colored quantum dots with tunable absorption/emission qualities can be used in tagging and tracking biological species, in anticounterfeiting applications to create special inks, in dyes and paints, in light displays, and in chemical sensing. Studies have shown that CdSe dots can be made to seep into cancerous tumors in the body. Then, when exposed to light, the particles glow—helping surgeons zero in on sick cells and leave the healthy cells alone.

Particles of titanium dioxide (TiO_2) and zinc oxide (ZnO)—both metal oxide semiconductors—have found use in sunscreens because they absorb ultraviolet radiation that damages the skin. Particles of both of these materials also scatter visible radiation, making them very good white pigments in paint, but this same property also turns skin white. Or at least it used to. While conventional sunscreens use particles measuring hundreds to thousands of nanometers in diameter, TiO_2 particles less than 50 nm in diameter are now available. Because they are smaller, these particles have larger band gaps, so they are transparent to visible light. This makes the sunscreen go on clear, although the particles still absorb the higher energy (shorter wavelength) ultraviolet radiation. Although such particles are already included in many sunscreens and cosmetics, their effect on the human body is still being studied. This raises concerns among those who feel this and other nanotechnology innovations should be better understood before becoming consumer products.

SOLAR PANELS—CATCHING MORE RAYS WITH QUANTUM DOTS

The Sun provides free photons from dawn to dusk. It is energy we should be able to use. However, we are still not very good at capturing photon energy. Solar cells that convert photons into electric energy have been around since 1954, but still have low efficiency (near 20%). This is not good enough to justify putting solar panels on every rooftop. The active material in most solar cells is a doped semiconductor such as silicon or gallium arsenide.

If an incoming photon's energy is equal to the band gap of the semiconductor, an electron is boosted from the valence band into the conduction band and contributes to the power output of the solar cell. If the photon's energy is less than the band gap, no electron can absorb it and the energy goes unused. Photons with energy greater than the band gap are absorbed but their extra energy (anything greater than the band gap) is lost as heat.

A typical band gap in a solar cell is 1.4 eV. Light from the Sun, on the other hand, is composed of a whole spectrum of energies. So, the solar cell acts as a lowpass filter, absorbing photons with greater than 1.4 eV, while photons with less than 1.4 eV are not absorbed. These lower-energy photons are either scattered or pass right through the cell.

There are semiconductors we could use that have lower band gaps (1 eV). So why not just use one of those and capture more photons? Because capturing more photons does not necessarily improve the efficiency. More photons are absorbed but there is a downside: we can only capture 1 eV per photon. So even more energy is wasted as heat.

One way to get better efficiency is to somehow make a solar cell with multiple band gaps. This may be possible using multiple materials. The high-band-gap material can grab the high-energy photons with minimal heat loss, while the low-band-gap material soaks up the remaining low-energy photons. Researchers know this approach can improve efficiency by 20%, but it has proven difficult to implement. For one thing, the crystal lattice structures of the various materials do not often match up very well, so such solar cells are troublesome to fabricate.

Enter quantum dots. We change their size and their band gap changes. So, if we were able to make a solar cell using layers containing different sizes of quantum dots, covering the whole spectrum of photon energies, we might truly drive up the efficiency (see Figure 8.7). Another possibility has been suggested by recent research with quantum dots in which a single high-energy photon is made to excite *multiple* electrons (instead of giving up the extraneous energy as heat). This phenomenon is called impact ionization.

These possible solutions do not necessarily get around the lattice mismatch issue or other material compatibility issues. And it may prove inefficient to get the excited electrons out of the quantum dots—thereby negating the improved absorbance efficiency. *Or . . .* quantum dots are exactly what is needed to finally get our photon's worth out of the Sun.

8.2.5.3 Lasers Based on Quantum Confinement

*L*ight *A*mplification by *S*timulated *E*mission of *R*adiation—that is what "laser" stands for. Stimulated emission occurs when a photon encounters an already excited electron and causes it to emit yet another photon with the same phase and direction. (The electron then drops back to its ground state.) The original photon is not absorbed and has therefore been amplified: there are two where there was once one. (Refer to Figure 8.2.) One condition under which stimulated emission can occur in a material is known as "population inversion." This is when the excited electrons outnumber the electrons in ground states. (I.e., more than half of all the electrons are excited.) Conventional lasers "pump" energy into the electrons using light or electricity or chemicals, but population inversion can also be achieved using quantum-confined systems. In fact, lasers are one of the most common

FIGURE 8.7   Quantum dots in a solar panel. A dot's size determines its band gap, which in turn determines how much energy the dot can absorb from a photon. The smaller the dot, the larger the band gap. If an incoming photon's energy equals the band gap, the dot absorbs all the energy. If the photon's energy is less than the band gap, the dot cannot absorb it. If the photon's energy is more than the band gap, it is absorbed but the extraneous energy is lost as heat.

applications for quantum confinement. Quantum dots, quantum wires, and quantum wells have all been used to make lasers. Let us take a closer look at the quantum well variety.

As we learned in Chapter 6, a quantum well is made by constraining electrons inside a region of minimal thickness. This can be done using a sandwich-type structure, with alternating layers of different semiconductors, such as gallium nitride (GaN) and indium-doped gallium nitride ($In_xGa_{x-1}N$). Doping involves adding a small amount of an impurity—indium metal, in this case—into a semiconductor crystal to make it a better conductor. Here, the amount of doping is indicated by x. In this way, each layer of doped GaN becomes a quantum well, electrically isolated by the less conductive GaN layers on either side of it.

The $In_xGa_{x-1}N$ layers are 3–4 nm thick. A power supply electrically connected to the quantum wells supplies electrons. Figure 8.8a shows a schematic of the main components of a quantum well laser. The $In_xGa_{x-1}N$ has a smaller band gap than pure GaN, giving the sandwiched layers an energy band profile similar to the one in Figure 8.8b.

Any electron that arrives in the conduction band of an isolated $In_xGa_{x-1}N$ layer is quickly trapped; it lacks the energy to surmount GaN's larger band gap and reach the conduction band, so it remains in the $In_xGa_{x-1}N$. Together, these trapped electrons assume quantized energy levels analogous to those found in a single atom. They behave like particles in a well (discussed in Chapter 3, Section 3.3.8). Just like the particles in the well, the trapped electrons can tunnel across the GaN barriers to reach similar energy levels in adjacent $In_xGa_{x-1}N$ wells.

Electrons usually spontaneously emit a photon within about 100 ns of becoming excited; however, in this case the excited electrons have another option: tunneling. Tunneling happens more quickly than spontaneous emission; so instead of emitting a photon, an excited electron just keeps "vanishing" and "reappearing" in an adjacent well. Before long, there is a huge buildup of excited electrons—the population inversion a laser needs.

Like a fugitive who cannot outrun the law forever, an excited electron cannot just keep tunneling indefinitely. Eventually, some of the excited electrons spontaneously emit photons, then drop back to ground states. These photons race through the well. Their EM fields perturb other excited electrons, stimulating emission of even more photons. This chain reaction produces many photons that reflect back and forth between mirrors on either end of the device, building up and emitting a coherent (single-frequency) photon stream through an aperture in one of the mirrors. This is the laser beam.

Quantum well lasers operate over a broad spectrum of wavelengths, from 400 nm to 1.5 μm, and are found in many devices, including CD players and laser printers. Because of their small wavelength, blue laser varieties enable high-density, optical information storage on DVDs and Blu-ray discs. Quantum wire lasers and quantum dot lasers are made in much the same way as quantum well lasers. (Quantum dot lasers are often made using thin films of quantum dots instead of the continuous thin films found in quantum wells.) Quantum wire lasers and quantum dot lasers can be switched on and off using less current, and in less time, than quantum well lasers. This offers possibilities for high-speed digital information transfer in telecommunications.

(a)

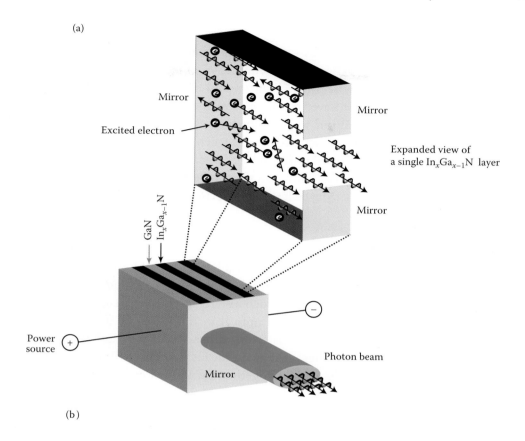

(b)

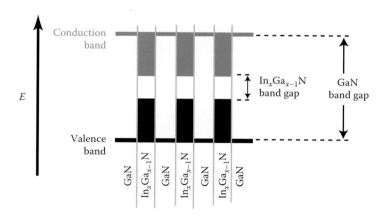

FIGURE 8.8 A quantum well laser. (a) In this simplified representation, a sandwich structure is formed by alternating layers of gallium nitride (GaN) and indium-doped gallium nitride ($In_xGa_{x-1}N$). A power supply provides electrons. Stimulated emission causes excited electrons trapped inside the $In_xGa_{x-1}N$ layer to emit photons that reflect between mirrors on either end of the device (as shown in the expanded view), until being emitted through an aperture as a laser beam. (b) The energy level diagram of the quantum well. Excited electrons in $In_xGa_{x-1}N$ layer are trapped because they lack the energy to overcome the larger band gap of GaN.

8.3 NEAR-FIELD LIGHT

Our eyes are great photon detectors—making light and sight practically one and the same—but only in a narrow range of wavelengths. We are unable to see the very small wavelengths of x-rays—otherwise x-ray "light" would represent the "visible" part of the EM spectrum. And although we have, over time, developed technologies to detect the other forms of EM radiation, from gamma waves to long waves, the wavelengths we can actually see with our eyes are the only wavelengths we use when we want to *look directly at something*. The problem is: the smallest wavelength humans can see is about 400 nm. As we will learn in the next section, this poses a limit on the smallest things we are able to make out with regular light microscopes. (Maybe it would be better if we could see x-rays.)

8.3.1 The Limits of Light: Conventional Optics

Optical telescopes and microscopes bend light to make small objects appear bigger, as shown in Figure 8.9. And it seems like if we could build the right microscope, we could shine light on a material and magnify it enough for our eyes to see atoms. Or, the other way around—that we could focus light to such a sharp point that we could illuminate a single atom. But we cannot. Conventional optical techniques have a lower limit. This limit goes by many names: Rayleigh criterion, diffraction limit, angular resolution, resolving power, precision, and others. The main idea is that visible light can be focused only so far. If two objects get close enough together, we will not be able to tell them apart using visible light.

   To explain how, let us conduct the virtual experiment shown in Figure 8.10. Here, we have a microscope that makes two scans in a row, from right to left, over a surface. During the first scan, hole 1 is open. As light shines up through the hole, photons reflect off the sides, causing diffraction and spreading the light out. When the microscope lens passes over the surface, the light is captured in a lens and routed to a light intensity meter (such as a photoelectric metal like we learned about in Chapter 3). This meter creates a graph of the light intensity at each point along the surface. We can see that the intensity peaks when the

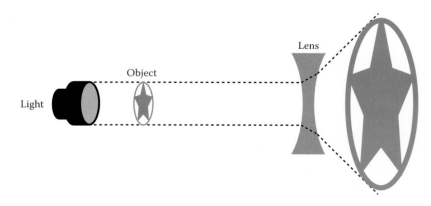

FIGURE 8.9   Microscope lenses bend light, making small objects appear large. The type of lens shown here is biconcave. The light spreads out as it passes through.

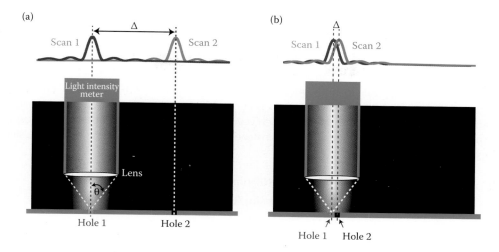

FIGURE 8.10 The resolution limit of a conventional optical microscope. The microscope performs two scans, from right to left. There is one hole in the surface, and light shines through it from underneath. Between scans, "hole 1" is plugged and a new "hole 2" is made at a distance, Δ, from its original location. As the microscope scans, it creates a graph of the light intensity. In (a), when we overlay the first scan (corresponding to hole 1) and the second scan (hole 2), both peaks are easily distinguished. But in (b), the hole barely moves. The peaks overlap. If Δ were any smaller, the microscope would not have the necessary resolution to tell the two peaks apart.

focal point of the lens is directly over the hole. We can also see some interference patterns in the graph from the photons reflecting off the sides of the hole. These patterns are due to the wave nature of light (see Chapter 3). After the microscope makes its first pass over the surface, the hole is moved a distance, Δ, from its original location to its new location at hole 2. The scan is repeated, with the microscope again registering the light intensity as it passes over the surface.

In Figure 8.10a, when we overlay the graph from scan 1 (corresponding to hole 1) and scan 2 (hole 2), we can easily distinguish the two peaks. In Figure 8.10b, the hole barely moves between scans. The peaks overlap. If they got any closer together, we would no longer be able to tell them apart. This is the resolution limit of a light microscope. It is the smallest distance, Δ, we can distinguish using an optical microscope, determined by

$$\Delta = \frac{1.22\lambda}{2n\sin\theta} \tag{8.10}$$

Here, λ is the wavelength of light used, n is the refractive index of the medium in which the lens operates, and θ is the collecting angle of the lens, as depicted in Figure 8.10. Although we have been discussing gathering light *into* a lens, this equation is also applicable to light focused *out* of a lens to make a small spot. In that case, Δ represents the smallest spot size the lens can make.

BACK-OF-THE-ENVELOPE 8.4

So what is the best resolution theoretically possible using an optical microscope? And what is the smallest spot of light we can make?

We obtain the answer to both questions by solving Equation 8.10. The best collecting angle we can expect to achieve with a lens is about 70°. Of course, we will want to use light with the shortest wavelength possible, which is violet ($\lambda = 400$ nm). If we operate in oil using a special immersion lens, then $n = 1.56$ (instead of operating in air, where $n = 1.0$). Using these parameters, we obtain

$$\Delta = \frac{1.22(400 \text{ nm})}{2(1.56)\,\sin(70°)} = 166 \text{ nm}$$

This is the theoretical limit. In practice, it is generally accepted that the best resolution is of the order of $\lambda/2$, which in this case equates to about 200 nm.

8.3.2 Near-Field Optical Microscopes

When we look at an object through a microscope, we see light. This light was either emitted by the object or emitted from elsewhere and scattered by the object, or some of both. It makes sense that no matter your distance from the object, the light you see will be the same. Which is to say that whether you stand 10 m or 1 m from a tree, you see the same tree. Of course, at 1 m you can see more detail—the spots on the leaves, the grittiness of the bark—but you can see those same details from 10 m with a good zoom lens. It makes sense that the tree (or, really, the light coming off the tree) is the same at 1 and 10 m. However, if you could somehow get within about 200 nm of the tree, the light would contain details you could not otherwise distinguish.

This is known as near-field light. The photons have just left the object and have not traveled even a single wavelength yet. The light at this distance is evanescent, transitory. It is part of a nonpropagating field, brimming with high-frequency spatial information. Near-field light, or "forbidden light" as it is sometimes called, decays exponentially. Within a single wavelength, it is completely gone. The highest amplitudes of near-field light occur within about 10 nm of the object. Far-field light, on the other hand, propagates across space and does not fade out. It goes until it hits something. To see near-field light coming off an object entails getting incredibly close to it, using tools and techniques enabled by nanotechnology. If you are too far away from an object, the highest-frequency spatial information about that object's shape has been filtered out.

To understand how this could be possible, consider what happens when you throw a rock into a pond (Figure 8.11). When the rock splashes in, it displaces water. In the region very near the splash, the ripples closely resemble the shape of the rock. As waves form and propagate outward from the splash, much of the high-frequency spatial information carried by the ripples is lost—a phenomenon called wave dispersion. Larger, low-frequency waves separate from and leave behind smaller, high-frequency waves. Small waves tend to lose energy and fade with distance, while long waves carry on. In oceans, long waves can

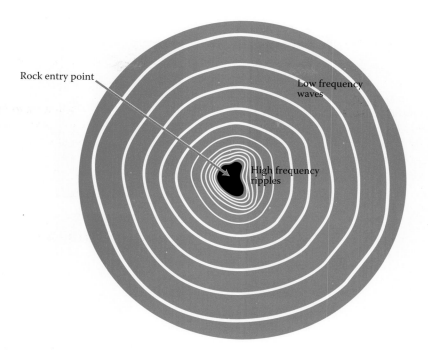

FIGURE 8.11   A simplistic analogy to near-field light. Here we see an overhead view of a rock dropped into a pond. Ripples near the entry point closely resemble the rock's shape. As waves form and propagate outward, the details of the shape are gradually lost. We can imagine near-field light like the ripples near the rock.

travel thousands of miles. In the pond, the large waves farther away from the splash are more or less circular, giving us a general idea of what the rock must have looked like, but not the details. The details are in the ripples.

The objective of near-field microscopy is to place a detector with a very small aperture within one wavelength of light from the object. In doing so, we can beat the limits imposed by traditional optical microscopes and see features smaller than 200 nm. The idea of obtaining ultra-high resolution took shape as early as 1928, but practical implementation proved difficult. After all, the techniques of nanotechnology had yet to be developed. It was impossible to build an aperture tiny enough, or to get it within 200 nm of a specimen.

Today there are numerous setups and modes of operation for near-field microscopes. Two common modes are depicted in Figure 8.12. One is collection mode, where the sample is illuminated from below and the transmitted light is collected in the near field by a probe scanning just nanometers from the sample. The other is illumination mode, where the probe illuminates the sample from just nanometers away, ensuring that only a few square nanometers of the sample are lit; the light is then collected and magnified by a lens below the sample. Because the probe needs to be in such close proximity to the sample, near-field microscopes are typically coupled to a scanning-probe microscope system with a feedback loop designed to track minute surface topography. (The scanning probe

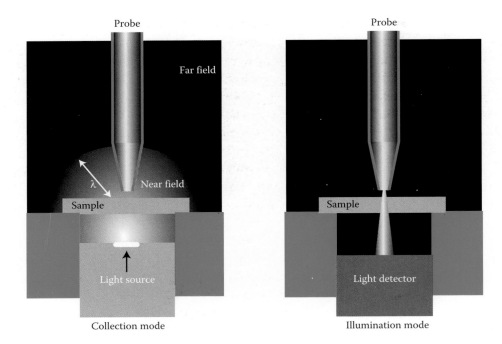

FIGURE 8.12 Two modes of near-field microscopy. In collection mode, a near-field probe collects light transmitted up through the sample. The distance from sample to probe is less than one wavelength, λ, of the light used. In illumination mode, the probe shines light on only a few square nanometers of the sample and this light is collected in a lens positioned underneath the sample.

microscope is discussed in Chapter 5, "Nanomechanics.") The probe is usually some type of optical fiber with a reflective metal coating, tapered at the tip to an aperture about 10–100 nm wide.

A typical near-field microscope will achieve resolution of less than 100 nm; some have achieved as low as 20 nm—an order of magnitude better than conventional optical microscopes. Near-field techniques have been used to examine quantum dots, thin films, human chromosomes, and even single fluorescent molecules. Near-field probes have even found use in lithography applications, where they are used to define structures smaller than the wavelength of light. The disadvantages of the near-field techniques are slow scanning speed and minimal working distance from the sample.

Another disadvantage is the depth of field, which is the distance in front of and behind the sample that appears in focus. As an example from photography, if you were to stand in the end zone of a football field and focus a camera on the 50-yard line, the 40-yard lines on either side of it might also still be in focus, while all the other lines are out of focus. In this case, the depth of field would be about 20 yards. Depth of field is proportional to the distance from the subject. In the near-field microscope's case, where the probe tip must be kept mere nanometers from the sample, the depth of field is a few nanometers or less, so only a very thin "slice" of the sample is focused at any one time.

8.4 OPTICAL TWEEZERS

The momentum of an object is classically given by the object's mass, m, multiplied by its velocity, v. A photon, as we understand it, has no mass, yet it can be shown that it *does* have momentum, $p = h/\lambda$, where h is Planck's constant and λ is the photon's wavelength (see Chapter 3). When light is reflected off an object, the photons change direction. For momentum to be conserved, the object must then move in the opposite direction. Of course, a photon's momentum is next to nothing. No building ever toppled west because the Sun rose in the east. However, if the object is small enough, light can move it. (Light can also move objects if the resistance to motion is small enough, as with the concept of solar sails—gigantic mirrors that would capture momentum from sunlight to propel spacecraft.)

Light has been used to push around all types of nanoscale matter: atoms, particles, viruses, bacteria, living cells, and DNA. One or more tightly focused laser beams is enough to exert picoNewtons of force and trap a nanoscale object in a particular place, with a resolution of 1 nm in three dimensions. This technique is commonly known as optical tweezers. It has been used to confine and move around different kinds of cells to sort them as well as to apply and measure small forces. In one application, tiny beads are attached to both ends of a DNA molecule. The optical tweezers push the beads apart, stretching out the molecule in order to measure its elasticity as well as the force required to alter or rupture it.

Although optical tweezers can be used to arrange structures on the nanoscale, it appears to be too slow a process for building any kind of larger, complicated structure.

8.5 PHOTONIC CRYSTALS: A BAND GAP FOR PHOTONS

Our understanding of the way electrons behave inside semiconductors has allowed us to make devices of unprecedented complexity and usefulness. However, photons have some advantages over electrons. For one, they are faster: photons move through a dielectric such as silicon dioxide at about 2×10^8 m/s, more than 100 times faster than electrons in a copper wire, which travel at about 1.6×10^6 m/s (for a wire 2 mm in diameter, carrying a 10 A current). (Note: This is an electron's average speed between collisions; the drift speed that makes electric current in this particular wire is a mere 250 µm/s because electrons zigzag instead of following a straight path down a wire.) The bandwidth, or usable frequency range, of dielectric materials is also larger than the bandwidth of electronic materials. For example, fiber-optic communication systems using dielectric materials and photons have bandwidths measured in Terahertz, as opposed to conventional electric systems (such as telephones), which are in the kilohertz range. In addition, photons do not interact as strongly as electrons and thus do not lose energy as easily.

Still, making a circuit that uses light instead of electricity engenders new challenges, such as finding ways to segregate, route, and control light—that is, taming the enigmatic photon. Photonic crystals offer the possibility of doing such things.

First envisaged in 1987, photonic crystals are a way of controlling photons like semiconductor devices control electrons. Semiconductor devices operate based on the electron energy band gap—the forbidden energies separating allowed energy bands. The analogy for photonic crystals is known as the photonic band gap.

Photonic crystals are made from periodic patterns of materials with different permittivities. When a photon traveling through one material encounters an interface with another material with a different permittivity, the photon is sometimes reflected. (This is Bragg diffraction and depicted in Figure 4.25.) This reflection depends on the way atoms in a material's crystal lattice are spaced out and organized. Photons of certain wavelengths will pass freely through the material, while photons within a specific range of wavelengths will be reflected. This represents the photonic band gap—a range of photon wavelengths (hence, frequencies) that are forbidden because they cannot penetrate into the materials inside a photonic crystal.

Most photonic crystals are man-made. (Opal is an example of a naturally occurring photonic crystal.) Typical spacing between material layers in a photonic crystal is similar to the wavelength of light, or about 500 nm. Fabrication of three-dimensional structures with features of this scale is difficult, although research and development in this area remains active. New fabrication techniques are enabling the creation of photonic crystals capable of organizing photons based on their energy, confining them to small cavities, or guiding them down wire-like conduits. Such options have the potential to enable the same kind of devices we make out of semiconductors—interconnects, modulators, filters, switches.

8.6 SUMMARY

We can make a very good guess as to how a material will interact with photons if we know how its electrons are organized. Conductors, semiconductors, and insulators each have unique electron energy band arrangements—most noticeably with regard to the band gap—giving all three of these materials unique photonic properties.

With conductors (metals), we treat the free electrons collectively, as a kind of plasma. The frequency at which this plasma oscillates determines what frequencies of EM radiation will be absorbed, reflected, and scattered. Particles measuring just a few tens of nanometers tend to absorb and scatter the most photons at a particular frequency. This frequency is inversely proportional to the size of the particle. With semiconductors, the size of the band gap determines what photon energies the material will absorb and emit. Band gap is inversely proportional to particle size. So in both cases (metals and semiconductors), we can "tune" the photonic properties of a nanomaterial, such as the colors it absorbs or emits, by altering its size. This ability enables numerous medical, sensing, energy conversion, and imaging opportunities. A quantum well laser is a good example of a device, found in myriad modern gadgets, that capitalizes on the photonic behavior of a semiconducting nanomaterial.

Three more nanophotonics-based technologies are near-field microscopes, optical tweezers, and photonic crystals.

Near-field light is a nonpropagating field that carries high-frequency spatial information. Using techniques enabled by nanotechnology, we are now able to bring ultra-small optical probes within just a few nanometers of an illuminated sample in order to capture this elusive, near-field light. In this way we can see features as small as 20 nm—10 times smaller than we can see using conventional light microscopes.

Optical tweezers use photons to push around nanoscale objects because photons have momentum. By directing a stream of photons at very small things—atoms, particles, viruses, bacteria, living cells, and DNA—we can position and manipulate them.

Photonic crystals control photons in a manner analogous to how semiconductor devices control electrons, only instead of an electronic band gap, there is a photonic band gap. These crystals are made from periodic patterns of materials with different permittivities, and make it possible to confine, guide, and filter photons.

HOMEWORK EXERCISES

8.1 Define nanophotonics.

8.2 What color is the light reflected off the glass of the Lycurgus Cup? What color is the glass when light is transmitted through it?

8.3 True or false? A material's optical properties are invariant with its size.

8.4 In what year did Einstein's published theories on the photon validate Planck's experimental data from 1900?

8.5 Which of the following statements is true?
   a. Photons have energy, but no momentum.
   b. Photons have momentum and energy, but no mass.
   c. Photons have velocity and momentum, but no energy.
   d. Photons have mass and velocity, but no momentum.

8.6 What electronic property of semiconductors and insulators determines how much energy they can absorb from a photon?

8.7 Use Table 6.1 and Equation 3.4 to show why a diamond is clear (i.e., unable to absorb visible light, the wavelengths of which range from 400 to 750 nm)?

8.8 Draw a picture showing a spontaneous emission in one electron triggering a stimulated emission in another electron. What happens, energy-wise, to each of the electrons after emission?

8.9 Which photon has more energy—an ultraviolet photon or a yellow photon?

8.10 True or false? An example of interband absorption is when a valence electron in a large block of sodium absorbs a photon and is boosted into the conduction band.

8.11 Use interband absorption to explain why glass is transparent.

8.12 a. Determine the permittivity (in Farads/meter) of quartz, water, and polyethylene.
   b. Which of these materials is the best dielectric?

8.13 True or false? The less polarizable the material, the stronger the electric field it can accommodate.

8.14 The ionosphere, which is the layer of atmosphere located in the region 50–500 km above the Earth, has a density of free electrons just like a metal.
   a. If the free electron density is 10^6 electrons per cubic centimeter, determine the plasma frequency of the ionosphere in hertz (Hz).
   b. A metal is transparent to radiation with frequencies higher than the free electron plasma frequency (this is usually in the ultraviolet range), and reflecting at lower wavelengths. Determine the wavelengths at which the ionosphere becomes transparent, and reflecting at lower frequencies.
   c. Is it transparent to at least some radio waves? (This is an important consideration for communication with satellites and spacecraft.)

8.15 In a metal particle, we must take into account the free electrons' collisions with _____.

8.16 The damping frequency of a metal's free electrons is determined by the inverse of the time between collisions for the free electrons.

    a. In general, how does minimizing the size of a metal particle affect the damping frequency? Why?

    b. Plot the damping frequency, $\gamma_{particle}$, as a function of the surface-to-volume ratio for a silver particle with a radius varying from 1 to 100 nm. The damping frequency of the bulk metal is $\gamma = 2.7 \times 10^{13}$ Hz. The Fermi velocity is $v_F = 1,400,000$ m/s.

    c. Is the plot of the curve linear or exponential?

8.17 The equation for determining the permittivity of the free electrons in a metal has both real and imaginary parts.

    a. Using values from Figure 8.5, create a plot of the real and imaginary parts of the permittivity of an aluminum particle 10 nm in diameter over wavelengths from 1 to 200 nm.

    b. At what wavelength does the real part transition from positive to negative?

    c. Over what range of wavelengths does the aluminum behave like a dielectric?

    d. What does the imaginary part of the permittivity indicate? Looking at the curve of the imaginary part, would you expect aluminum to be opaque or transparent at wavelengths below 40 nm?

8.18 Decreasing the size of a metal particle tends to cause which two things to happen to the extinction peak?

8.19 True or false? Both semiconductors and conductors have a band gap.

8.20 True or false? Band gap is inversely proportional to particle size.

8.21 A quantum dot has a band gap of 5 eV. Can it emit visible light?

8.22 Solar radiation in the wavelength range of 290–400 nm burns our skin.

    a. What range of the EM spectrum do these wavelengths correspond to?

    b. A batch of particles with band gaps ranging from 3.0 to 3.5 eV is prepared for use in sunscreen. What percentage of the harmful wavelengths can this range of particles absorb?

8.23 True or false? A solar cell material with a smaller band gap can absorb a broader range of photon energies, yet this does not necessarily improve the solar cell's efficiency.

8.24 How does doping gallium nitride (GaN) with indium metal affect the band gap?

8.25 Define the electronic condition necessary for a laser to begin working.

8.26 Typically, excited electrons are quick to lose their extra energy by emitting a photon. What faster phenomenon do the excited electrons "opt" for instead in a quantum well laser?

8.27 True or false? Conventional optical microscopes do not resolve features any smaller than about 200 nm because that is the smallest wavelength humans can see.

8.28 In photolithography, visible light is focused at a thin film of photosensitive polymer on a surface. This can make the exposed portions of the polymer

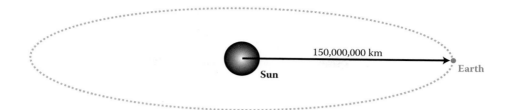

FIGURE 8.13   For Homework Exercise 8.31.

harden. In this way we can define raised features on a surface. What is the thinnest polymer line we could create using this process if conducted in air, using a lens with a collecting angle of 50°?

8.29 At what distance from an illuminated surface does a probe need to be to captured near-field light?

8.30 Near-field probe tips have apertures measuring about _____ in diameter.

8.31 The luminosity of the sun is 3.827×10^{26} W (1 W = 1 J/s). This is the approximate power it emits over a range of photon wavelengths, in all directions, out into the solar system. The distance from the Earth to the Sun is approximately 150,000,000 km (Figure 8.13). The Earth's diameter is 12,756 km.

a. Assuming no photons are absorbed or scattered in the space between the Sun and the Earth, how much of this power is incident on the Earth's surface?

b. We assume that 40% of the Sun's total luminosity is emitted at a wavelength of 550 nm. (This happens to be yellowish light.) What is the energy (in Joules) of a single photon with this wavelength?

c. How many photons with this wavelength are incident on the Earth every second?

d. What is the momentum carried by a second's worth of these 550 nm-wavelength, earth-bound photons?

e. The heaviest train ever pulled weighed 99,790,321 kg. It had 5648 wheels, measured over 7 km long, and traveled across part of Western Australia on June 21, 2001. How fast would this train have to travel (in m/s) to have momentum equal to the photons you calculated in part (d)?

RECOMMENDATIONS FOR FURTHER READING

1. C. Bohren and D. Huffman. 1998. *Absorption and Scattering of Light by Small Particles.* Wiley-VCH.
2. M. Grundmann. 2002. *Nano-Optoelectronics: Concepts, Physics, and Devices.* Springer.
3. P. Prasad. 2004. *Nanophotonics.* Wiley-Interscience.
4. J. P. Fillard. 1996. *Near Field Optics and Nanoscopy.* World Scientific.
5. J. Simmons and K. Potter. 1999. *Optical Materials.* Academic Press.
6. V. Markel and T. George. 2000. *Optics of Nanostructured Materials.* Wiley-Interscience.
7. R. Menzel. 2001. *Photonics: Linear and Nonlinear Interactions of Laser Light and Matter.* Springer.
8. C. F. Klingshirn. 2006. *Semiconductor Optics.* Springer.

CHAPTER 9

Nanoscale Fluid Mechanics

9.1 BACKGROUND: BECOMING FLUENT IN FLUIDS

Now it is time to speak the language of fluid mechanics. Fluid mechanics is the study of fluids in motion or at rest. From the point of view of conventional fluid mechanics, matter is either a fluid or a solid. A fluid is a substance that will deform continuously under shear stress, and a solid is a substance that will not. That is, a fluid will not support shear stress, but will flow and take the shape of its container, as shown in Figure 9.1. This section will derive the equations used to describe fluid motion, which includes the behavior of a fluid at the nanoscale.

9.1.1 Treating a Fluid the Way It Should Be Treated: The Concept of a Continuum

Unlike solids, the molecules in a fluid are not fixed to a lattice—they are free to move relative to each other. These molecules are more widely spaced out in a gas than in a liquid, the distance between them being much larger as compared to their molecular diameter. Brownian motion, or the random motions of the molecules due to their individual thermal energies, makes all these molecules move all over the place—unlike a solid. Therefore, it is not easy to track every single molecule in fluids: a mere picogram of water contains 10^{10} H_2O molecules. As a practical alternative, scientists have come up with the concept of a *continuum*. What this means is that we assume that the fluid properties are the same throughout a certain volume of fluid. The continuum concept also assumes that the fluid is infinitely divisible, in other words, its variation in properties is so smooth that we can use differential calculus to analyze the flow. To understand this further, let us think about a specific fluid property: density.

What is the density of a fluid? It is the total amount of mass in a given volume. If the volume is on the molecular scale, the addition or removal of a single molecule can change the total mass significantly, and with it the density. So we cannot really assign a precise meaning to density unless we have a volume large enough that the gain or loss of a few molecules will not significantly affect the total mass. It is at this volume where we can say we have a continuum fluid. For example, air can be considered a continuum at volumes greater than about 10^{-21} m^3 at room temperature and pressure since we can assume that the density does not

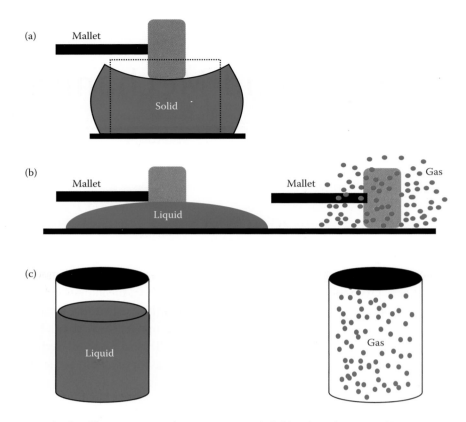

FIGURE 9.1   Fluid will not support shear stress. A solid, like that shown in (a), can support shear stress after deforming to a given shape (in this case because a rubber mallet is dropped on it). However, a fluid, like the liquid or gas in (b) cannot support shear stress but will flow away from, and around, the rubber mallet. Because fluids flow, they can take the shape of their containers, as shown in (c).

fluctuate at this volume. The length scales required for this assumption to hold are different for the flows of liquids and gases. For gases, the appropriate length scale is typically the mean free path of the gas. In liquids, molecules are tightly packed and we use the characteristic size of the molecule. In general, the continuum approximation holds for lengths greater than 40 nm in liquids (i.e., a volume of 40 × 40 × 40 nm) and 1000 nm in gases.

At the nanoscale, the continuum approximation may not work due to the small length scales. However, as you read further on in this chapter, you will see that this approximation actually works quite well for several applications. To alert you to the fact that we may not always be able to use this approximation, we will also touch briefly on other ways to analyze fluid dynamics at the molecular level (called molecular dynamics)—where we take on the great challenge of tracking each molecule individually.

9.1.1.1 Fluid Motion, Continuum Style: The Navier–Stokes Equations

In this section we will present equations that describe fluid motion using continuum theory. First, recall in Chapter 5, "Nanomechanics," where we learned the basic equations of motion.

Most notably, we learned that the sum of all the forces acting on an object, ΣF, equals the object's acceleration, a, multiplied by its mass, m (or $\Sigma F = ma$). This famous equation of motion is really all we need to describe any type of fluid motion as well. Using a molecular dynamics approach, we use this equation to analyze the motion of individual molecules and atoms as they flow. But when we are dealing with too many molecules to analyze individually, we have to bring in the continuum concept. Applying the continuum theory to the famous equations of motion, we can derive the also-famous Navier–Stokes equations.

These Navier–Stokes equations were first developed by Claude-Louis Navier in 1822, and have been studied and used extensively since then to describe fluid motion. They have been used to develop accurate models for breathing, blood flow, swimming, pumps, fans, turbines, airplanes, ships, rivers, windmills, pipes, missiles, icebergs, engines, filters, jets, sprinklers, and yes, nanofluidics. If you go to a library and look up fluid mechanics and you will find hundreds of books dedicated to these equations. So before we delve into them, we just want you to keep in mind that we will be looking at these equations at a very superficial level.

The Navier–Stokes equations boil down to $F = ma$. After all, we are just describing the equations of motion for a fluid. However, since we are dealing with a continuum fluid, it makes more sense to deal with fluid *density* instead of fluid *mass*, because it is difficult to track all the mass (i.e., individual particles) of a fluid flowing. This is not too much of a change since density is just mass per unit volume, but is a big change conceptually because we can deal with a continuum, not individual particles. So if we divide Newton's law of motion by volume V, we obtain the following:

$$\rho \vec{a} = \sum \frac{\vec{F}}{V} \tag{9.1}$$

Here, ρ is the density of the fluid (kg/m³) and a is the acceleration of the fluid (m/s²). The right-hand side of the equation represents the forces per unit volume acting on the fluid (N/m³). From Chapter 5, we know that acceleration is the differential change in velocity with respect to time, so the equation becomes

$$\rho \frac{D\vec{v}}{Dt} = \sum \frac{\vec{F}}{V} \tag{9.2}$$

Here, v is the velocity of the fluid and t is the time. In this equation, first notice that there are arrows on top of the symbols representing force and velocity. This means that these variables are vectors; and have an associated magnitude and direction. Notice also that the derivative of velocity with respect to time looks a little bit different than in Chapter 5. This is because we are taking a slightly different kind of derivative. We are taking a derivative of the velocity of the continuum fluid, not just an individual fluid particle. This derivative (also called a substantial convective, or total derivative) in two dimensions is defined as

$$\frac{D*}{Dt} = \frac{\partial *}{\partial t} + \frac{\partial x}{\partial t}\frac{\partial *}{\partial x} + \frac{\partial y}{\partial t}\frac{\partial *}{\partial y} = \frac{\partial *}{\partial t} + v_x \frac{\partial *}{\partial x} + v_y \frac{\partial *}{\partial y} \tag{9.3}$$

Here, * refers to the variable that we are taking the total derivative of. Note that the first term is a partial derivative of the variable with respect to time, and the next two terms are partial derivatives of the variables with respect to space in two dimensions. Since the fluid is always moving under shear stress, the derivative of any fluid property must also include the velocity of the fluid in each direction, which we capture in Equation 9.3, since $\partial x/\partial t$ is the velocity of the fluid in the x-direction (which can also be written as vx), and $\partial y/\partial t$ is the velocity of the fluid in the y-direction (which can also be written as vy). So, the derivative of velocity in two dimensions using Equation 9.3 becomes

$$\frac{D\vec{v}}{Dt} = \frac{\partial \vec{v}}{\partial t} + \frac{\partial x}{\partial t}\frac{\partial \vec{v}}{\partial x} + \frac{\partial y}{\partial t}\frac{\partial \vec{v}}{\partial y} = \frac{\partial \vec{v}}{\partial t} + v_x\frac{\partial \vec{v}}{\partial x} + v_y\frac{\partial \vec{v}}{\partial y}$$

Here, $\vec{v} = v_x$ (velocity in the x-direction) $+ v_y$ (velocity in the y-direction).

Now, applying this derivative definition to Equation 9.2, we can write the following (for two dimensions):

$$\rho\left(\frac{\partial v_x}{\partial t} + v_x\frac{\partial v_x}{\partial x} + v_y\frac{\partial v_x}{\partial y}\right) = \sum\frac{\vec{F}_x}{V} \tag{9.4a}$$

$$\rho\left(\frac{\partial v_y}{\partial t} + v_x\frac{\partial v_y}{\partial x} + v_y\frac{\partial v_y}{\partial y}\right) = \sum\frac{\vec{F}_y}{V} \tag{9.4b}$$

Equation 9.4a describes motion in the x-direction, whereas Equation 9.4b describes fluid motion in the y-direction.

Now we are ready to start looking at all the forces per unit volume that affect a fluid. These forces can either be surface forces or body forces. Surface forces are those that act on the surface of the fluid, such as shear stress and pressure. Body forces are forces that act on the entire volume of the fluid, such as gravity, electric fields, and magnetic fields.

BACK-OF-THE-ENVELOPE 9.1

Fluid flows in the x-direction through a converging duct as shown in Figure 9.2a. After putting a velocity probe into the fluid at points (A) and (B), we measure the velocities at these points in the duct to be $v_a = 1$ mm/s and $v_b = 5$ mm/s in the x-direction only. Estimate the fluid acceleration (in the x-direction) between these two points if the distance between these two points is 5 mm.

Since this flow is only in the x-direction, we are just looking at the change of velocity over time in the x-direction. Employing Equation 9.3, we obtain the following equation for acceleration in the x-direction:

$$a_x = \frac{Dv_x}{Dt} = \frac{\partial v_x}{\partial t} + \frac{\partial x}{\partial t}\frac{\partial v_x}{\partial x} + \frac{\partial y}{\partial t}\frac{\partial v_x}{\partial y}$$

(Note that this is the same as Equation 9.3, where we have replaced the *s with v_x.)

Since there is only flow in the x-direction, $\partial y/\partial t$ is zero, making the entire third term on the right-hand side of the equation equal to zero. In addition, the flow is steady, meaning that it is not time-varying; and at any point in time, the velocity at points A and B will be the same. This means that the first term on the right-hand side of the equation is also zero. What remains is the following:

$$a_x = \frac{\partial x}{\partial t} \frac{\partial \vec{v}}{\partial x}$$

We can estimate $\partial x/\partial t$ as the average velocity in the x-direction, $\vec{v}_{avg}$, which is simply

$$v_{x\text{-avg}} = \frac{(v_a + v_b)}{2}$$

Therefore, the acceleration in the x-direction is

$$a_x = \frac{\partial x}{\partial t} \frac{\partial v}{\partial x} = \frac{1}{2}(v_a + v_b)\frac{v_b - v_a}{\Delta x}$$

$$a_x = \frac{1}{2}\left(1 \times 10^{-3}\,\text{m/s} + 5 \times 10^{-3}\,\text{m/s}\right)\left(\frac{5 \times 10^{-3}\,\text{m/s} - 1 \times 10^{-3}\,\text{m/s}}{5 \times 10^{-3}\,\text{m}}\right) = 2.4 \times 10^{-3}\,\text{m/s}^2$$

9.1.1.1.1 Surface Forces on a Fluid: Pressure and Shear Stress   The surface forces that act on a fluid are the most notable forces in fluid mechanics and these lead to the terms that make the Navier–Stokes equations look different from Newton's Second Law of motion. First, there are a few key assumptions that we will make in order to characterize these forces. These assumptions include the following:

1. The fluid is continuous, and the shear stress (force per unit area) is directly proportional to the rate of change in velocity between layers in the fluid.

2. The fluid is isotropic, meaning that all the fluid properties are independent of direction (in other words, we can use any coordinate system and always get the same result).

3. When the rate of change of velocity between layers is zero, there is no shear force on the fluid. In this case, the only force acting on the fluid is pressure.

If the fluid fits these three assumptions, it is called a Newtonian fluid, and we can use the Navier–Stokes equations to describe its motion. In addition to water, most gases (including air), oils, solvents, and many biological fluids are Newtonian fluids. Any discontinuous fluid that would flow out of a jar in clumps like peanut butter cannot be considered a

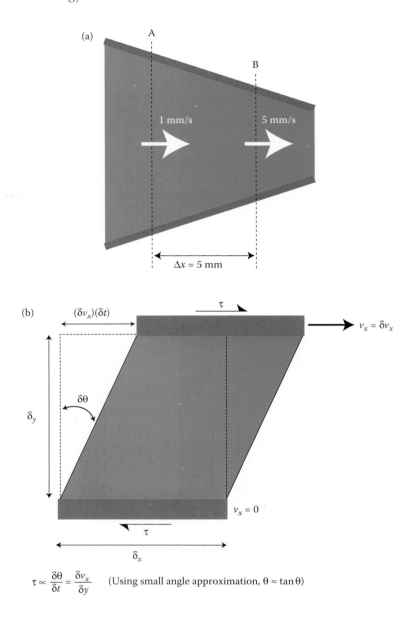

FIGURE 9.2 Fluid flow in a converging duct. A converging duct is schematically portrayed in (a). The velocity at point A is 1 mm/s and the velocity at point B is 5 mm/s. Part (b) shows an example of shear stress, τ, on a small section of the fluid in this duct. The shear stress is related to the velocity gradient (the difference in velocity over distance), $\tau = dv_x d_y$.

Newtonian fluid. Additionally, some biological fluids, such as blood with a lot of clots, may not be considered Newtonian fluids.

The other part of assumption 1 is that the shear stress on the fluid is proportional to the change in velocity between layers in the fluid. If you are shearing the fluid, you are moving the fluid in a certain direction. For example, if you have a fluid moving in a duct like that in Figure 9.2a or in another channel, the shear stress τ, on the fluid is proportional to the velocity in the following manner:

$$\tau = \mu \frac{dv_x}{dy} \tag{9.5}$$

Here, τ is the surface shear stress acting on the fluid (N/m^2), μ is a proportionality constant ($N\ s/m^2$), and v_x is the fluid flow velocity in the x-direction (m/s). This is shown for a small section of fluid in Figure 9.2b. Note that there is only one term here because the flow is in the x-direction only. The proportionality constant between stress and the change in velocity divided by the change in distance is known as the coefficient of viscosity μ. All Newtonian fluids have a coefficient of viscosity, which is a property of the fluid just like density. Table 9.1 lists the coefficient of viscosity for some typical fluids at 25°C.

It is the relationship between stress and the rate of change in velocity that gives us the pressure and shear forces in the Navier–Stokes equations. For example, in the x-direction, the pressure and shear forces per unit volume in a two-dimensional flow are the following:

$$\sum \frac{\vec{F}_{pressure+shear\_x\text{-}dir}}{V} = -\frac{\partial P}{\partial x} + \frac{\partial}{\partial x}\left(2\mu\frac{\partial v_x}{\partial x} + \lambda\left(\frac{\partial v_x}{\partial x} + \frac{\partial v_y}{\partial y}\right)\right) + \frac{\partial}{\partial y}\left(\mu\left(\frac{\partial v_x}{\partial x} + \frac{\partial v_y}{\partial x}\right)\right) \tag{9.6}$$

Here, P is the pressure acting *on* the fluid (that is why it is negative), μ is the coefficient of dynamic viscosity, and λ is the coefficient of bulk viscosity. Your first reaction may be: Ack! Look at all those terms! (Or: All right, thanks for the details!) Remember that this is only for force in the x-direction in a two-dimensional flow. We have two more dimensions to go! What did we just get ourselves into? Well, it turns out that if you multiply through all the derivatives (not shown as a matter of simplicity), you can completely cancel out all the terms with the coefficient of bulk viscosity and also some of the dynamic viscosity terms using the conservation of mass relationship for an incompressible fluid. This law states that all the mass going into a system must equal all the mass going out of a system;

TABLE 9.1   Coefficient of Viscosity for Typical Newtonian Fluids

Liquid at 25°C	Viscosity ($\times 10^{-3}$ Pa s)
Acetone	0.306
Methanol	0.544
Water	0.89
Ethanol	1.074
Mercury	1.526
Olive oil	81
Molasses	870
Pitch	2.3×10^{11}

thus, if a fluid is incompressible, like most liquids, then this equation can be simplified to the following (*x*-direction, two-dimensional case):

$$\sum \frac{\vec{F}_{\text{pressure+shear\_x-dir}}}{V} = -\frac{\partial P}{\partial x} + \mu \left(\frac{\partial^2 v_x}{\partial x^2} + \frac{\partial^2 v_x}{\partial y^2} \right) \tag{9.7}$$

The first term should be straightforward; it is just the pressure acting in the *x*-direction. For the *y*-direction equation, this term would be $\partial P/\partial y$. (Notice that it is a partial differential because we are only taking the derivative in one direction, the *y*-direction.) The second term deals with the shear stress.

9.1.1.1.2 Body Forces on a Fluid: Gravity and Electric Fields   We will now consider body forces, which are those that act upon the entire volume of the fluid. The two forces that are important for our discussion are gravitational and electric forces. As we know, the force due to gravity is $F_{\text{gravity}} = mg$, where *m* is the mass of an object and *g* is the acceleration due to gravity. Therefore, the gravity force per unit volume just becomes

$$\sum \frac{\vec{F}_{\text{gravity}}}{V} = \rho \vec{g} \tag{9.8}$$

Here, ρ is the density of the fluid. Next, we want to also consider an electric force on a fluid that contains ions (like water) due to an electric field. This force is known as a Lorentz body force term and is of the form:

$$\sum \frac{\vec{F}_{\text{electric}}}{V} = \rho_{\text{E}} \vec{E} \tag{9.9}$$

Here, ρ_{E} is charge per unit volume (C/m³) and *E* is the electric field (V/m).

Combining Equations 9.4 through 9.9, we can write the full, simplified Navier–Stokes equations for a two-dimensional flow:

$$\rho \left(\frac{\partial \vec{v}_x}{\partial t} + v_x \frac{\partial \vec{v}_x}{\partial x} + v_y \frac{\partial \vec{v}_x}{\partial y} \right) = -\frac{\partial P}{\partial x} + \mu \left(\frac{\partial^2 v_x}{\partial x^2} + \frac{\partial^2 v_x}{\partial y^2} \right) + \rho g_x + \rho_{\text{E}} E_x \tag{9.10a}$$

$$\rho \left(\frac{\partial \vec{v}_y}{\partial t} + v_x \frac{\partial \vec{v}_y}{\partial x} + v_y \frac{\partial \vec{v}_y}{\partial y} \right) = -\frac{\partial P}{\partial y} + \mu \left(\frac{\partial^2 v_y}{\partial x^2} + \frac{\partial^2 v_y}{\partial y^2} \right) + \rho g_y + \rho_{\text{E}} E_y \tag{9.10b}$$

Let us now look at these equations a little bit and figure out what each term means. The left-hand side of the equation is the "*ma*" part of *F* = *ma*. The right-hand side of the equation deals with all the forces that are imparted on the fluid, be it viscous forces, pressure forces, or body forces such as gravity or an electric force. It is important to understand what each of these terms mean since later on in the chapter we will be able to neglect these terms, depending on what kind of force is dominating in our flow situation.

9.1.1.2 Fluid Motion: Molecular Dynamics Style

All the equations that we just derived apply to continuum flow. But what if the fluids we are dealing with are flowing in channels that have smaller dimensions than the continuum approximation allows? Then we must write individual equations of motion for every atom. Additionally, every atom will have three equations associated with it, due to the three different dimensions to analyze. And, if that is not enough, each of these equations may also include models of chemical reactions and interactions between molecules. As you can guess, this gets very complicated and cumbersome.

As we see from the calculation in Back-of-the-Envelope 9.2, there are an exorbitant number of equations associated with even a very small amount of fluid if we wish to analyze its motion atom by atom. Assuming that a good computer can solve about a thousand equations per second, it would take 4.7 trillion years to solve all the equations of motion of the atoms in the water inside the 1-mm-long nanotube. As you can see, this is hardly feasible. (Note that supercomputers these days can solve equations much, much faster, but simulation times are still on the order of weeks.)

Additionally, molecular simulations are even more difficult due to very strong, specific chemical interactions between molecules, and the appearance of larger-scale structures in the flow, which are difficult to model. All these reasons make the molecular dynamics approach for fluidic systems very unfavorable. However, to mitigate these obstacles, new models are being developed that combine classical mechanics with molecular dynamics.

BACK-OF-THE-ENVELOPE 9.2

How many individual calculations must we perform to analyze the movement of water in a carbon nanotubes with a radius, r, of 30 nm and a length, l, of 1 mm?

The volume of the nanotube that we are analyzing is

$$V_{tube} = 2\pi r l = 2\pi \left(30 \times 10^{-9} m\right)\left(1 \times 10^{-3} m\right) = 1.89 \times 10^{-10} m^3$$

Now let us assume that a water molecule is spherical, with a radius of about 0.14 nm. Therefore, the volume that an individual water molecule occupies is

$$V_{water} = \frac{4\pi r^3}{3} = \frac{(4)(\pi)(1.4 \times 10^{-10} m)^3}{3} = 1.15 \times 10^{-29} m^3$$

So the number of water molecules in the nanotube that we would have to analyze would be

$$\frac{V_{tube}}{V_{water}} = \frac{1.89 \times 10^{-10} m^3}{1.15 \times 10^{-29} m^3} = 1.64 \times 10^{19} \text{ molecules}$$

And since each molecule has three atoms, and each atom has three degrees of freedom, the number of equations that we would need to solve is

$$\#\, equations = (1.64 \times 10^{19})(3)(3) = 1.5 \times 10^{20} \text{ equations}$$

A major goal is to bridge the gap between continuum and molecular scale modeling by incorporation of noncontinuum nanoscale physics into legacy continuum mechanics codes that have developed over decades. This approach is particularly beneficial in dealing with systems too large to permit direct molecular dynamics simulations. Although we would like to be able to model fluid flow using more molecular models, this research is still very much in its infancy, and continuum equations seem to work at the nanoscale for most nanoscale fluid mechanics applications we will discuss here. So for now, we will keep looking at fluids through the continuum eyeglass, although we need to understand that this approximation is limited, and may change.

9.2 FLUIDS AT THE NANOSCALE: MAJOR CONCEPTS

Now that we have all the equations to describe fluid motion, what makes fluid flow at the nanoscale unique? First of all, it matters whether the "fluid" is a liquid or a gas. In gases, the mean free path can be as large as 1 μm in some cases, which is much longer than it is in liquids. If we confine these gases to channels smaller than a micrometer, continuum theory will definitely break down, and we are back to the challenge of modeling with molecular dynamics—with the additional challenge being that the density can vary greatly because gases are compressible fluids. However, there are not many applications for nanoscale gas flows, and therefore we will focus on liquid flow for the remainder of this chapter.

The most notable property of flow at nanoscale dimensions is the high surface-area-to-volume ratio of the fluid conduits. This leads to effects such as adsorption of molecules onto the nanochannel surface, surface roughness effects and changes in effective viscosity. These and other interesting nanoscale effects lead to unique phenomena and new applications, such as the ability to examine the motion of large molecules (like DNA) on an individual basis, and perform useful tasks based on this motion. This section will prepare us to use the tools we developed in the previous section to describe not only bulk fluid flow in a nanochannel, but also flow of nanoscale objects, such as particles and biomolecules. We will learn about the Reynolds number, a useful nondimensional parameter that helps us gain a physical understanding of flow at the nanoscale. Next, we will learn about surface effects, which tend to have important consequences at the nanoscale because of the large surface-to-volume ratios of nanochannels. Finally, we will discuss molecular diffusion, an important topic regarding movement of individual particles or molecules in small-scale flow.

9.2.1 Swimming in Molasses: Life at Low Reynolds Numbers

Do airplanes and bugs use different mechanisms to fly? Do bacteria swim the way we (humans) swim? Do a football and a peanut act differently when thrown in air? The answers to all these questions can be solved by examining the different kinds of forces acting on the object. Airplanes flying, humans swimming, and footballs spiraling—all of these have *inertia*—mass that tends to resist acceleration or deceleration. They have momentum—mass multiplied by velocity. However, bugs, bacteria, and peanuts have such small masses, and therefore so little momentum, that the force caused by the viscosity of the fluid medium represents a bigger force to be reckoned with. How do we get a grasp on this concept? We use the *Reynolds number*.

9.2.1.1 Reynolds Number

The Reynolds number is a nondimensional number, which means it has no units. It is the ratio of momentum to viscosity (gooeyness):

$$\text{Re} = \frac{\rho v D}{\mu} = \frac{\text{Inertial Forces}}{\text{Viscous Forces}} \qquad (9.11)$$

Here, ρ is the density of the fluid (kg/m³), v is a characteristic velocity (m/s), D is the characteristic dimension of the system (m), and μ is the coefficient of dynamic viscosity (N s/m²). The Reynolds number is named after Osborne Reynolds (1842–1912), a British engineer who first proposed the number in 1883. If the Reynolds number is high, then inertia (momentum) dominate. If it is really high, the flow is turbulent. Moderate Reynolds numbers imply laminar (smooth) flow. If the Reynolds number is very low, typically less than 1, then viscous forces dominate, and this is known as creeping, or Stokes flow. Table 9.2 shows some typical viscosities for different fluidic systems.

BACK-OF-THE-ENVELOPE 9.3

What is the Reynolds number that characterizes a human (2 m long) swimming 1 m/s through water? What about a bacterium (1 μm long) swimming through the same water at 30 μm/s? What would it feel like for the human to swim in the flow that the bacterium feels?

Using Equation 9.11 and the properties of water ($\rho = 1000$ kg/m³, $\mu = 0.89 \times 10^{-3}$ N s/m²), the Reynolds number for the person is

$$\text{Re} = \frac{\rho v D}{\mu} = \frac{(1000 \text{ kg/m}^3)(1 \text{ m/s})(2 \text{ m})}{0.89 \times 10^{-3} \text{ N s/m}^2} = 2.3 \times 10^6$$

As for the bacterium, the Reynolds number is

$$\text{Re} = \frac{\rho v D}{\mu} = \frac{(1000 \text{ kg/m}^3)(30 \times 10^{-6} \text{ m/s})(1 \times 10^{-6} \text{ m})}{0.89 \times 10^{-3} \text{ N s/m}^2} = 3.4 \times 10^{-5}$$

So what would it feel like for a human to swim like a bacterium? Well, we would have to match the bacterium's Reynolds number. And there are 11 orders of magnitude separating the two Reynolds numbers. Thus, for a human to have the same Reynolds number as the bacterium, we have to adjust the velocity, the viscosity, and/or the density. Let us slow down the swim to a reasonable 0.5 m/s and try to find a viscosity that would give us the same Reynolds number, using the density of water and then double-checking with a corrected density once an estimate has been made. The viscosity that we would need to match our bacterium's Reynolds number is

$$\mu = \frac{\rho v D}{\text{Re}} = \frac{(1000 \text{ kg/m}^3)(0.5 \text{ m/s})(2 \text{ m})}{3.4 \times 10^{-5}} = 3 \times 10^7 \text{ Ns/m}^2$$

That is a very high viscosity. It is almost impossible to achieve with common fluids, but from the table of viscosities we see that we need a fluid that is more viscous than molasses and almost as viscous as pitch—so for an illustration that might be understandable to most, a bacterium swimming in water is like a human swimming in molasses, or something even more gooey (see Figure 9.3).

The density of molasses is about twice that of water and the density of pitch about the same as water, so our approximation using the density of water is within an order of magnitude and good enough for this illustration. Now let us think about this: what is it like to swim in molasses? Well, the viscosity is so high that when you try to swim with a freestyle stroke, the molasses goes back to its original place before you have moved forward at all. When you are swimming in water, you are displacing the water and using the inertia of the water to propel you forward. (You can think of throwing packets of water behind you to propel you forward—accelerating the mass of water in one direction results in a force on you in the other direction.) However, in molasses, you cannot "throw" enough fluid fast enough, so you do not generate the force you need to move forward. How do you move? How do bacteria move? It turns out that you need to use a nonreciprocating motion, like a corkscrew, that travels only in one direction. This way, you can slowly churn your way through the molasses. So as you can see, viscous forces are very important when you decrease size significantly (or increase viscosity). Life at low Reynolds numbers is tough.

9.2.2 Surface Charges and the Electrical Double Layer

In nanometer-sized channels, the surface-to-volume ratio is very large, which means that the surfaces will have a significant effect on fluidic motion. In general, most surfaces acquire a surface electric charge, especially when in contact with an ionic (charged) fluid (also known as an *electrolyte*). Water can be considered an electrolyte, because there are H_3O^+ ions and OH^- ions in water. However, more traditional electrolytes are salt solutions, such as NaCl or KCl dissolved in water, or buffered solutions, such as borate or phosphate buffers. The most common way for fluidic channels to acquire surface charge is the protonation/deprotonation of surface chemical groups from the surface due to the ionic charges in the electrolyte. Deprotonation refers to the surface giving up a hydrogen ion (with a +1 charge) to the solution, making the surface negative, whereas protonation of the surface does the opposite, leaving the surface with a net positive charge. Other ways for

TABLE 9.2 Typical Viscosities for Different Systems

Liquid at 25°C	Viscosity ($\times 10^{-3}$ Pa s)
Acetone	0.306
Methanol	0.544
Water	0.89 (i.e., water is 0.89×10^{-3} Pa s)
Ethanol	1.074
Mercury	1.526
Olive oil	81
Molasses	870
Pitch	2.3×10^8

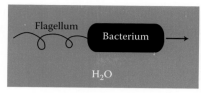
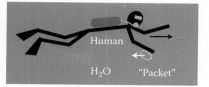

Reynolds number = 3.4×10^{-5} Reynolds number = 2.3×10^{6}

FIGURE 9.3   Swimming in molasses. A bacterium swimming in water is equivalent to a human swimming in molasses or something even more gooey, when you compare the Reynolds number of the two systems. When swimming slowly in molasses, the normal swimming motions you make with your arm will not propel you forward because by the time you finish your stroke, viscosity will make the fluid go back to where it started and you go nowhere. When swimming in water, you can "throw packets" of water in one direction to create a propulsion force in the opposite direction. The way bacteria move is by "churning" their way through liquid with the corkscrew-type motion of flagella.

surfaces to acquire charge include ion adsorption onto the surface from an electrolyte, ionic dissolution from the surface to the electrolyte, and deprotonation/ionization of surface groups on the surface.

The charges at the interface between solids and liquids allow interesting electrokinetic phenomena to occur, including the movement of fluid. This section will summarize the basic concepts behind electrokinetics to give us the tools we need to examine electrokinetic flow in nanometer-scale channels.

9.2.2.1 Surface Charges at Interfaces

First, let us discuss why there are charges at an interface between a solid channel surface and a fluid. Some mechanisms for charge separation at an interface between any two materials include

1. The electrons in both materials have greater affinity for one material than the other. For example, electrons between a semiconductor and conductor preferring the conductor.

2. The ions in both materials have greater affinity toward one material than the other. Example: ions in an electrolyte attracted to a solid charged surface; or the ion in a lattice attracted to the electrolyte.

3. Ionization of molecules at the solid surface. This is usually what happens at a solid–liquid interface, and is the mechanism we will focus on for the remainder of this section.

So what exactly does ionization of molecules at the surface mean? Well, let us assume that we have a fluidic channel etched in a piece of pure glass. The molecular formula of glass is SiO_2, but at the surface there are silanol (SiOH) groups. When a liquid comes into contact with the glass, protons (H^+ ions) tend to be donated into the solution and thus make the glass negatively charged. The equilibrium reaction associated with this deprotanation can be represented as

$$SiOH \rightarrow SiO^- + H^+$$

Just how many protons are donated into a solution depends on the acidity of the solution. (We measure acidity on a pH scale, with a pH of 7 being neutral, less than 7 being acidic, and greater than 7 being basic; a solution of pH = 8.2 will have a hydrogen ion [H^+] concentration of $10^{-8.2}$ mol/L.) If the solution contains a lot of positive ions in it (meaning it is acidic), the glass may not donate many protons since the solution is already very positive. However, if the solution is very basic (more negative ions), then the surface groups of the glass will tend to donate most of their protons, and the glass surface will acquire a greater negative charge. In general, glass is a strong proton donor, so it is therefore classified as an acid. In practice, the full deprotonation of a glass surface is achieved for pH values greater than about 9. Figure 9.4 shows the states of a glass surface (a) when dry, (b) in contact with an acid, and (c) in contact with a strong base. Remember that it is the deprotonation of the silanol groups that determines the surface charge on the glass.

9.2.2.2 Gouy–Chapman–Stern Model and Electrical Double Layer

The surface charge between a solid and a liquid causes ions in the solution with opposite charge compared to the surface charge (called *counter-ions*) to be attracted toward the surface, and ions of like charge (*co-ions*) to be repelled from the surface. This attraction and repulsion, when combined with the random thermal motion (Brownian diffusion) of the ions, creates two layers next to the surface: a fixed layer of ions called the Stern layer, and a diffuse, or mobile layer of ions further away from the surface. These two layers together are called the electrical double layer (EDL). This model is schematically shown in Figure 9.5.

The Stern layer is a layer of ions from the fluid that is attracted to the wall so much that they are essentially stuck to the wall (adsorbed). We can think of these ions as part of the glass itself. However, the layer does not completely shield off the wall charge. More ions from the solution shield this wall charge; but these ions do not get "stuck." Instead, these ions stay free due to random Brownian motion and diffusion that kicks them around and keeps them from sticking to the wall. This layer of mobile ions is called the diffuse layer (also known as the Gouy–Chapman layer). The interface between the Stern and Gouy–Chapman layer is called the shear plane, and it is from this plane that we say that the fluid/ions start to move (when a force is applied). In addition, we define the zeta potential, ζ, as the voltage potential at the shear plane—which is a potential that we can measure experimentally unlike the potential at the surface. This is because the adsorbed ions in the Stern layer make measuring the naked wall potential very difficult. And we need the potential to figure out how strongly the wall is attracting the ions. Figure 9.5 shows where the zeta potential is measured.

The next important thing to know about this model is how big the "diffuse" region is, or how far it extends from the surface into the solution. The electric field is very strong near the surface and the ions are greatly attracted to this surface. However, near the outer edge

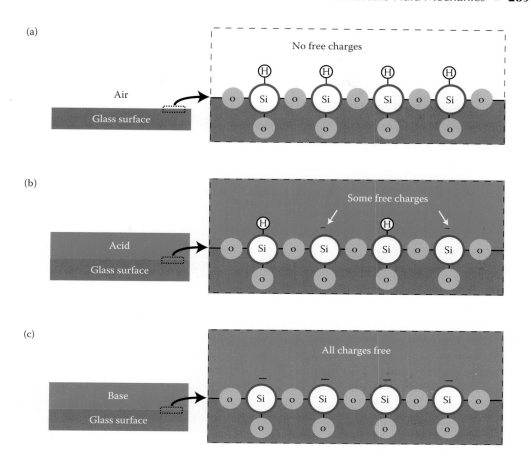

(a) No free charges

Air

Glass surface

(b) Some free charges

Acid

Glass surface

(c) All charges free

Base

Glass surface

FIGURE 9.4   Charges on a glass surface. This schematic shows the amount of charge on a glass surface (a) when dry, (b) in contact with an acid, and (c) in contact with a base. When dry, the surface does not give up any protons (hydrogen ions); however, when in contact with a weak acid, some of the protons from the glass are donated into the solution. When in contact with a strong base, the glass is almost completely ionized—giving up all of its free protons to the solution.

of this diffuse "cloud," the electric field is weak and the ions may have enough thermal energy to escape the electrostatic potential. (For each ion, this thermal energy is roughly equal to $k_B T$, where k_B is Boltzmann's constant, 1.38×10^{-23} J/K, and T is the temperature in Kelvin. For more details, see Chapter 7, Section 7.2.1.) Therefore, we can define the edge of this double layer at a place where the potential energy is approximately equal to the thermal energy of the counter-ions. To know exactly where this edge is, we solve for the potential distribution and screening charge density. According to the diffuse model, this ionic screening charge density decays from the surface following a potential distribution as the Boltzmann distribution, and is written as follows:

$$n_i = n_{i\infty} \exp\left(-\frac{z_i e \psi}{k_B T}\right) \quad (9.12)$$

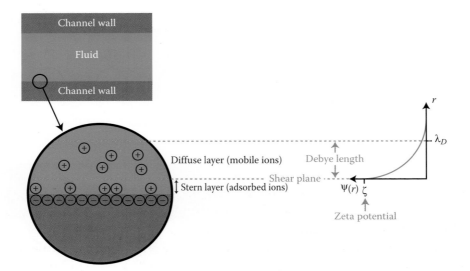

FIGURE 9.5 Gouy–Chapman–Stern model. This schematic shows the diffuse double layer region along with a Stern layer, or region of adsorbed ions on the surface of a channel. Because these ions do not move when an electric field is applied, the shear plane is assumed to be at the interface between the Stern layer and the diffuse layer. The diffuse layer of ions can move under an applied electric field, and this is what pulls fluid through the channel. Also shown in this figure is the potential distribution, marking the length of the diffuse region, and the Debye length λ_D Note that this potential distribution goes to zero near the center of the channel (depending on how large the channel is and how large the Debye length is). The potential at the shear plane is known as the zeta potential ζ.

Here, n_i is the number density (#/m³) of a specific kind of ion, i, $n_{i\,\infty}$ is the value of the number density of that kind of ion in the bulk solution, far away from the wall, z_i is the charge of the ion, e is the charge of an electron, k_B is Boltzmann's constant, T is the temperature, and ψ is the potential distribution associated with the surface charge.

The actual potential distribution due to the electric double layer is generally highly nonlinear and difficult to solve for exactly. However, we can make an assumption (popularly known as the Debye–Huckel Approximation):

$$\frac{z_i e \psi}{k_B T} \ll 1$$

And with this assumption, we can solve for the potential to be

$$\frac{\psi}{\zeta} = \exp\left(\frac{-y}{\lambda_D}\right) \tag{9.13}$$

Here, boundary condition, ζ is the zeta potential (V), y is the distance away from the wall (m), and finally we have a characteristic length of the double layer, λ_D, which is called the Debye length and is defined as

$$\lambda_D \equiv \left(\frac{\varepsilon k_B T}{e^2 \sum_{i=1}^{N} z_i n_{\infty,i}} \right)^{1/2} \tag{9.14}$$

Here, ε is the electric permittivity of the fluid.

From Equation 9.13, you can see that the approximate distribution is exponential. It is actually this distribution that is shown in Figure 9.5. The Debye–Huckel Approximation is generally valid for wall potentials less than 50 mV.

If the electrolyte is symmetric, meaning that both the positive and negative ions in the solution have the same charge, then $z_+ = -z_- = z$ (such as with NaCl, which is Na = +1 and Cl = −1, so $z_+ = 1$, $z_- = -1$, and $z_+ = -z_- = 1$). In this case, we can further simplify Equation 9.14 for the Debye length:

$$\lambda_D \equiv \left(\frac{\varepsilon k_B T}{2e^2 z^2 n_\infty} \right)^{1/2} \tag{9.15}$$

Remember that the number density of a particular ion can be related to the molar concentration of that ion through the following equation:

$$n_i = N_A c_{i,\infty} \tag{9.16}$$

Where n_i is the number density of an ion (#/m³), N_A is Avogadro's number (6.022×10^{23} #/mol), and $c_{i,\infty}$ is the molar concentration of that ion in mol/m³. (Note that molar concentration is often given in units of "molar," where 1 molar = 1 M = 1 mol/L = 1000 mol/m³. It is also sometimes given in units of millimolar, mM, which is 10^{-3} molar. So, a solution that is 5 mM NaCl is 5×10^{-3} mol/L.)

As we can see in Back-of-the-Envelope 9.4, as the concentration of the electrolyte decreases, the Debye length increases. The reason is that there are more ions in the fluid to shield the wall charge. For a univalent electrolyte (an electrolyte is only able to make one chemical bond), the Debye lengths are usually between 1 and 100 nm. Therefore, the Debye layer thicknesses can be comparable to the dimension of a nanochannel.

9.2.2.3 Electrokinetic Phenomena

Now for the reason we introduced surface charges and the electric double layer in the first place: they enable *electrokinetic flow* and, specifically, *electroosmosis*.

There are four major kinds of electrokinetic flow:

1. *Electrophoresis.* This kind of flow is concerned with the motion of charged bodies inside a fluid when an electric field is applied. One common example of electrophoresis is DNA gel electrophoresis, where negatively charged DNA migrates toward a positive electrode.

BACK-OF-THE-ENVELOPE 9.4

What is the Debye length of a microfluidic system with 10 mM NaCl salt solution dissolved in water? What about for 10 μM?

Since NaCl is a symmetric electrolyte, we can use Equations 9.15 and 9.16 to solve for the Debye length. In this case, $z_+ = -z = z = 1$, and we can use the permittivity of water (which is its relative permittivity, $\varepsilon_r = 78.3$, multiplied by the permittivity of free space, $\varepsilon_o = 8.85 \times 10^{-12}$ C²/J/m) as that of the liquid, since NaCl is very dilute compared to the water. Therefore, the equation becomes

$$\lambda_D = \left(\frac{\varepsilon k_B T}{2e^2 z^2 N_A c_{i,\infty}} \right)^{1/2}$$

$$= \left(\frac{(78.3)(8.85 \times 10^{-12} \text{ C}^2/\text{J}^{-1}\text{m}^{-1})(1.38 \times 10^{-23} \text{ J/K})(293 \text{ J})}{2(1.602 \times 10^{-19}\text{C})^2 (1)^2 (6.023 \times 10^{23}\text{mol}^{-1})(10^{-3} \text{ mol/L})(1000 \text{ L/m}^3)} \right) = 3 \text{ nm}$$

And for 10 μM

$$\lambda_D = \left(\frac{\varepsilon k T}{2e^2 z^2 N_A c_{i,\infty}} \right)^{1/2}$$

$$= \left(\frac{(78.3)(8.85 \times 10^{-12} \text{ C}^2 \text{J}^{-1}\text{m}^{-1})(1.38 \times 10^{-23} \text{ J/K})(293 \text{ J})}{2(1.602 \times 10^{-19}\text{C})^2 (1)^2 (6.023 \times 10^{23}\text{mol}^{-1})(10^{-6}\text{mol/L})(1000 \text{ L/m}^3)} \right) = 95 \text{ nm}$$

---

2. *Electroosmosis.* This kind of flow is due to the motion of electrolyte liquids with respect to a surface when an electric field is applied. The liquid moves because of the electric double layer. This type of motion is the most interesting at the nanoscale and will be discussed in Section 9.3.

3. *Streaming potential.* This kind of flow occurs when a liquid is forced along a charged surface, and gives rise to an electric field (the opposite of electroosmosis).

4. *Sedimentation potential.* This kind of flow occurs when charged particles are moving relative to a stationary liquid, creating an electric field (the opposite of electrophoresis).

Electroosmosis and electrophoresis are shown in Figure 9.6.

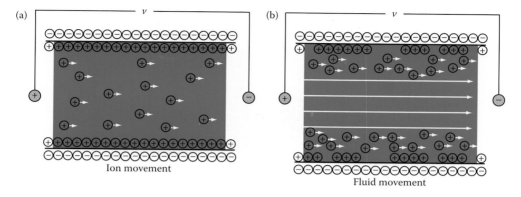

FIGURE 9.6   Summary of electrokinetic flow phenomena. This schematic shows the two major types of electrokinetic flows: (a) electrophoresis, where there is motion of charged bodies due to an applied axial electric field; and (b) electroosmosis, the flow due to motion of electrolyte liquids with respect to a surface under an applied axial electric field. This electroosmotic motion is caused by (1) the attraction of the charged diffusion layer toward the walls of the channel, (2) the force of the applied axial electric field on the diffusion layer, and (3) the shear force on the fluid in the middle of the channel caused by the motion of the diffusion layer at the channel wall. Note: These schematics only show net charge, there are many other ions in the bulk fluid that cancel each other's charge.

9.2.3 Small Particles in Small Flows: Molecular Diffusion

When thinking about nanofluidics, you can either think of fluid movement in a nanometer-sized channel or tube, or you can think of nanometer-sized particles moving through a fluid. Although most of this chapter focuses on bulk fluid movement inside channels, we are going to spend a little time on the movement of particles in fluid and the flow around them. This is applicable for movement of biomolecules, quantum dots, small ions, or other types of particles that are all of nanometer-scale dimensions. However, before deriving the flow patterns around these particles, there is one property that is especially important at the nanoscale, and this is the diffusion coefficient d.

When you put a drop of blue dye into a jar of clear water, what happens? As you can see from Figure 9.7, the drop of dye diffuses into the water, and lightens in color over time. Although the local concentration in and near the drop of dye is decreasing over time, the overall amount of dye in the container is the same. This is analogous to heat being transferred from a hot to a cold body. As we learned in Chapter 7, heat can be transferred from a hot body to a cold body by collisions that transfer kinetic energy from fast-moving to slow-moving molecules. Thermal processes are also responsible for the spreading of dye in water by a process known as diffusion. Diffusion can be described mathematically using either continuum theory or molecular theory (also known as Brownian motion). We come to the same result either way. Here, we are going to explain the phenomenon in terms of concentration gradients and continuum equations.

The equation to describe the diffusion of one material into another is called Fick's First Law, and states that the mass flux of a diffusion material is proportional to the

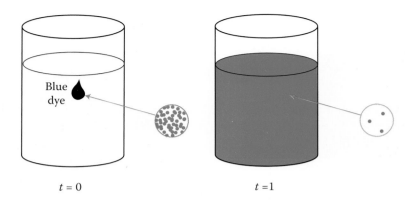

$t = 0$ $t = 1$

FIGURE 9.7   Diffusion of molecules in jar. When blue dye is dropped into a jar of water at time zero, the dye diffuses into the water. As the dye spreads out, the local concentration in and near the drop of dye decreases, although the total amount of dye in the jar remains the same.

concentration gradient in a given direction. Thus, for blue dye that is diffusing in the x-direction in water, this mathematically means

$$J = -d\frac{\partial c}{\partial x} \qquad (9.17)$$

Here, J is the mass flux per unit area (mol/m² s), d is the diffusion coefficient (m²/s), and c is the concentration (mol/m³). This equation is valid for any solute c diffusing into a solvent that is much more concentrated than the solute. In the case of the particles that we are examining, this is almost always the case because most solvents are very concentrated. For example, water is 55 M (55 mol/L), and particle concentrations can be on the order of nanomolar or picomolar. If the dilute assumption is not valid, it is necessary to use a more rigorous diffusion equation (not covered here).

BACK-OF-THE-ENVELOPE 9.5

Suppose we have a channel that is 30-mm long and has a 100 nm × 100 nm cross-section area. At the inlet of the channel, there is a high level of CO_2 gas, at a concentration of 34 mg/m³. But at the outlet of the channel the CO_2 level is monitored and kept at 10 mg/m³. If the diffusivity of CO_2 is 0.5 m²/min, what is the CO_2 mass flux toward the end of the channel?

Introducing the area, A, into Equation 9.17, we know that the total mass flux will be equal:

$$JA = -dA\frac{\partial c}{\partial x}$$

And we obtain

$$-dA\frac{\partial c}{\partial x} = \left(0.5\,m^2/min\right)\left(\frac{1min}{60\,s}\right)\left(100 \times 10^{-9}\,m\right)^2\left(\frac{\left(34\,mg/m^3 - 10\,mg/m^3\right)}{30 \times 10^{-3}\,m}\right) = 6.67 \times 10^{-14}\,mg/s$$

Fick's Second Law describes the diffusion of particles as a function of time, which, in the x-direction is

$$\frac{\partial c}{\partial t} = d\,\frac{\partial^2 c}{\partial x^2} \tag{9.18}$$

Here, $\partial c/\partial t$ is the rate of change of concentration over time (the change of mass stored in a system) and the term on the right side of the equation, $d\,(\partial^2 c/\partial x^2)$, is the diffusive flux of a species into and out of a system. This equation is also known as the diffusion equation.

The solution of this differential diffusion equation in one dimension (x) gives us the concentration of the diffusing solute in both space (x-dimension) and time, t:

$$C(x,t) = \frac{M}{\sqrt{4\pi\,dt}}\,e^{-x^2/4dt} \tag{9.19}$$

Here, M is the total concentration of the diffusing solute (mol) and d is the diffusivity. This solution can also be derived using molecular theories and considering a probability distribution of a large number of particles. Figure 9.8 shows the solution to this equation for blue dye molecules diffusing in the x-direction over time. Since the solution is a Gaussian distribution that extends to infinity, it is convenient to define a parameter that gives us a realistic estimate of how much the molecules have diffused over a certain time. This is generally done by assigning a standard deviation σ to the solution function, which, for flow in the x-direction is

$$\sigma = \sqrt{2dt} \tag{9.20}$$

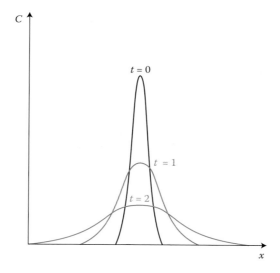

FIGURE 9.8   One-dimensional (1-D) diffusion equation solution. This graph shows the solution to the diffusion equation in one dimension for a point source. Over time, the Gaussian profile spreads out, although the area under the curve always remains the same.

Here, d is diffusivity and t is time. Note that this result also applies to Brownian motion, where the movement of each particle is random and does not depend on its previous motion.

SIMILARITIES AMONG THE TRANSPORT OF MASS, MOMENTUM, HEAT, AND CURRENT

Looking at the equations for the transport of mass, momentum, heat, and current side by side (Table 9.3), we see that for all cases the transport is proportional to a descriptive constant multiplied by the rate of change of one parameter with respect to distance in the medium. In the table, q'' is the heat flux, k is the thermal conductivity, T is temperature, I is current (electron flux), σ is conductivity, and ϕ is potential. In all of these cases, fluxes are along gradients and typically are directly proportional to the magnitude of the gradient.

BACK-OF-THE-ENVELOPE 9.6

A quantum dot that is 30 nm in diameter is entering a 200-nm tall channel. Assuming the diffusion coefficient, d, is 4×10^{-8} m²/s, approximately how long would it take for the quantum dot to hit the wall of the channel? What if the channel was 2 mm tall?

From Equation 9.20, we know that the time for diffusion, t_{diff}, is

$$t_{diff} = \frac{D^2}{2d} \tag{9.21}$$

Here we have replaced σ by D, a characteristic dimension.

Assuming that the dot originated at the center of the channel, the characteristic time for a quantum dot to hit a 200-nm tall channel wall would be

$$t_{diff} = \frac{\left(100 \times 10^{-9}\,\text{m}\right)^2}{2\left(4 \times 10^{-8}\,\text{m}^2/\text{s}\right)} = 1.25 \times 10^{-7}\,\text{s}$$

That is a mere 125 nanoseconds! Now for a 2-mm tall channel, the characteristic time is

$$t_{diff} = \frac{\left(1 \times 10^{-3}\,\text{m}\right)^2}{2\left(4 \times 10^{-8}\,\text{m}^2/\text{s}\right)} = 1.25 \times 10^5\,\text{m} = 12.5\,\text{s}$$

That is 100 million times longer.

From Back-of-the-Envelope 9.6 we can see that it does not take too long for particles to travel far through diffusion at the nanoscale. But what if these molecules are flowing through a channel at the same time that they are diffusing in the channel? Does the advection (horizontal flow/transport) of the particles due to the velocity profiles in the channel play a more important role than the diffusion of the particles due to their random thermal energy? What about surface effects such as the surface potential? Does this effect

TABLE 9.3 Comparison of Fluxes

Kind of Transport	Transport Relation	1-D Formulation
Mass	Fick's Law	$J = -d\dfrac{\partial c}{\partial x}$
Momentum	Newton's Law of Fluids	$\tau = \mu\dfrac{\partial v_x}{\partial y}$
Heat	Fourier's Law	$q'' = -k\dfrac{\partial T}{\partial x}$
Current	Ohm's Law	$I = -\sigma\dfrac{\partial \phi}{\partial x}$

play a significant role? It turns out that they are all important in different cases—and we will examine them further in the next section when we solve for the diffusion equation with convection and electromigration.

9.3 HOW FLUIDS FLOW AT THE NANOSCALE

We have arrived: we finally have all tools we need to start examining flow at the nanoscale. The first major concept we learned is that at nanoscale dimensions the Reynolds number is very low. This means that inertial forces are not very important. In fact, viscous forces are so dominant that the inertial forces in the Navier–Stokes equation can be neglected. This simplifies things for nanofluidic flow. Therefore, Equation 9.10 becomes

$$0 = -\frac{1}{\rho}\frac{\partial P}{\partial x} + \mu\frac{1}{\rho}\left(\frac{\partial^2 v_x}{\partial x^2} + \frac{\partial^2 v_x}{\partial y^2}\right) + g_x + \frac{1}{\rho}\rho_E E_x \qquad (9.22a)$$

$$0 = -\frac{1}{\rho}\frac{\partial P}{\partial y} + \mu\frac{1}{\rho}\left(\frac{\partial^2 v_y}{\partial x^2} + \frac{\partial^2 v_y}{\partial y^2}\right) + g_y + \frac{1}{\rho}\rho_E E_y \qquad (9.22b)$$

Remember that P is the pressure acting on the fluid (N/m^2), μ is the coefficient of dynamic viscosity (Pa s), ρ is the density of the fluid (kg/m^3), g is the acceleration of gravity (m/s^2), ρ_E is charge per unit volume (C/m^3), and E is the electric field (V/m). These equations are valid for Reynolds numbers much smaller than 1, which will be the case in almost any micro- or nanofluidic flow, since the characteristic dimension will always be orders of magnitude smaller than the viscosity. (To achieve a Reynolds number above 1, the fluid may have to travel at the speed of sound. See the Homework Exercises for more details.)

In this section, we will first solve for pressure, gravity, and electrically driven flow in long, wide, and thin channels, since we can derive exact solutions for these cases. Next, we will discuss flows of nanoparticles inside channels, and look at the interesting effects that can happen at a nanoscale with both fluid particles and bulk flow.

9.3.1 Pressure-Driven Flow

To solve for the velocity profile and flow rates of a fluid in a nanochannel due to pressure, we need to solve Equation 9.22. To make the math simpler, let us assume that the channel is much wider than it is tall, so that we can pretend it is infinite in the z-direction, making the problem a two-dimensional problem. In this case, the equation set is

$$0 = -\frac{\partial P}{\partial x} + \mu\left(\frac{\partial^2 v_x}{\partial x^2} + \frac{\partial^2 v_x}{\partial y^2}\right) \tag{9.23a}$$

$$0 = -\frac{\partial P}{\partial y} + \mu\left(\frac{\partial^2 v_y}{\partial x^2} + \frac{\partial^2 v_y}{\partial y^2}\right) \tag{9.23b}$$

First, let us think about the physical problem that we are trying to solve. We are applying a pressure gradient in the x-direction to move fluid in the x-direction. Therefore, we will have parallel flow in the x-direction, and should have no flow in the y-direction. In addition, the velocity will vary in the y-direction, depending on the distance from the channel wall. That is, the velocity cannot depend on the x-coordinate. Since we are looking at a long, straight channel with constant cross-section, if our velocity depended on x, we would have a huge traffic jam. With this knowledge, our complex Navier–Stokes equations boil down even further, to the following:

$$0 = -\frac{\partial P}{\partial x} + \mu\left(\frac{\partial^2 v_x}{\partial y^2}\right) \tag{9.24a}$$

$$0 = -\frac{\partial P}{\partial y} \tag{9.24b}$$

Now it is time to solve Equation 9.24 to find the velocity profile in the nanochannel. First, let us rearrange the equation:

$$\frac{\partial^2 v_x}{\partial y^2} = \frac{1}{\mu}\frac{\partial P}{\partial x}$$

Next, let us integrate with respect to y:

$$\int\frac{\partial^2 v_x}{\partial y^2}\,dy = \int\frac{1}{\mu}\frac{\partial P}{\partial x}\,dy \rightarrow \frac{\partial v_x}{\partial y} = \frac{1}{\mu}\frac{\partial P}{\partial x}y + C_1$$

Integrating again, we obtain:

$$\int\frac{\partial v_x}{\partial y}\,dy = \int\left(\frac{1}{\mu}\frac{\partial P}{\partial x}y + C_1\right)\partial y \rightarrow v_x = \frac{1}{2\mu}\frac{\partial P}{\partial x}y^2 + C_1 y + C_2 \tag{9.25}$$

Now, to solve for the constants C_1 and C_2, we need to apply boundary conditions. For this flow, we know that the flow velocity at the walls must be zero. This is because the solid is not moving at all; so once we get really close to the wall, we cannot have flow. This boundary condition is a very important one in fluid mechanics and is known as the "no-slip" condition. It means fluid cannot "slip" at a wall—that the closest layer of molecules must be completely stuck to the wall.

This boundary condition translates to the following in mathematical terms: the velocity v_x, must equal 0 at $y = 0$ and at $y = h$ (the top and bottom of the channel cross section). Therefore,

$$0 = C_2 \ (v_x = 0 \ \text{when} \ y = 0) \tag{9.26a}$$

$$0 = \frac{1}{2\mu}\frac{\partial P}{\partial x}h^2 + C_1 h\left(v_x = 0 \ \text{when} \ y = h\right) \rightarrow C_1 = -\frac{1}{2\mu}\frac{\partial P}{\partial x}h \tag{9.26b}$$

Thus, combining Equation 9.26 with Equation 9.25 results in the final solution for the velocity profile in a channel of height h (and width, $w \gg h$):

$$v_x = -\frac{1}{2\mu}\frac{\partial P}{\partial x}\left(hy - y^2\right) \tag{9.27}$$

Equation 9.27 gives the equation for the velocity of fluid in a nanochannel for a given pressure gradient. Figure 9.9 shows this velocity profile. The profile is parabolic, with the flow in the center of the channel much faster than the flow at the walls. To solve for the overall flow rate per unit depth, q, we need to integrate this solution over the height:

$$q = \int v_x \ dy$$

$$q = \int_0^h -\frac{1}{2\mu}\frac{\partial P}{\partial x}\left(hy - y^2\right)dy = -\frac{1}{2\mu}\frac{\partial P}{\partial x}\left|\frac{hy^2}{2} - \frac{y^3}{3}\right|_0^h$$

$$q = -\frac{h^3}{12\mu}\frac{dP}{dx} \tag{9.28}$$

Here, q is the overall flow rate per unit depth, h is the height of the channel (m), μ is the coefficient of dynamic viscosity (Pa s), and P is the pressure (N/m^2).

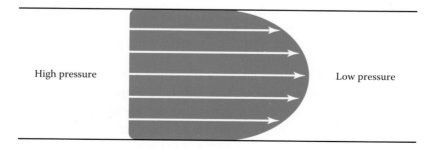

FIGURE 9.9 Pressure-driven flow. This figure shows the velocity profile of fluid in a channel with pressure as a driving force. The profile is parabolic, with the flow in the center of the channel much faster than the flow at the walls.

To solve for the average velocity v_{ave}, we can just divide the flow rate by h:

$$v_{ave} = -\frac{h^2}{12\mu}\frac{dP}{dx}$$

BACK-OF-THE-ENVELOPE 9.7

Let us look at the difference in pressure it will take to drive water in a nanochannel versus a macrochannel at the same velocity. Assuming that we want to drive our flow at an average of 1 m/s, what is the difference in pressure between a 100-nm high channel versus a 1-m high channel?

The pressure gradient dP/dx in the 100-nm high channel will be

$$-\frac{\partial P}{\partial x} = v_{ave}\frac{12\mu}{h^2} = \frac{12\left(1.7 \times 10^{-3}\,N s/m^2\right)}{\left(100 \times 10^{-9}\,m\right)^2} = 2 \times 10^{12}\,Pa/m$$

The pressure gradient in a 1-m high channel will be

$$-\frac{\partial P}{\partial x} = v_{ave}\frac{12\mu}{h^2} = \frac{12\left(1.7 \times 10^{-3}\,N s/m^{-2}\right)}{\left(1m\right)^2} = 0.02\,Pa/m$$

That is a 14 order-of-magnitude difference in pressure between the macrochannel and the nanochannel.

9.3.2 Gravity-Driven Flow

We can do the same exact thing we did above for pressure-driven flows to solve for gravity-driven flows (i.e., simplify to a two-dimensional problem). In this case, the Stokes flow equations simplify to

$$0 = \mu\left(\frac{\partial^2 v_x}{\partial x^2} + \frac{\partial^2 v_x}{\partial y^2}\right) \tag{9.29a}$$

$$0 = \rho g_y + \mu\left(\frac{\partial^2 v_y}{\partial x^2} + \frac{\partial^2 v_y}{\partial y^2}\right) \tag{9.29b}$$

Again, μ is the coefficient of dynamic viscosity (Pa s), ρ is the density of the fluid (kg/m^3), and g is the acceleration of gravity (m/s^2). (Remember that gravity only acts in the y-direction.) Since flow will only be in the y-direction this time, we can use the same procedure (but in a different direction) to solve for the velocity profile and flow rates. We come up with the following relations (which you will derive yourself in the Homework Exercises):

$$v_y = \frac{\rho g}{2\mu}\left(wx - x^2\right) \tag{9.30}$$

$$Q = \frac{\rho g w^3}{12\mu} \tag{9.31}$$

$$v_{ave} = \frac{\rho g w^2}{12\mu} \tag{9.32}$$

Here, w is the width of the channel. Remember that the flow will be downward since gravity acts in the negative y-direction.

BACK-OF-THE-ENVELOPE 9.8

Compare the difference in velocities between water falling through a 100-mm wide channel and 100-nm wide channel due to gravity.

Using Equation 9.32, the velocity in the 100-nm wide channel will be

$$v_{ave} = \frac{\rho g w^2}{12\mu} = \frac{\left(1000\,\text{kg/m}\right)\left(9.8\,\text{m/s}^2\right)\left(100 \times 10^{-9}\,\text{m}\right)^2}{12\left(1.7 \times 10^{-3}\,\text{Ns/m}^2\right)} = 4.8 \times 10^{-9}\,\text{m/s} = 4.8\,\text{nm/s}$$

The velocity in 100-mm wide channel will be

$$v_{ave} = \frac{\rho g w^2}{12\mu} = \frac{\left(1000\,\text{kg/m}^3\right)\left(9.8\,\text{m/s}^2\right)\left(100 \times 10^{-3}\,\text{m}\right)^2}{12\left(1.7 \times 10^{-3}\,\text{Ns/m}^2\right)} = 4800\,\text{m/s}$$

The velocity in the 100-nm wide channel is negligible compared to the velocity in the 100-mm wide channel.

9.3.3 Electroosmosis

From the two previous examples we found that we need an incredible amount of pressure to drive fluid in nanometer dimensions. We also showed that gravity does not have a huge effect on nanoscale flows. So how do we drive fluids in nanochannels? One way is through an electric force, and the corresponding flow is called electroosmosis (described briefly in Chapter 9, Section 9.2.2.3). We now come back to examine it in detail, and solve for the velocity profile and flow rates. We will discover that this type of flow works very well on the nanoscale—much better than it does on a macroscale, actually.

Going back to Equation 9.22, if we are only dealing with electrical forces, the equation set simplifies to

$$0 = \mu\left(\frac{\partial^2 v_x}{\partial x^2} + \frac{\partial^2 v_x}{\partial y^2}\right) + \rho_E E_x \tag{9.33a}$$

$$0 = \mu\left(\frac{\partial^2 v_y}{\partial x^2} + \frac{\partial^2 v_y}{\partial y^2}\right) + \rho_E E_y \tag{9.33b}$$

Again, μ is the coefficient of dynamic viscosity (Pa s), ρ_E is charge per unit volume (C/m³), and E is the electric field (V/m). As in the other two cases (pressure- and gravity-driven flows), we have parallel flow, so we can eliminate a few more terms. In addition, we are only applying an electric field in the x-direction, so our equation set simplifies to

$$0 = \mu\left(\frac{\partial^2 v_x}{\partial y^2}\right) + \rho_E E_x \tag{9.34a}$$

$$0 = \rho_E E_y \tag{9.34b}$$

Now, before solving Equation 9.34a, we need to figure out what the volumetric charge density, ρ_E, is. We know that the only excess charges in the fluid come from the electric double layer, and we already solved for the potential distribution in this double layer (Equation 9.13). Using the conservation of charge equation, which states that the volumetric charge density divided by the electrical permittivity equals the second derivative of the potential, we can write ρ_E as

$$\rho_E = -\varepsilon \frac{d^2 \psi}{dy^2} \tag{9.35}$$

Here, ψ is the potential associated with the electric double layer (V) and ε is the electric permittivity of the fluid (F/m). Substituting this back into Equation 9.34 and rearranging, we have the following:

$$\frac{\partial^2 v_x}{\partial y^2} = \frac{\varepsilon E_x}{\mu} \frac{d^2 \psi}{dy^2} \tag{9.36}$$

Now all we have to do is integrate twice with respect to y and apply boundary conditions. The two boundary conditions are: (1) the potential at the shear plane (remember that the shear plane is where we say the fluid moves, not the wall) is equal to the zeta potential ζ; and (2) the flow is symmetric, so at the center of the channel $d\psi/dy = 0$. With these boundary conditions, we obtain the following result for the velocity profile in electro-osmotic flow:

$$v_x = \frac{\varepsilon E_x \zeta}{\mu}\left(1 - \frac{\psi(y)}{\zeta}\right) \tag{9.37}$$

If the thickness of the EDL is much smaller than either this channel width or channel height, then the potential is close to zero everywhere (except right next to the channel walls). In this case, Equation 9.37 simplifies to

$$v_x = \frac{\varepsilon E_x \zeta}{\mu} \tag{9.38}$$

This is known as the Helmholtz–Smoluchowski equation for electroosmotic (EO) flow with a thin EDL. This velocity profile is just a flat profile, as seen in Figure 9.10a. However, for a thicker EDL, the velocity profile looks more parabolic, like that shown in Figure 9.10b. EO flow is used in many microchannel applications all the way down to the nanoscale—with the difference in velocity profile making a difference in some cases. But as you will see in Back-of-the-Envelope 9.9, these flows are not actually feasible at the macroscale.

BACK-OF-THE-ENVELOPE 9.9

What electric fields are needed to drive water through a 100-nm high nanochannel versus a 100-mm high macrochannel at 0.5 mm/s? Assume the double layers are thin (~3 nm), the zeta potential $\zeta = -100$ mV, the relative permittivity of water $\varepsilon_r = 78.3$ (permittivity ε equals relative permittivity multiplied by the permittivity of free space, $\varepsilon_o = 8.85 \times 10^{-12}$ C²/J/m).

Since we have a thin EDL, we can use the Helmholtz–Smoluchowski equation to solve for the electric field E_x:

$$E_x = \frac{\mu v_x}{\varepsilon \zeta}$$

Looking at this equation, we see that there is no dependence on channel height—which means the electric field would be the same for a nanochannel and a macrochannel. This electric field would be

$$E_x = \frac{\left(1.7 \times 10^{-3}\,\text{Ns/m}^2\right)\left(0.5 \times 10^{-3}\,\text{m/s}\right)}{(78.3)\left(8.85 \times 10^{-12}\,\text{C}^2/\text{J/m}\right)\left(100 \times 10^{-3}\,\text{V}\right)} = 12{,}000\ \text{V/m} = 12\,\text{kV/m}$$

That is a strong electric field. And although it is certainly possible to generate 12 kV/m, what we need to worry about is current. If we have too much current, then we have a phenomenon known as Joule heating, where the current will heat up the liquid. In addition, too much current will make the system dangerous to operate. So what is the difference in current at the nanoscale and the macroscale?

An equation for current flow is

$$I_{\text{total}} = \sigma_{\text{bulk}} E_x A$$

Here, I is the total current, σ_{bulk} is the conductivity of the solution (or concentration of ions in the fluid), and A is the cross-sectional area. Since the conductivity of the solution is the same in both cases, the current will scale with the cross-sectional area of the channel. What this means is for a conductivity of 100 μS/m in the nanochannel:

$$I_{\text{total}} = \left(100 \times 10^{-6}\,\text{S/m}\right)\left(12{,}000\ \text{V/m}\right)\left(100 \times 10^{-9}\,\text{m}\right)^2 = 1.2 \times 10^{-14}\,\text{amps} = 12\,\text{fA}$$

The macrochannel current would be

$$I_{\text{total}} = \left(100 \times 10^{-6}\,\text{S/m}\right)\left(12{,}000\ \text{V/m}\right)\left(100 \times 10^{-3}\,\text{m}\right)^2 = 0.012\ \text{amps} = 12\,\text{mA}$$

At 12 kV, a current of 12 mA could kill you. This high current is not only impractical in real systems, but it would cause Joule heating, which would heat up the flow and cause boiling. Therefore, electroosmotic flow is not feasible on the macroscale (see Figure 9.11). However, the nanoscale channel's 12 fA is a safe current that will not cause noticeable Joule heating.

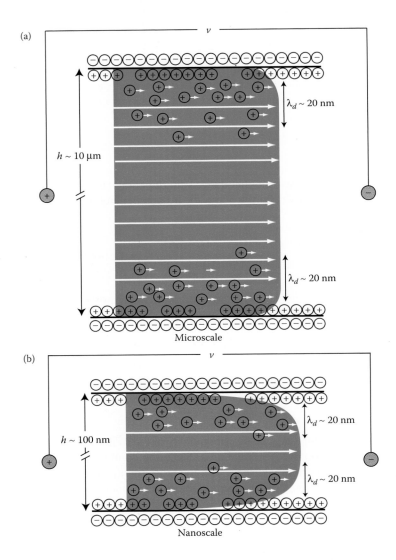

FIGURE 9.10 Electroosmotic (EO) flow profiles. When the electric double layer is very thin compared to the channel dimension, the velocity profile for EO flow is simply a plug profile, as shown in (a). However, at the nanoscale, where the EDL thickness is larger compared to the channel dimension, the profile looks more and more parabolic (although one should think of it more as two exponential distributions overlapping), as shown in (b).

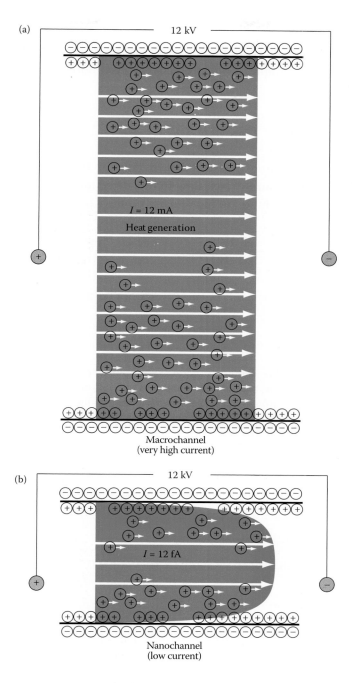

FIGURE 9.11   Macro- versus nanoscale electroosmotic flow. (a) Electroosmotic flow is not used in macrochannels because the current would be about 1 trillion times higher than it is for a nanoscale channel, and this would heat up the flow through the movement of the many ions (Joule heating). In addition, it would be a safety hazard. A current of 12 mA at 12 kV is sufficient to kill a person. (b) In a nanochannel, flow can happen at very low current and with virtually no Joule heating due to the smaller number of ions.

9.3.4 Superposition of Flows

Since our solutions to equations (Equations 9.22a and 9.22b) are linear, we can actually combine solutions. For example, if our problem involved both pressure and gravity, our solution would be

$$v_y = -\frac{1}{2\mu}\frac{\partial P}{\partial y}(hx - x^2) + \frac{\rho g}{2\mu}(hx - x^2) \tag{9.39a}$$

Note that we are assuming pressure-driven flow in the y-direction in this case. Additionally, if we had pressure-driven flow in one direction and electroosmotic flow in the opposite direction, the solution to the velocity profile would simply be

$$v_x = -\frac{1}{2\mu}\frac{\partial P}{\partial x}(hy - y^2) - \frac{\varepsilon E \zeta}{\mu}\left(1 - \frac{\psi(y)}{\zeta}\right) \tag{9.39b}$$

Figure 9.12 shows schematically how these flows can be superposed.

BACK-OF-THE-ENVELOPE 9.10

A 30-μm tall glass channel is standing upright on a table. The capillary is filled with water, and an electric field of 1 kV is applied from the top to the bottom of the channel. Assume that the zeta potential $\zeta = -100$ mV. What is the flow rate through the channel?

In this case, we have both gravity-driven and electroosmostic flow, so the flow rate would just be a superposition of the two flow rates:

$$q = \int_0^h v_x\,dy = \frac{\rho g w^3}{12\mu} - \frac{\varepsilon E \zeta h}{\mu}$$

$$= \frac{(1000\,\text{kg/m}^3)(9.8\,\text{m/s}^2)(30 \times 10^{-6}\,\text{m})^3}{(12)(1.7 \times 10^{-3}\,\text{Ns/m}^2)}$$

$$- \frac{(78.3)(8.85 \times 10^{-12}\,\text{C/J/m})(1 \times 10^3\,\text{V/m})(100 \times 10^{-3}\,\text{V})(30 \times 10^{-6}\,\text{m})^2}{1.7 \times 10^{-3}\,\text{Ns/m}^2}$$

$$= 1.3 \times 10^{-8}\,\text{m}^2/\text{s}$$

(Note: A volumetric flow rate would be given in units of cubic meters per second [m³/s]; here, however, because we are only working in two dimensions, the equation yields a "flow rate per unit depth," expressed in square meters per second [m²/s].)

9.3.5 Ions and Macromolecules Moving through a Channel

Now that we have finally solved for the velocity profile within the channel, we are going to discuss what happens when you put ions and other macromolecules within the channel. This is actually where a lot of interesting phenomena start to happen—such as interactions

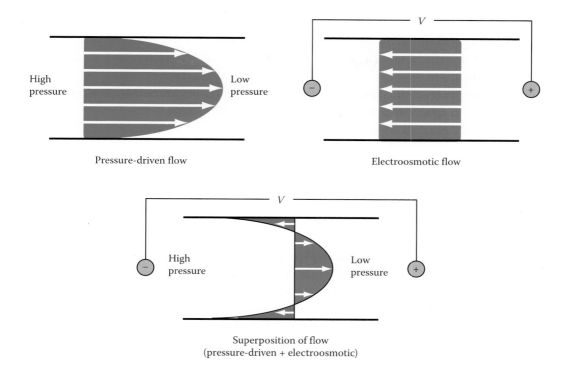

Pressure-driven flow

Electroosmotic flow

Superposition of flow
(pressure-driven + electroosmotic)

FIGURE 9.12 Superposition of flow example. Because Equations 9.22a and 9.22b are linear, pressure-driven flow, gravity-driven flow, and electroosmotic flow can all be superposed to generate a solution with combined flows. For example, if we have a pressure-driven flow in one direction and electroosmotic plug flow in the other, the solution will be a simple addition of both velocity profiles.

between the particles and the wall and different sorts of particle motion due to different fields. We start this section by describing known solutions for low-Reynolds-number flow around a particle with just pressure-driven flow, and then discuss phenomena involving surface charges and electrokinetics.

9.3.5.1 Stokes Flow around a Particle

Low-Reynolds-number, pressure-driven flow is typically known as creeping flow, or Stokes flow. The equation set used to describe Stokes flow is the same as Equation 9.23, shown again here for reference:

$$0 = -\frac{\partial P}{\partial x} + \mu \left(\frac{\partial^2 v_x}{\partial x^2} + \frac{\partial^2 v_x}{\partial y^2} \right) \tag{9.23a}$$

$$0 = -\frac{\partial P}{\partial y} + \mu \left(\frac{\partial^2 v_y}{\partial x^2} + \frac{\partial^2 v_y}{\partial y^2} \right) \tag{9.23b}$$

Remember that P is the pressure acting on the fluid (N/m^2), and μ is the coefficient of dynamic viscosity (Pa s). Stokes developed these equations, which have subsequently been solved exactly for various types of flow. These flow types include fully developed duct flow (which we derived ourselves in the previous section for two-dimensional parallel flow), flow around immersed bodies, flow in narrow but variable passages, and flow through porous media.

The velocity profiles for flow around a spherical particle can be obtained by solving Equation 9.22 in spherical coordinates. The solution is as follows:

$$v_r = v\cos\theta\left(1+\frac{a^3}{2r^3}-\frac{3a}{2r}\right) \tag{9.40a}$$

$$v_\theta = v\sin\theta\left(-1+\frac{a^3}{4r^3}+\frac{3a}{4r}\right) \tag{9.40b}$$

Here, v is the velocity a good distance upstream (or downstream) of the particle, v_r (m/s) is the radial velocity of the fluid (the particle is stationary), v_θ (m/s) is the angular velocity of the fluid, a is the radius of the particle, r is the radial position, and τ is the angular position. Radial position is the length of a line drawn from the center of the particle to the location of interest; angular position is the angle of this radial line.

Figure 9.13a shows Stokes flow around a spherical particle using Equation 9.40. The interesting thing to note from this figure is that the flow displays perfect fore/aft symmetry. That is, the flow looks exactly the same before and after it encounters the particle—which

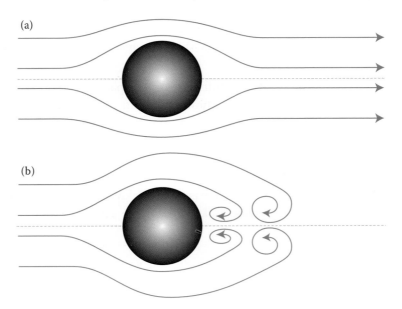

FIGURE 9.13 Stokes flow over a particle. Part (a) shows the velocity profiles for low Reynolds number, pressure-driven flow (Stokes flow) around a particle. The velocity profiles show fore/aft symmetry. This contrasts with the velocity profiles at much higher Reynolds numbers, as shown in part (b), where eddies shed off the particle and the profile aft of the particle looks very different from profile at the fore.

is very different from the high-Reynolds-number case (Figure 9.13b), where inertial forces are dominant and cause eddies to shed off the spherical particle.

From these solutions, we can also calculate the pressure, stress, and force on the particle due to the fluidic motion. The pressure distribution is

$$P(r,\theta) = P_\infty - \frac{3\mu a v}{2r^2}\cos\theta \qquad (9.41)$$

This equation gives the pressure distribution $P(r,t)$ around a spherical particle of radius a (m), in a fluid with viscosity μ (Pa s), at a velocity v (m/s), where the pressure far away from the object is P_∞ (N/m²). The shear stress distribution is

$$\tau(r,\theta) = -\frac{\mu v \sin\theta}{r}\left(\frac{3a^3}{2r^3}\right) \qquad (9.42)$$

This equation gives the shear stress distribution, $\tau(r, \theta)$, around a spherical particle of radius a, in a fluid with viscosity μ, at a velocity v. We can then integrate both the pressure and shear stress distributions to find the force on the particle:

$$F = 6\pi\mu a v \qquad (9.43)$$

And this equation gives the force F needed to move a particle of radius a through a fluid with viscosity μ (Pa s) at a velocity v (m/s).

BACK-OF-THE-ENVELOPE 9.11

A little boy at the beach just threw a bunch of sand into your glass of water. What is the settling velocity of the sand in the water? We can assume the sand density to be about 4000 kg/m³.

Let us assume that there is one-dimensional flow downward in the water. Also assume that there is a uniform sand grain size (let us say it is about 10 μm in radius), so we will have a uniform settling velocity. Now we can look at a force balance on the sand in the glass of water (see Figure 9.14).

Since the sum of forces must equal zero, we have the following:

$\Sigma F = 0 = -$Viscous drag forces $-$ Buoyancy of liquid $+$ Gravitational force on particle

The viscous drag force (described by Equation 9.43) is the force that must be overcome to move the particle through the liquid:

$$F_v = 6\pi\mu a v_s$$

In this case, v_s is the settling velocity of the particle.
The buoyancy force of the liquid is just

$$F_b = \rho_f V_p g$$

Here, force F_b due to buoyancy is equal to the density of the fluid ρ_f times the volume of the particle V_p times the gravitational acceleration g.

And the gravitational force on the particle is

$$F_g = \rho_p V_p g$$

The force due to gravity F_g is equal to the density of the particle ρ_p times the volume of the particle V_p, times the gravitational acceleration g.

Combining, we obtain

$$0 = -F_v - F_b + F_g = -6\pi\mu a v_s - \rho_f V_p g + \rho_p V_p g$$

And solving for v_s, we find that the equation for the settling velocity is

$$v_s = \frac{2}{9} \frac{a^2 \rho_f}{\mu} \left(\frac{\rho_p}{\rho_f} - 1 \right) g$$

(Note that this is a version of the equation we used in Back-of-the-Envelope 2.6 in Chapter 2.) This means that for our particles of sand, we will have a settling velocity of approximately

$$v_s = \frac{2}{9} \frac{\left(10 \times 10^{-6}\,\text{m}\right)^2 \left(1000\,\text{kg/m}^3\right)}{1.7 \times 10^{-3}\,\text{Ns/m}^2} \left(\frac{4000\,\text{kg/m}^3}{1000\,\text{kg/m}^3} - 1 \right)\left(9.8\,\text{m/s}^2\right) = 3.84 \times 10^{-4}\,\text{m/s}$$

That is, 384 μm/s. If the glass is 25 cm tall, how long would it take the sand to reach the bottom of the glass?

$$\Delta t = \frac{\Delta d}{v} = \frac{25 \times 10^{-2}\,\text{m}}{384 \times 10^{-6}\,\text{m/s}} = 651\,\text{s}$$

So it takes nearly 11 min for the sand to settle to the bottom of the glass.

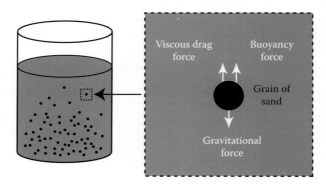

FIGURE 9.14 Sand settling in a glass. This picture accompanies Back-of-the-Envelope 9.11, where sand is settling in a 25-cm tall glass of water. In (b), a force diagram is shown for one grain of sand, with buoyancy and viscous drag forces acting in the positive y-direction, and gravity acting in the negative y-direction.

9.3.5.2 The Convection–Diffusion–Electromigration Equation: Nanochannel Electrophoresis

We also know that a particle has an inherent diffusivity—meaning that it will move on its own due to random thermal motion. So what happens when we add a flow field around these particles? Do they still diffuse? Or does fluid flow take over? Also, what if the particle is charged and you apply an electric field? Even if there is no flow in the channel and no surface charge on the channel wall, the particle will be attracted to the oppositely charged electrode via electrophoresis, as we learned earlier. And how does this tie into diffusion and convection?

Well, the answers to all of these questions reside in the convection–diffusion–electromigration equation, which is derived from the conservation of mass equation. We now have two more fluxes that are moving mass into and out of the system; these fluxes are convection, due to the bulk flow, and electromigration, due to the charged ions moving in the flow. These fluxes (in a two-dimensional nanochannel system) can be represented as follows:

$$\text{Convective flux} = cv_x \tag{9.44}$$

Here, c is the concentration (mol) and v_x is the bulk velocity (m/s).

$$\text{Diffusive flux} = -d\frac{\partial c}{\partial x} \tag{9.45}$$

This we already learned above; recall that d is the diffusion coefficient (m^2/s).

$$\text{Electromigration flux} = -cvzFE_x - cvzF\frac{\partial \psi}{\partial y} \tag{9.46}$$

Here, c is concentration (mol), F is Faraday's constant ($F = 96,485$ C/mol), E_x is the applied axial electric field (V/m), z is the charge number of the ion, ψ is the potential due to the electric double layer (in the y-direction)(v), and v is the electrophoretic mobility. The electrophoretic mobility of an ion is defined as the velocity an ion will travel under an applied electric field. Mobility is actually a property of an ion and can be related to diffusivity d in many cases with the following relation:

$$d = vRT \tag{9.47}$$

Here, R is the universal gas constant (8.314 J/K mol) and T is the temperature. The electrophoretic mobility actually takes into account the movement of ions due to an applied axial field. This is the first term of the electromigration flux of Equation 9.46. The second term deals with the transverse electromigration of the ions due to the EDLs. As the EDLs get thicker, this flux will have a greater effect on the ions.

Note that all these fluxes are valid only if we are dealing with dilute concentrations of particles; that is, the concentration of the solute molecules is much less than that of the solvent. The total flux with convection, diffusion, and electromigration is

$$J = cv_x - cvzFE_x - cvzF\frac{\partial\psi}{\partial y} - d\frac{\partial c}{\partial x}$$

These fluxes can be put directly into the diffusion equation, making the diffusion equation (now called the convective, diffusive, electromigration equation) equal to

$$\frac{\partial c}{\partial t} + \underbrace{v_x\frac{\partial c}{\partial x}}_{\text{advection}} = \underbrace{vzFE_x\frac{\partial c}{\partial x}}_{\text{electrophoresis}} + \underbrace{vzF\frac{\partial\psi}{\partial y}\frac{\partial c}{\partial y} + vzFc\frac{\partial^2\psi}{\partial y^2}}_{\text{electromigration}} + \underbrace{d\frac{\partial^2 c}{\partial x^2}}_{\text{diffusion}}$$

Solving this equation will allow us to solve for the movement of ions and particles in a nanochannel. However, before trying to tackle this equation, let us look at each of the terms and see how each term affects the velocity profile.

The term on the left-hand side is called the advection (flow/transport). When this term dominates, it means that the band of molecules flowing down the system will look like the bulk velocity profile of the fluid (solved with the Navier–Stokes equations). Figure 9.15a shows the advection-dominating case for a parabolic flow profile. The first term on the right-hand side is the electrophoresis migration term, which tells us how fast the charged particle will move on its own in an electric field. If this term dominates, the particles will flow at their electrophoretic velocity. The next term is the transverse electromigration term, which tells us how much the molecules will move toward the wall, due to the surface charges on the wall. If this term dominates, charged particles will either be attracted or repelled from the wall, as shown in Figure 9.15b. And finally, the last term is diffusion. If diffusion dominates, then we will just get a plug that keeps increasing axially, as we see in Figure 9.15c.

So what happens at the nanoscale? Which term "wins?" To find out, let us look to the time scales for each of these fluxes:

$$t_{\text{diff}} \sim \frac{\lambda_D^2}{d}$$

$$t_{\text{electromigration}} \sim \frac{1}{\zeta*}\frac{\lambda_D^2}{d}$$

$$t_{\text{advection}} \sim \frac{\sigma_0}{v}$$

Here, λ_D is the Debye length (m), d is the diffusivity (m²/s), $\zeta*$ is a nondimensional zeta potential (v), σ_0 is a characteristic plug length of ions in a channel, and v is a characteristic velocity of a pressure-driven or electroosmotic flow (m/s). Characteristic values for each of these terms are listed in Table 9.4.

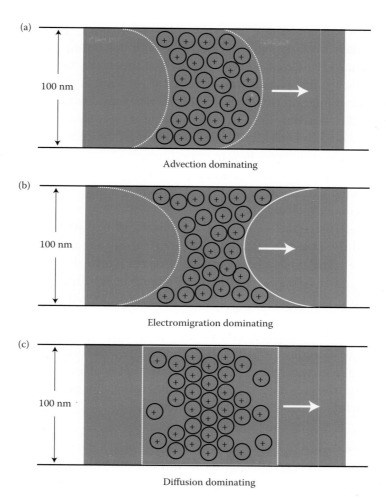

FIGURE 9.15 Concentration distributions of ions in a nanochannel for different cases of flow. When advection dominates and the velocity profile is parabolic, the ion distribution in the channel will look like that shown in (a). However, if transverse electromigration due to the double layer potential dominates, the ions will be distributed in higher concentration near the walls, as shown in (b). Finally, if diffusion dominates, we will not see the effects of advection or electromigration, and the concentration profile will be symmetric about the center line, as shown in (c).

TABLE 9.4 Characteristic Values of Properties at the Nanoscale

Property	Characteristic Value
λ_D	10 nm
d	1×10^{-10} m^2/s
ζ	1
σ	50 μm
ν	100 μm/s

For typical nanoscale systems, ζ^* is on the order of 1, making $t_{diff} \sim t_{electromigration}$. Using the rest of the values from the table, we find that both t_{diff} and $t_{electromigration}$ are typically much smaller than $t_{advection}$. We therefore assume that the transverse concentration distribution at any point in time is quasi-steady. What this means is that in a typical system, diffusion and electromigration win. So if the particles are charged, then they will be attracted toward the wall. But wait a minute: looking at the velocity profile near the wall, we see that the velocity is typically much lower than in the center of the channel. (See Section 9.3.3 about electroosmotic flow in nanochannels.) What this means is that we will have a parabolic-type flow profile, and charged species in the flow are going to be affected by the flow profile as well as the transverse electromigration. To explain this situation a little better, let us take a look at Figure 9.16.

At the microscale, when the electric double layers are thin, two ions with the same electrophoretic mobility but different charge will move at the same speed. However, at the nanoscale, these electric double layers are thick compared to the channel dimension—meaning that there will be a distribution of ions along the channel because of the "pull" from the walls. This "pull" combined with the parabolic profile will make the ions move at different speeds, depending on their charge. More negatively charged ions will want to be near the center of the flow, and will therefore go faster; whereas less negative and/or more positive ions will be closer to the wall and go slower. It is this unique phenomenon at the nanoscale (also called nanochannel electrophoresis) that enables an entirely new way to separate species from each other—and this has applications for bioanalytical separations, which we will learn more about in the next section.

9.3.5.3 Macromolecules in a Nanofluidic Channel

In nanoscale channels, it is possible for fluid molecules to be almost the same size as the channel diameter (as seen in Figure 9.17). We may encounter this scenario when dealing with molecular filtering applications, or even when trying to model some biological systems, like transport through cell membranes or biomolecules traveling in gels. (In the past 20–30 years, there has been a great deal of progress in our ability to analyze and separate biomolecules quickly, inexpensively, and accurately, and this has often been done using nanoporous structures such as gels.) In these systems, the molecules are hindered because the channel sizes are small compared to their diameter, so they end up hitting the channel walls, a phenomenon known as steric hindrance. The diffusion coefficient of a solute molecule with radius, a, through a channel with radius, r, has been approximated as the following:

$$\frac{d}{d_0} = \left(1 - \frac{a}{r}\right)^2 \left[1 - 2.104\left(\frac{a}{r}\right) + 2\left(\frac{a}{r}\right)^3 - 0.95\left(\frac{a}{r}\right)^5\right] \tag{9.48}$$

Here, d_0 is the diffusion coefficient that the macromolecule would have in free solution (or in bigger channels). The first term on the right-hand side of the equation represents this steric hindrance effect caused by the pore.

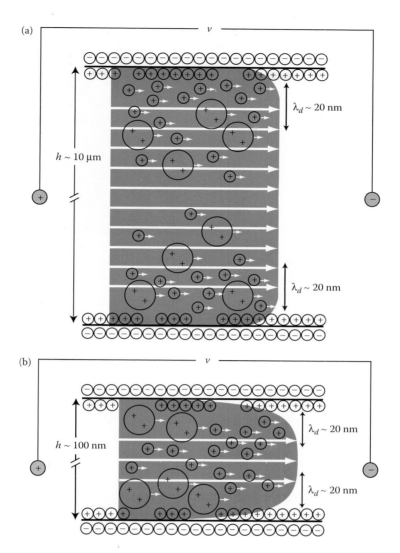

FIGURE 9.16 Nanochannel electrophoresis, illustrating the difference between electrophoresis in nanochannels and microchannels. In microchannels, two analytes with the same mobility but different charge will not separate, as seen in (a). However, in a nanochannel (b), these two analytes will separate due to the "pull" they feel from the transverse potential distribution, coupled with the nonuniform velocity profile. In this case, a faster-moving +1 ion will separate from a slower-moving +2 ion.

9.4 APPLICATIONS OF NANOFLUIDICS

Let us take a closer look at a few technologies that capitalize on the unique properties of nanoscale flow. There are many new enabling technologies at the nanoscale, especially in the biological field. These include engineering devices for biological applications, such as nanoexplorers, cell manipulators, and electrokinetic flow in chip-based devices for detection of chemical and biological species in which the transport process is carried out by the methods that we have discussed in this chapter. Understanding the transport processes

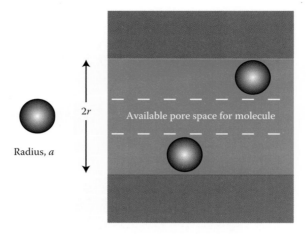

FIGURE 9.17 Macromolecule in a nanochannel. This diagram demonstrates the steric hindrance effect. The available pore space for the molecule with radius, a, in a channel with radius, r, will be limited due to the steric hindrance effect (interference effect).

relevant to biomolecular devices can greatly increase the probability of finding target molecules and identifying important biological processes at the subcellular level. Other applications include membrane filtration with nanochannels, nanoair vehicles, nanojets for inkjet printing, nanoscale fuel cells, electroosmotic pumps, drug-delivery systems, and systems for disease treatment and prevention. In this section, we will discuss a couple of the more mature nanofluidic technologies that exist, including nanofluidics for bioanalytical applications and electroosmotic pumps for pressure and cooling applications.

9.4.1 Analysis of Biomolecules: An End to Painful Doctor Visits?

Earlier in the chapter we found how nanofluidic technology enables the separation of ions in a unique way. We also learned that macromolecules diffuse through nanochannels differently than larger channels due to steric hindrance. At the nanoscale, we can use these two effects to create nanofluidic channels that can separate biomolecules from each other based on charge and size. The larger the molecule, the slower it will travel through a channel because of steric hindrance. And we know that in a channel with a negatively charged surface, the negative ions will travel faster than the positive ions. Therefore, we can use both phenomena to our advantage to create unique electrokinetic separation devices.

Figure 9.18 shows a schematic of a possible separation device to separate biomolecules in nanofluidic channels. There is a detector downstream of the separation channel that allows for the detection of bands of molecules passing by. This detector can either detect a fluorescent tag attached to the molecule, or conductivity, or charge of the molecules as they pass by. The resulting data that are collected are referred to as an electropherogram, which shows the intensity of the collected signal over time. Figure 9.18b shows an example electropherogram of DNA molecules and two markers separating out in a nanofluidic channel. One important thing to note is that the analysis of DNA in the case of nanofluidic flow can be achieved in less than 2 min; this is significant because current DNA analyses take

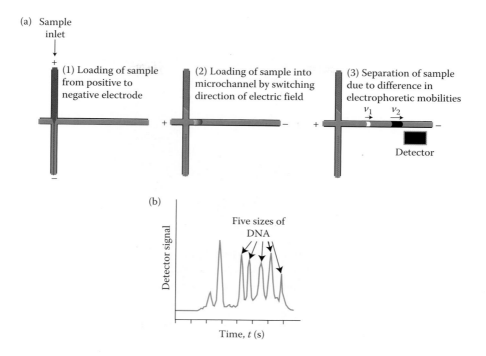

FIGURE 9.18 DNA separation/analysis chip in action. (a) Concept for a nanofluidic biomolecule separation device. On the chip are fluidic inlets and a device for detecting separated species. (b) Demonstration data showing the separation of two marker dyes (left peaks) from five different sizes of DNA molecule, each size due to the number of base-pairs the molecule contained (1, 10, 25, 50, or 100).

overnight. Imagine going to the doctor and knowing all your lab results within a couple minutes of taking a blood sample.

This brings us to another important point—namely, the point of a needle. Nanofluidic channels allow for much less sample to be extracted for analysis. Instead of the doctor taking an entire vial of blood to test, the use of nanofluidic devices necessitates that only a very small fraction of blood needs to be used—as little as a thousandth or a millionth of a drop. So the needles used to extract blood can be much, much smaller—so small, in fact, that the needle can go in between the pain sensors in the skin, so that we will not be able to feel anything while blood is being drawn. Such ~1-μm needles have been developed using microelectromechanical systems (MEMS) fabrication technologies and are currently in use for in-home glucose tests. Imagine combining them with nanofluidic bioanalytical devices: no more painful doctor's visits.

As nanofluidic channels get smaller and smaller, the number of biomolecules that we detect with a scanner will become lower and lower. Researchers have already reached the realm of single-molecule detection, and nanofluidics allows for understanding the transport of the molecule to the specific detection location. That is, understanding nanofluidic phenomena is very important in designing bioanalytical systems capable of detecting very low-volume, low-abundance biomolecules.

9.4.2 EO Pumps: Cooling Off Computer Chips

Electroosmotic (EO) flow in nanochannels can be used for creating pumps with no moving parts. EO pumps are purely driven by electric fields, due to the electric double layer region discussed earlier. Pressures can be generated as high as 200 atm using electroosmotic

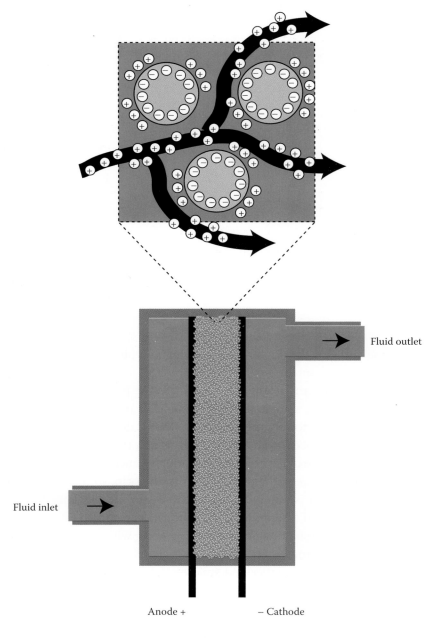

FIGURE 9.19 EO pump: a simplified schematic of an electroosmotic pump, where the electrolyte is pulled through a porous glass structure and forced out of the pump by an applied electric field.

pumps, and typical flow rates are in the range of 10–1000 nL/s. When the pump is used to move flow (as opposed to generating pressure), flow rates up to 50 cm³/min can be achieved. An EO pump consists of a glass capillary packed with small glass particles to create a porous glass structure that acts like many nanofluidic flow channels in parallel. A schematic of an EO pump is shown in Figure 9.19.

9.4.3 Other Applications

Electroosmotic flow depends on the surface potential of the microfluidic and nanofluidic channels. Applying an external field can control the surface potential of a fluidic channel. So imagine applying an electric field directly on the channel walls and changing the surface potential from negative to positive—all of a sudden we change the magnitude and sign of the EO flow. This is analogous to a field-effect transistor (FET), although instead of moving electrons, we are controlling the movement of fluidic ions. Due to this direct analogy, this emerging technology has been dubbed a FlowFET.

BACK-OF-THE-ENVELOPE 9.12

Consider an EO pump like the one shown in Figure 9.19. Assume that the nanochannels are about 500 nm in height, h, and there is no flow. For a potential of 5 kV and a zeta potential, $\zeta = 60$ mV, how much pressure could this system generate?

Using the superposition principle, we know that the velocity profile for electroosmotic flow with pressure driven-flow is

$$v_x = -\frac{1}{2\mu}\frac{\partial P}{\partial x}\left(hy - y^2\right) - \frac{\varepsilon E \zeta}{\mu}\left(1 - \frac{\psi(y)}{\zeta}\right)$$

For a conservative estimate, let us assume the double layer is thin and take the area integral of both sides to obtain the flow rate:

$$q = \int_0^h v_x dy = -\frac{h^3}{12\mu}\frac{dP}{dx} - \frac{\varepsilon E \zeta h}{\mu}$$

Since, $q = 0$, we can rearrange the above equation to find that

$$\frac{dP}{dx} = -\frac{12\varepsilon E \zeta}{h^2}$$

Thus, we see that if h is small, electroosmotic flows can be used to generate very high pressures. For $E = 5$ kV, $\zeta = 60$ mV, and $h = 100$ nm, the pressure generated is

$$\frac{\partial P}{\partial x} = -\frac{12(78.3)\left(8.85 \times 10^{-12}\,\mathrm{C/J/m}\right)\left(5000\,\mathrm{V/m}\right)\left(60 \times 10^{-3}\,\mathrm{V}\right)}{\left(500 \times 10^{-9}\,\mathrm{m}\right)^2} = 997{,}852\,\mathrm{N/m^3}$$

This pressure equates to about 100 atm = 1500 psi. High-pressure generation is a very useful application for many bioanalytical systems, as well as for new emerging technologies, such as microthrusters and rockets.

However, in addition to generating high pressure, these electroosmotic pumps can be used to pump fluids from one place to another. Examples include pumping methanol and water into micro fuel cells, or pumping water in a closed-loop channel for laptop cooling applications.

9.5 SUMMARY

Nanoscale fluid flow behavior can be quite accurately described using the concept of a continuum if the mean free path of the fluid molecules is smaller than the dimension of the channel—as is often the case, especially with liquids. Otherwise, calculation-heavy molecular dynamics simulations are required. To estimate the flow properties at the nanoscale, we use a handy, nondimensional number called the Reynolds number. Nanoscale fluid flow is heavily dependent on the surface properties of the channel in which the fluid flows since the surface-to-volume ratio in these systems is so high. Surface charges give rise to an electric double layer, which enables different types of electrokinetic flow. Therefore, at a nanoscale, fluids can flow not only because of pressure differences, as is the case at larger scales, but also because of electrokinetic interactions.

Continuum theory can predict the movement of small molecules in nanochannels, using Stokes flow. The applications of these sorts of flows include the analysis of biomolecules, electroosmotic pumps for cooling and pumping purposes, and devices like the FlowFET.

HOMEWORK EXERCISES

9.1 Describe the concept of a continuum.

9.2 Air is flowing through a 40-nm radius carbon nanotube. Can we use the Navier–Stokes equation to analyze this flow? What about if it were in a 4-m radius pipe?

9.3 Determine whether the following items can be considered a Newtonian fluid:
   a. Banana chocolate chip pancake batter
   b. WD-40
   c. Saliva
   d. Helium gas
   e. Fresh milk

9.4 Write the Navier–Stokes equation in the x-direction if the only forces acting on the fluid are pressure forces, viscous forces, and an electric force.

9.5 True or false? The Navier–Stokes equations treat each molecule of liquid individually, tracking each one as it moves through the fluidic system.

9.6 What are the inertial terms in the Navier–Stokes equations? When can they be completely neglected? Why?

9.7 Reynolds number is the ratio of inertial forces to viscous forces. Can you think of another nondimensional grouping that may be able to tell us the ratio between gravitational forces and viscous forces?

9.8 Compute the Reynolds number for each case:
   a. A major league fastball (100 mph)
   b. A 10-μm diameter protozoan flowing at 100 μm/s in water
   c. A blade of grass floating in a pond at 1 mm/s

9.9 Compute the Reynolds numbers of
   a. A particle 1 μm in diameter moving 0.1 m/s in water (viscosity of 1×10^{-3} Pa s; density of 1000 kg/m³).
   b Flow over the 5-meter wing chord of an F16A Fighting Falcon traveling 500 m/s in air (viscosity at 12,000 ft: 1.619×10^{-5} Pa s; density at 12,000 ft: 0.737 kg/m³).

9.10 True or false? The Reynolds number determines similarity between velocity flow profiles.

9.11 What are mechanisms for charge separation at an interface?

9.12 In order for glass to fully deprotonate, the pH of the solution must be higher than what value?

9.13 What is the electric double layer?

9.14 Calculate the electric double layer length for pure water at pH 7.

9.15 True or false? Zeta potential is the potential at the surface of a solid surface.

9.16 What are the major types of electrokinetic flow phenomena?

9.17 If you put a drop of 5-mM blue dye in a very large jar, what is the equation describing the one-dimensional concentration distribution of dye after 5 min? Assume that the diffusion coefficient of the dye is $5 \times 10^{-8} m^2/s$.

9.18 If you want to make sure that a specific protein molecule with diffusion coefficient 3.2×10^{-6} m²/s will diffuse to a target in less than 5 s, what is the largest dimension that you can make the system?

9.19 Calculate the flow rate per unit depth for air that is driven by 50 atm per unit meter of pressure in a nanofluidic channel that is 100 nm × 100 nm in cross-sectional area.

9.20 What is the shear stress for pressure-driven flow in a nanochannel? How about gravity-driven flow?

9.21 Derive the equations of gravity-driven motion in a two-dimensional nanochannel. Assume that the nanochannel is much wider than it is deep.

9.22 If we have 5 kV to drive fluid at 100 μm/s, is it better to use one big channel or multiple nanochannels? Why?

9.23 True or false? We cannot simply add our solutions of pressure-driven flow and gravity-driven flow in a channel together to find the velocity distribution, but must derive everything from the original Navier–Stokes equations.

9.24 Calculate the amount of pressure required to negate the gravity force in a vertical channel that is
   a. 40 nm wide.
   b. 4 mm wide (Hint: Flow rate = 0 in this case).

9.25 Using the Stokes flow solution, compute the following for a cell 10 μm in radius cell moving at 100 μm/s in the x-direction:
   a. The Reynolds number.
   b. The drag force.
   c. The velocity of the fluid 1 μm away from the cell.

d. The velocity of the fluid 1 mm away from the cell in the y-direction. Use the viscosity of water for your answer.

9.26 Draw the diffusion of a band of colored solute in a channel for the following situations (refer to Figure 9.15):

a. Diffusion > Advection, Electromigration > Diffusion.

b. Advection > Diffusion, Electromigration.

9.27 True or false? The thicker the double layer, the better separation there will be between two differently charged species. (Think carefully about this one— what happens when the double layers get really, really big?)

9.28 Using the typical values in Table 9.4, calculate the electromigrative, diffusive, and advective time scales for nanochannel electrophoresis. What must be the typical velocity in order to get an advective time scale on the order of diffusion and electromigration? Is this reasonable?

9.29 What is the diffusion coefficient of a DNA molecule (diameter = 2 nm) trying to serpentine its way through a 5-nm radius channel? The diffusion coefficient of this particular DNA in free solution is 3×10^{-8} m^2/s.

9.30 Draw an electropherogram for the following cases:

a. Two species with the same mobility but different charge in a microchannel

b. Two species with the same mobility but different charge in a nanochannel

c. Two species with different mobilities but the same charge in a microchannel

d. Two species with different mobilities but the same charge in a nanochannel

9.31 Derive the equation that solves for current through an electroosmotic pump that is generating a certain pressure but has no flow.

9.32 Explain a FlowFET device.

RECOMMENDATIONS FOR FURTHER READING

1. E. M. Purcell. 1977. Life at low Reynolds number. *American Journal of Physics*, 45:3.

2. N.-T. Nguyen and S. T. Wereley. 2006. *Fundamentals and Applications of Microfluidics*, 2nd Edition. Artech Print on Demand.

3. R. F. Probstein. 1994. *Physicochemical Hydrodynamics*. Wiley-Interscience.

4. F. M. White. 2006. *Viscous Fluid Flow*. McGraw-Hill.

5. M. T. Napoli, S. Pennathur, and J. C. T. Eijkel. 2010. Nanofluidic technology for biomolecule applications: A critical review. *Lab on a Chip*, 10:957.

6. D. Huber, M. Markel, S. Pennathur, and K. Patel. 2009. Oligonucleotide hybridization and free-solution electrokinetic separation in a nanofluidic device. *Lab on a Chip*, 9(20):2933.

7. S. Pennathur and J. G. Santiago. 2005. Electrokinetic transport in nanochannels: 2. Experiments. *Analytical Chemistry*, 77(21):6782.

8. T. Squires and S. Quake. 2005. Microfluidics: Fluid physics at the nanoliter scale. *Reviews of Modern Physics*, 77:978.

9. B. Kirby. 2010. *Physics of Micro- and Nano-scale Fluid Mechanics*. Cambridge University Press.

CHAPTER **10**

Nanobiotechnology

10.1 BACKGROUND: OUR WORLD IN A CELL

If a curious passerby were traveling the universe and stopped in on us to collect a single sample—one souvenir that would represent our planet—what would we give? Would we pick something alive? Would we be able to give him something that contains a message, some bit of fundamental information? Perhaps an example of our most highly developed technology, something to represent the pinnacle of our planet's progress up until now? It turns out there is something that might very well cover all these options—a living cell. It certainly would not take up much trunk space, as we can see in Figure 10.1. Measuring just a few micrometers long, an *Escherichia coli* (*E. coli*) bacterium in particular makes an ideal baton to hand off to the alien.

Why?

First, it is alive. And, like the majority of living things on Earth, it is a single-celled organism. Inside it are instructions, written in the oldest and most universal language on Earth—DNA—for making a copy of itself. Finally, the "technology" that exists in just a single cell such as *E. coli* rivals or beats anything mankind has ever developed.

It is also a good example of man's understanding of his world. We know more about the tiny, rod-shaped *E. coli* than we do about any other living organism, even humans. (Note: there are hundreds of strains of *E. coli* that live symbiotically in our gut, although one strain gets most of the press: *E. coli O157:H7*, an uncommon strain that can make us sick if we eat an infected raw hamburger or bad bag of spinach, for example.) We sequenced *E. coli*'s genome—all 4,639,221 nucleotide pairs—by 1997. We did not know our own genome until 2003. *E. coli* is considered a model organism because so much of how it operates is universal. Everything we learn about it sheds light on other aspects of biology. Most of what we know about the basic workings of life itself, such as how DNA works and how proteins get put together, we know from studying *E. coli*.

However, *E. coli* is just one of a mind-boggling 100 million total species estimated to currently live on Earth. This diverse and well-established living world has billions of successful implementations of nanotechnology. Our own bodies, arguably the most complex

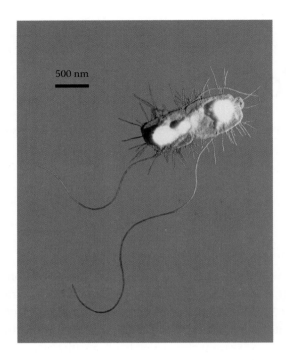

FIGURE 10.1 The *Escherichia coli* (*E. coli*) bacterium. This image of the well-understood, single-celled organism was captured by an SEM. Protruding from the cell are two flagella, which are corkscrew-shaped appendages that rotate at about 6000 rev/s, propelling the bacterium through the water, and multiple fimbriae, which are smaller appendages used to adhere to other cells. When the flagella stop rotating, the bacterium drifts less than the width of an atom because inertia plays such a minimal role, compared to viscous forces, at that scale. (Remember swimming in molasses in Section 9.2.1?) (Image copyright by Dr. Fred Hossler/Visuals Unlimited.)

things we know of, operate at a molecular level, with nanoscale organic machines storing and using energy, reproducing, protecting themselves, and controlling movement.

Nanobiotechnology is nanotechnology that is present in living things. It is the molecular machinery of organisms. These machines are not of our design but we can certainly find uses for them. Because of their nanometer sizes and functionality, we classify them not only as examples of nanotechnology, but also phenomenal examples. Even if life just went on and we never figured out a way to use DNA as a computer or use ATP synthase to propel a drug-delivery device through our arteries, these machines still qualify as nanotechnology in our book.

This marks a divergence from prior chapters. Usually, we are discussing things that have been made very small in order to take advantage of new physical properties at that scale—man-made things. Here, we are discussing things that already exist, that work wonderfully. Living things. Things we can start using today, without having to go to the trouble of inventing them. Many professionals call nanobiotechnology a "bottom-up" approach, as opposed to a "top-down" approach where a technology is built from bigger structures. There are many examples of "top-down" engineering in the other chapters of this book;

however, in this chapter we will discuss building things from molecules and using machines that currently exist on a nanoscale to help us construct useful nano-machines.

In this chapter, we will discuss a highly essential existing set of biological nanotechnologies with broad applications for mankind. More specifically, we will discuss the machinery of a living cell, which contains four important nanoscale components— sugars, fatty acids, amino acids, and nucleotides. All of these components can be found in just about any living cell on Earth, not just the *E. coli* bacterium. With knowledge of the long-established mechanisms in cells, engineers can leapfrog intricate development steps and find ways to use the *existing* machinery of life for new purposes, using the "bottom-up" approach. Effective and highly efficient, this machinery encompasses everything nanotechnology aspires to achieve—or just inherit. However, before jumping into the machinery of the cell, we will first discuss a few aspects of how biology is a little bit different at a nanoscale level.

10.2 INTRODUCTION: HOW BIOLOGY "FEELS" AT THE NANOMETER SCALE

Feelings are fuzzy. They can be irrational, indescribable sensations more likely found in the gut than the head. Engineers and scientists are often expected to purge feelings in favor of reason—hard data outweighing hunches. A similar kind of prejudice infects those of us accustomed to dealing with the perfect corners, blocky shapes, and solid materials typical in engineering. Bridges, integrated circuits, and internal combustion engines—these things are hard and linear. And so there may be a subtle distrust of the squishy world of biology. Riddled with carbon, not steel. Everything soaked in water. Right angles are few and far between. Perhaps this prejudice, like any other, stems from unfamiliarity. Maybe your organic chemistry is a little rusty; or maybe not: perhaps you relish the miraculous messiness of all things "bio," knowing full well that what seems like disorder is actually complexity. Either way, we are diving in. Prepare to get wet.

10.2.1 Biological Shapes at the Nanoscale: Carbon and Water Are the Essential Tools

The biomolecular machines inside a cell take on shapes better recreated in Play-Doh™ than Lego™ blocks. These intricate, bulbous forms can change their shapes with the addition or subtraction of energy. The shapes of the machines determine their interactions with each other, which in turn determine what fundamental biological processes will occur, such as energy conversion, reproduction, or movement. One of the most important things involved in these processes is water.

Why is water so important to living cells?

First, all living things are made up of at least 70% water, which means everything is always wet. So it makes sense that the major driving forces for molecular interaction exploit water. Water is what ensures that biomolecular machines are interacting with each other. It is a polar molecule, and electrically charged molecules, as well as other polar molecules such as alcohols, sugars, and molecules with nitrogen- and oxygen-rich regions, are all attracted to water's polar regions (see Figure 10.2). Molecules that are attracted to water are

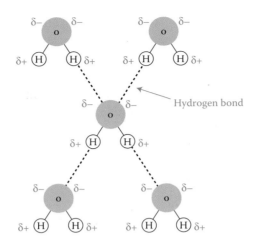

FIGURE 10.2   A group of five water molecules held together transiently by hydrogen bonding. The positive and negative regions of water (denoted by δ^+ and δ^-) show that water is polar and can attract positive and negative molecules via two positive regions (on the two hydrogen atoms) and two negative regions (on the one oxygen atom). Here we see it attracting itself through hydrogen bonding. Molecules attracted to water are known as hydrophilic ("water loving") and readily participate in hydrogen bonding interactions and dissolve easily into water.

called hydrophilic ("water loving"). They naturally participate in hydrogen bonding interactions with the water molecules and therefore dissolve easily into aqueous solutions.

Almost all the molecules in a cell contain carbon, water being the primary exception. Carbon atoms have four vacancies in their outer shells and can thus form four covalent bonds. This, and the fact that carbon can bond to itself to form the chains, branched trees, and rings shown in Figure 10.3, causes complex and gigantic organic molecules.

However, the carbon-rich area of a molecule tends to be hydrophobic ("water fearing") and will not readily form hydrogen bonds with water. Molecules containing carbon-rich areas therefore contort so as to pull these areas inward, thus causing biological molecular machines to fold into their unique shapes—shapes that endow such molecules with very specific chemical functionalities. The forms assumed by DNA, RNA, and most proteins are such that hydrophilic regions remain on the outside, in contact with water, while hydrophobic regions bunch into the core.

10.2.2 Inertia and Gravity Are Insignificant: The Swimming Bacterium

We learned in Chapter 2 about scaling laws, and in Chapter 9 we learned that at the nanoscale, inertia and gravity make virtually no difference. To see this in action, let us go back to a swimming bacteria cell.

So, according to our calculation, a bacterium swimming through water comes to a stop in a distance less than the diameter of a hydrogen atom. The motion of things in and around a cell is governed not as much by momentum as by physical interactions with other things. Attractive forces, such as van der Waals forces, and viscous forces between small objects are much stronger than the force of gravity and inertia at that scale.

BACK-OF-THE-ENVELOPE 10.1

Bacteria swim through water at about 30 μm/s, using corkscrew-shaped, flagella-like propellers (see Figure 10.1). We will examine a single bacterium and assume it to be relatively spherical, in order to make an estimate using Stokes' Law:

$$F = 6\pi\mu r_p v \qquad (10.1)$$

This equation gives the force, F, needed to move a particle (in this case, the bacterium) of radius, r_p, through a fluid with viscosity, μ, at a velocity, v. For water at room temperature, $\mu = 0.001$ N s/m². If we assign the bacterium a radius of 0.5 μm, we obtain

$$F = 6\pi\left(0.001\,\mathrm{Ns/m^2}\right)\left(0.5 \times 10^{-6}\,\mathrm{m}\right)\left(30 \times 10^{-6}\,\mathrm{m/s}\right) = 2.8 \times 10^{-13}\,\mathrm{N}$$

This is the force the bacterium must generate with its flagella to move forward at 30 μm/s. Because force equals mass m times acceleration a, we can estimate the rate at which the bacterium (weighing approximately 1 pg) would stop if it suddenly stopped moving its flagella:

$$a = \frac{F}{m} = \frac{2.8 \times 10^{-13}\,\mathrm{N}}{1 \times 10^{-15}\,\mathrm{kg}} = 280\,\mathrm{m/s^2}$$

That is, it decelerates at 280 m/s². Using well-known equations of motion, we can determine the time, t, it takes the bacterium to stop completely (or, more specifically, undergo a change in velocity, Δv, from 30 μm/s to zero):

$$t = \frac{\Delta v}{a} = \frac{30 \times 10^{-6}\,\mathrm{m/s}}{280\,\mathrm{m/s^2}} = 1 \times 10^{-7}\,\mathrm{s}$$

It stops in merely 0.1 μs. How about the distance, x, it travels while coasting to a halt? That can be determined by

$$x = vt - \frac{1}{2}at^2 = \left(30 \times 10^{-6}\,\mathrm{m/s}\right)\left(1 \times 10^{-7}\,\mathrm{s}\right) - \frac{1}{2}\left(280\,\mathrm{m/s^2}\right)\left(1 \times 10^{-7}\,\mathrm{s}\right)^2 = 2 \times 10^{-12}\,\mathrm{m}$$

Thus, this simplified model shows that the bacterium stops within less than the width of a single atom. Forget stopping on a dime: scale-wise, this is analogous to a jumbo jet stopping on a bacterium.

As a result, all of these molecules and machines and cell parts are in constant motion, being pushed and pulled around in quick, random trajectories determined by their diffusion coefficient. This is also called Brownian motion. As we learned in Chapter 9, the diffusion coefficient can be derived from Brownian motion. We will learn more about this motion in the next section.

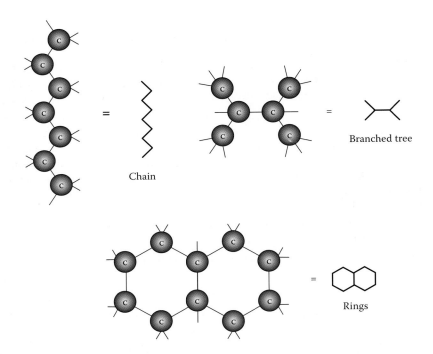

FIGURE 10.3   Carbon skeletons. Carbon plays an important role in living cells because it creates strong covalent bonds with itself and forms chains, branched trees, and rings. The bond structure and common symbols of these three forms are shown here.

10.2.3 Random Thermal Motion

In 1827, the biologist Robert Brown noticed that if you look at pollen grains in water through a microscope, the pollen jiggles. Although he could not figure out the cause of the jiggling, he named it after himself: Brownian motion. He actually assumed (incorrectly) that since the pollen was moving, it must be alive.

It would be Albert Einstein who later came up with the correct explanation for this movement in one of his three seminal papers of 1905 (see the sidebar "1905: The Big Year for Big Al").

1905: THE BIG YEAR FOR BIG AL

In 1905, Albert Einstein was 26-years-old. He was young and broke, so had to work full-time at a patent office in Bern. He was a newlywed and was writing his PhD thesis at the University of Zurich. However, in addition to working full-time, having a family, and finishing his PhD thesis, he somehow found time to publish three of the most seminal papers in history—all in the same year and all about completely different phenomena.

In the first paper, Einstein used Max Planck's quantum hypothesis to describe visible electromagnetic radiation, in other words, light. He claimed that light could be imagined to consist of discrete bundles, or packets, of radiation, as we learned about in Chapter 3,

"Introduction to Nanoscale Physics." This paper formed the basis for much of quantum mechanics (described in Chapters 3 and 5).

In the second of his three papers, Einstein proposed what is now known as the theory of relativity, where the famous equation, $E = mc^2$, comes from. Einstein figured out how, in a certain manner of speaking, mass and energy were equivalent. He was not the first to propose all the elements that went into the special theory of relativity but his contribution lies in having unified important parts of classical mechanics and Maxwellian electrodynamics.

The third of Einstein's seminal papers of 1905 dealt with statistical mechanics. He calculated the average trajectory of a microscopic particle buffeted by random collisions with molecules in a fluid or in a gas. He predicted that the thermal energy of particles was responsible for this "jiggling" motion, the apparently erratic movement of pollen in fluids, which had been noted first by the British botanist Robert Brown. Einstein's paper provided convincing evidence for the physical existence of atom-sized molecules, which had already received much theoretical discussion.

What a year!

Einstein realized that the jiggling was due to molecules of water hitting the tiny pollen grains, like players kicking the ball in a game of soccer. The pollen grains were visible but the water molecules were not, so it looked like the grains were bouncing around on their own. He figured out that the Brownian motion of the molecules within a cell is driven by thermal energy, and that this is how you can derive the diffusion coefficient of molecules. (We learned about the diffusion coefficient in Chapter 9, "Nanoscale Fluid Mechanics.") This energy is always up for grabs in cells, which tend to operate best (and therefore live) at temperatures of about 37°C. Inside every cell is a crowd of churning, bumping, battering nanoscale substances. A few examples of nanoscale species in a cell include proteins, water molecules, ions, sugars, and lipids. We will discuss the main ones in this chapter. Every interaction between different species in a cell leads to a bond of some kind. Maybe a fleeting van der Waals attraction, an ionic bond, or a hydrogen bond; maybe a lasting covalent bond.

Molecules inside cells come in contact with one another more often than we might expect. The cell is a chemical reactor. Thousands upon thousands of reactions are occurring at all times, and these reactions need parts and materials—like the assembly line of a car. But instead of linear, conveyor-belt type delivery mechanisms, the cell often relies on the fact that, eventually, the parts, materials, and tools will simply bump into one another.

And the wait is not long. The diffusion time, or mixing time as we call it here, t_{mix}, is a way of characterizing how quickly a molecule released at one point inside a cell can be found at any other location within the cell's volume. (We discussed diffusion in Chapter 9.) We can determine the mixing time, t_{mix}, using this relationship:

$$t_{mix} = \frac{D^2}{d} \tag{10.2}$$

Here, D is the characteristic dimension of the cell (such as its diameter) (m) and d is the diffusion coefficient of the molecule (m²/s). Typical mixing times inside a micrometer-sized cell are a second or less.

BACK-OF-THE-ENVELOPE 10.2

A sugar molecule with a diffusion coefficient in water of $d = 10^{-10}$ m²/s is released inside a cell 2 μm in diameter. How much time passes before it is probable that the same molecule can be at any location inside the cell?

$$t_{mix} = \frac{D^2}{d} = \frac{\left(2 \times 10^{-6}\ m\right)^2}{10^{-10}\ m^2/s} = 0.04\ s$$

A second enlightening parameter is the traffic time $t_{traffic}$, which estimates the time for two molecules inside a cell to meet if they start off separated by a distance equal to the cell's characteristic dimension D. It is given by

$$t_{traffic} = \frac{D^3}{dr} \tag{10.3}$$

The diffusion coefficient is the sum of the diffusion coefficients for each molecule, $d = d_1 + d_2$, and r is the sum of the molecules' radii, $r = r_1 + r_2$.

BACK-OF-THE-ENVELOPE 10.3

A pair of enzyme molecules, 12 and 10 nm in diameter, respectively, each have a diffusion coefficient in water of $d = 10^{-10}$ m²/s. How often do these two molecules collide with one another inside a cell 2 μm in diameter?

This can be determined by the traffic time:

$$t_{traffic} = \frac{D^3}{dr} = \frac{\left(2 \times 10^{-6}\ m\right)^3}{\left(10^{-10}\ m^2/s\right)\left(12 \times 10^{-9} + 10 \times 10^{-9}\right)} = 0.4\ s$$

These two molecules, minute in comparison to the cell they occupy, still manage to meet one another more than twice every second.

Note that both t_{mix} and $t_{traffic}$ are strongly dependent on a cell's characteristic dimension—scaling with D^2 and D^3, respectively). For a cell 1 μm in diameter, $t_{traffic}$ will be close to a second; for a 10-μm cell, it will be minutes; for a 100-μm cell, it will be hours. In the human body, where a neuron cell can be a meter long, relying on Brownian motion alone would mean it would take about a thousand years to get one small protein from one end of the neuron to the other. (Kinesin, a molecular motor, is used instead. It transports the protein in about a week. We will discuss molecular motors later in this chapter.)

However, we can see the remarkable effectiveness of Brownian motion for delivering molecules to where they need to be very quickly inside small cells. These cells solve their logistics problems by having molecules go everywhere all the time, with the end result being that the molecules pass through where they need to be very often.

10.3 THE MACHINERY OF THE CELL

The warm, dense, wiggling stuff inside a cell is akin to machinery. It is humming 24 hours a day, burning energy to carry out jobs. Most of this machinery lasts only a short time. A year would be an especially long life for anything inside a cell. A few seconds is more typical. These machines take form in an instant, interact with their chemical environment, complete a task, and then break up to supply the raw materials for the next machine.

A typical cell contains thousands of different kinds of molecules. Despite their apparent diversity, most of these molecules are chemically related. Almost every cellular molecule belongs to one of four molecular families: sugars, fatty acids, amino acids, and nucleotides. Cells use these four families to build all manner of structures and machinery. There is absolutely no reason we cannot use them too, and develop our own technologies, from the bottom up.

Icons representing each of the four families are introduced in Figure 10.4. In the next sections, we will learn details about the form and function of each family and take a closer look at some specific examples of cellular machinery. It will be like peering under the hood of a car, where we see whirring fans and pumping pistons. We see hoses and wires running

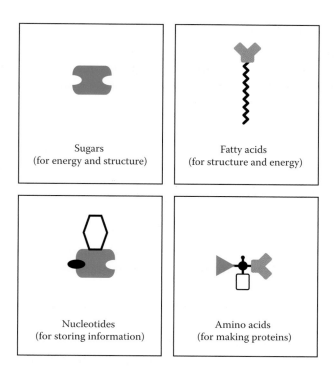

Sugars
(for energy and structure)

Fatty acids
(for structure and energy)

Nucleotides
(for storing information)

Amino acids
(for making proteins)

FIGURE 10.4 Icons representing the four molecular families common to living cells. Cells use molecules from these families for fuel, for archiving information, and to build larger, more complex cellular machinery and structures.

everywhere. Sure, all this stuff is great for making a car go—but could we not rip some of it out and use it to make a chainsaw? Or a refrigerator? A boat?

Now consider the *E. coli* bacterium. It is a fully operational submarine. Can we not find a way to reuse its parts to build useful nanoscale machines?

10.3.1 Sugars Are Used for Energy (but also Structure)

Any dieter with a sweet tooth probably knows them as "carbs." Sugars, and the molecules made out of sugars, are called carbohydrates, the simplest of which have the general chemical formula $(CH_2O)_n$, where n is typically 3, 4, 5, 6, 7, or 8. Sugars are our bodies' key energy source, providing 4 kilocalories (kcal) of energy per gram. (Note that 1 food calorie equals 1 kcal; also, for comparison, gasoline—which is also rich in carbon and hydrogen—contains about 11.5 kilocalories per gram (kcal/g).) Unlike most molecules that assume one particular shape for a given chemical formula, a single type of sugar can take on a variety of molecular orientations—which in turn make for a variety of very different functions for the same type of molecule.

Sugars serve two main roles: (1) cells can either break them down one at a time to use as energy for their machinery, or (2) build them up by the thousands into gigantic polymers for mechanical support. Glucose is a prime example of a sugar that plays both roles.

10.3.1.1 Glucose

In animals, plants, fungi, and bacteria, glucose's main purpose is to be consumed. It circulates through our bodies, taking energy to our cells—hence its other common name, "blood sugar." The formula of glucose is $C_6H_{12}O_6$ and, so like all carbohydrates, it can be made from simple ingredients; during photosynthesis, plants make glucose out of just water (H_2O) and carbon dioxide (CO_2), using the energy they get from absorbed photons.

When cells need energy, they break glucose back down into its constituent parts in a process known as glycolysis, after the Greek words *glukus*, meaning "sweet," and *lusis*, meaning "rupture." This process yields water, carbon dioxide, and the all-important adenosine triphosphate (or ATP). (Cells do not use glucose directly for energy just as cars do not use crude oil; they use ATP, a specialized molecule discussed in greater detail later in this chapter.) Glycolysis is a complex, 10-stage reaction that, in its early stages burns two ATP molecules. This energy "investment" pays dividends in the latter stages, which produce four ATPs—a net gain of two ATPs for every glucose molecule broken down.

A sugar molecule can exist in one of two forms: (1) ring or (2) open chain. These two forms of glucose are shown in Figure 10.5. In water, most of the glucose molecules exist as rings. This is because the aldehyde group ($_H > C = O$) at the top of the chain reacts with a hydroxyl (–O–H) group on the same glucose molecule, closing off the chain into a ring. We can identify the carbon atom of the original aldehyde as the only one in the ring bonded to two oxygen atoms. Once the ring is formed, this same carbon atom is able to create an additional link to a hydroxyl-bonded carbon on a nearby sugar molecule. Glucose alone is known as a "monosaccharide." When it bonds to another sugar, it becomes a "disaccharide." Add another sugar, it is a "trisaccharide," then "tetrasaccharide," and so on—all the way up to the gigantic "polysaccharides."

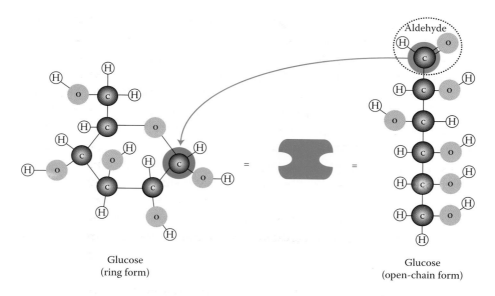

Glucose
(ring form)

Glucose
(open-chain form)

FIGURE 10.5   The sugar, glucose. Although it has the formula $C_6H_{12}O_6$, glucose's oxygen (O), carbon (C), hydrogen (H), and nitrogen (N) atoms can be covalently bonded in numerous ways. Two main forms, the ring and the open chain, are shown here using ball-and-stick models. Each stick represents a bond; each ball an atom. In water, the aldehyde group ($_H > C = O$) at the top of the chain reacts with a hydroxyl (–OH) group further down the chain, forming a bond that closes the chain into a ring. The carbon from the aldehyde becomes the only one in the ring bonded to two oxygen atoms. We use the icon in the middle to broadly represent all sugars.

Because sugars such as glucose tend to be covered in hydroxyl groups, there are numerous ways for the rings to link up (polymerize), creating polysaccharides of all shapes and sizes. Some polysaccharides are long, strong, linear chains, such as the cellulose found in plant cell walls. Cellulose is the most abundant organic chemical on Earth. Formed from thousands of linked glucose rings, cellulose fibers endow wood, paper, cotton, and other plant structures with strength. Other polysaccharides form branched structures, such as that of glycogen. Glycogen is a mega-molecule, each one made from about 10,000 to 100,000 glucose rings. It is the long-term storage place for sugar energy in animals and humans. The branched structure has many free ends that can be quickly detached when energy is needed. Both cellulose and glycogen are shown in Figure 10.6.

10.3.2 Fatty Acids Are Used for Structure (but also Energy)

Fatty acids are long, skinny molecules also made from carbon, hydrogen, and oxygen. As we see in Figure 10.7, these molecules have a long hydrocarbon chain for a tail and a carboxyl (–COOH) group for a head. In solution, this carboxyl loses its hydrogen atom and becomes ionized (–COO⁻). Do you see the oxygen atom shown in Figure 10.7 with just a single bond to the carbon atom? That oxygen still has one valence electron to spare, making it chemically reactive and likely to form bonds with other molecules, including other fatty acids. Nearly all the fatty acids in a cell are covalently bonded to other molecules via

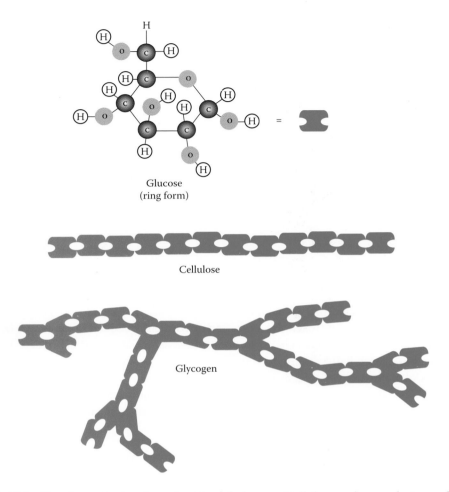

FIGURE 10.6 Two important polysaccharides of glucose. Cellulose, a linear chain made from thousands to tens of thousands of glucose rings, is Earth's most abundant organic substance. Glycogen, the long-term storage form of glucose in animals and humans, has a branched structure made from tens to hundreds of thousands of glucose rings. Its numerous free ends can be easily detached when sugar energy becomes necessary.

this head group. Also, this head group is polar—meaning it has an unequal distribution of electrons, making one side more positive, the other more negative. When polar objects are near one another, they favor an orientation where the positive portion of one object aligns with the negative portion of its neighbor, similar to a pair of bar magnets. This makes the head group a willing participant in the constant hydrogen bonding interactions inside water and other polar liquids. For this reason, the head of a fatty acid is hydrophilic ("water loving"). The carbon-rich tail meanwhile is nonpolar, making it hydrophobic ("water fearing").

A molecule with a hydrophobic tail and a hydrophilic head is a known as a surfactant. In water, the tails bunch together to keep "dry," letting their heads get "wet." So the mere presence of water can make fatty acids self-assemble into interesting structures engineers

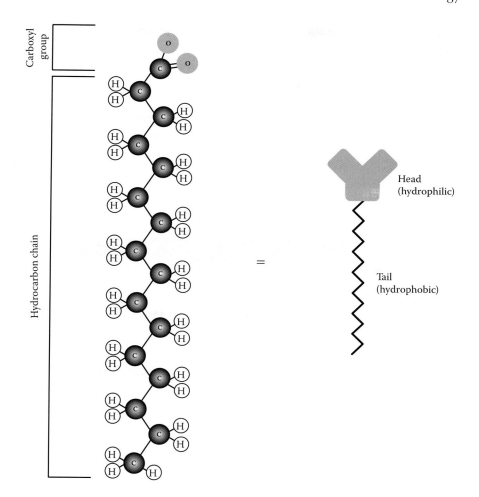

FIGURE 10.7 A fatty acid. Fatty acids have two parts—a hydrophobic hydrocarbon chain of carbon (C) and hydrogen (H) atoms, and a hydrophilic carboxyl group with two oxygen (O) atoms. The fatty acid shown here is palmitic acid, $C_{16}H_{32}O_2$, in its ionized form, which has carboxylic acid as the carboxyl group. We represent all fatty acids with the simplified icon on the right. The length of the hydrocarbon chain varies by fatty acid. Palmitic acid is found in many oils, fats, and soaps.

might be able to use. These structures are classified as micelles and can be globular fat droplets, with diameters of just a few nanometers up to a few hundred nanometers, or spread thinly, as a surface film or fatty bilayer a few nanometers thick. (These are discussed in Chapter 4, Section 4.4.6.3 and shown in Figure 4.23.)

The main role for fatty acids is structural. Cells rely on lipid bilayers to build membranes. A membrane is a thin molecular sheet, about 5–10 nm in width, that envelops every cell; numerous other membranes are found on the inside of cells also, where they compartmentalize and protect cellular organs and machinery. We will discuss the cellular lipid bilayer in more detail in the next section.

TABLE 10.1   Common Fatty Acids That Occur Naturally in Fats and Oils

Fatty Acid	Formula	Selected Sources
Butyric acid	$C_4H_8O_2$	Butterfat
Caproic acid	$C_6H_{12}O_2$	Butterfat
Caprylic acid	$C_8H_{16}O_2$	Coconut oil, palm oil, breast milk
Capric acid	$C_{10}H_{20}O_2$	Coconut oil, palm oil, breast milk, whale oil
Lauric acid	$C_{12}H_{24}O_2$	Coconut oil, palm oil, cinnamon oil
Myristic acid	$C_{14}H_{28}O_2$	Coconut oil, palm oil, nutmeg oil, whale oil
Palmitic acid	$C_{16}H_{32}O_2$	Butterfat, palm oil, animal fat
Palmitoleic acid	$C_{16}H_{30}O_2$	Animal fats
Stearic acid	$C_{18}H_{36}O_2$	Animal fats
Oleic acid	$C_{18}H_{34}O_2$	Grape seed oil, olive oil
Arachidic acid	$C_{20}H_{40}O_2$	Peanut oil, fish oil

Cells also utilize fatty acids as a concentrated energy reserve. One gram of fatty acids, when broken down, provides about six times more energy than the same mass of glucose. Yet any marathoner will tell you that energy is released from fats at about half the rate of sugars (via glycolysis), making carbohydrates the first choice for fast fuel. "Hitting the wall" at the end of a long workout means you have used up your glucose reserves and are running—painfully—on slow-burning fat.

There are hundreds of different fatty acids. They vary only in the length and bonding structure of their hydrocarbon chains. The fat in meat, butter, and cream is made from fatty acids, as are the oils derived from plants, including corn oil and olive oil. A list of some common fatty acids and their sources is provided in Table 10.1.

10.3.2.1 Phospholipids

A broader term for fatty acids and the larger molecules made out of fatty acids is *lipids*, a class of molecules that includes fats. Lipids account for about half the mass of most animal cell membranes (the remainder being mostly protein). A phospholipid is shown in Figure 10.8a. It is formed from a pair of fatty acids that are bonded to glycerol, which in turn bonds to a phosphate group (hence the name *phospho*lipids). Bonded to the top of the phospholipid molecule is a small, hydrophilic molecule such as choline. The tails are usually 14–24 carbon atoms long, and each one can be a different length. As we can see, the basic form is the same one we have been discussing: a hydrophilic head and a hydrophobic tail. This means that water causes spontaneous self-assembly of phospholipids.

The form they assume is that of a bilayer, like the one shown in Figure 10.8b. The bilayer is the preferred structure because phospholipids are relatively cylindrical shapes—as opposed to conical, in which case the assumed shape would be spherical. (Refer to Chapter 4, "Nanomaterials," Section 4.4.6.3 about micelles.) This bilayer holds together cellular membranes. The phospholipids pack together with a density of about 5 million molecules per square micrometer of bilayer (counting both layers).

As long as there is water around, bilayer assembly is automatic and constant. This endows membranes with an incredible gift—self-healing. A small rip in the bilayer exposes an edge of the bilayer to water. The hydrophobic hydrocarbon chains hidden inside the bilayer

(a)

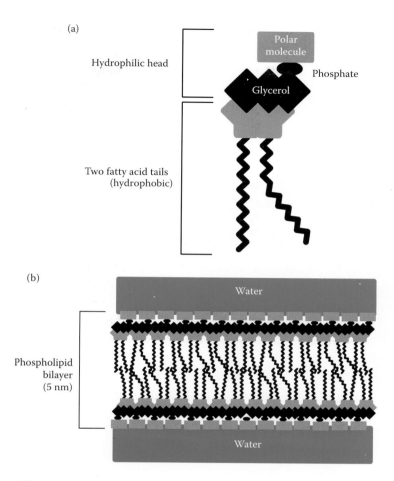

Hydrophilic head

Polar molecule

Phosphate

Glycerol

Two fatty acid tails
(hydrophobic)

(b)

Water

Phospholipid
bilayer
(5 nm)

Water

FIGURE 10.8 Cell membranes are made using phospholipids. The individual phospholipid molecule in (a) has hydrophobic fatty acid tails, which can be bent or kinked, depending on the fatty acid, and a hydrophilic head. The head is made from a glycerol molecule, a phosphate molecule, and a small polar molecule, such as choline. (b) In water, phospholipids pack together in a sandwich structure to keep water off their tails. The resulting phospholipid bilayer is the structural basis of cell membranes.

suddenly get "wet." Because they cannot form hydrogen bonds with the hydrocarbons, nearby water molecules are forced to reorganize so as to repel the chains. This requires energy, and natural systems always avoid the use of extraneous energy.

Imagine a hanging colony of bats sleeping contentedly in a dark cave—when suddenly a cave wall rips away and sunlight comes rushing in. The bats awaken and flap around furiously until they find a new place to escape the harsh light of day. So also do the hydrocarbon chains come alive and seek out new hiding places. The molecules jiggle about, forming new bonds and therefore resealing tears in the membrane, all just to be dry again. Consider the importance of this phenomenon: because a flat, planar bilayer has exposed edges that are hydrophobic, it tends to close in on itself, forming a sealed pouch—as shown in Figure 10.9. If such pouches did not form, neither would cells. Life on Earth would be a vast, messy

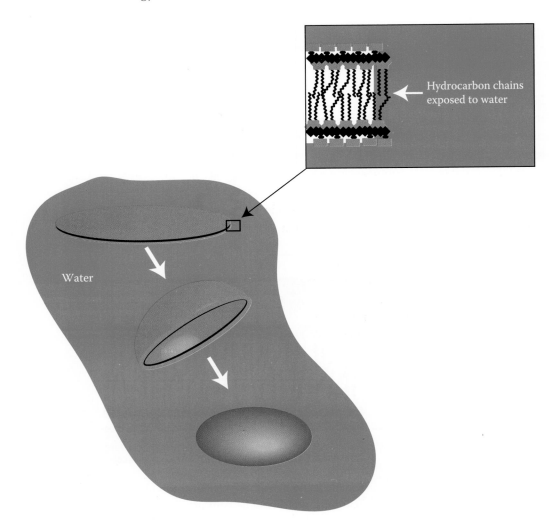

Hydrocarbon chains
exposed to water

Water

FIGURE 10.9 A phospholipid bilayer closing upon itself to form a sealed pouch. Because the edges of the bilayer have exposed hydrocarbon chains that are hydrophobic, the planar bilayer automatically closes upon itself to form a sealed compartment. Large compartments are what enclose cells; tinier pouches enclose the machinery within cells.

soup. Consider also the nanoscale engineering possibilities for a material capable of fixing itself and creating protective compartments around sensitive machinery. Recently, scientists have tested lipid membranes as possible ways to carry drugs for drug delivery inside humans.

Phospholipid membranes are stable, and yet fluidic, structures. A given phospholipid on one side of a bilayer swaps places with one of its neighbors about 10 million times per second. However, the same phospholipid will "flip-flop" to the other side of the bilayer less than once a month. The lateral diffusion coefficient, d, for phospholipids within one side of a bilayer is estimated to be about 1 $\mu m^2/s$, meaning that the average phospholipid molecule diffuses from one end of, say, a large bacterial cell to the other in about a second. This fluidity makes membranes flexible. Disk-shaped red blood cells in our bodies routinely contort enough to squeeze through capillaries half their diameter.

10.3.3 Nucleotides Are Used to Store Information and Carry Chemical Energy

Nucleotides are another carbon-based set of molecules prevalent in cells, mainly there to store information and carry chemical energy. All nucleotides have three things in common: (1) a nitrogen-containing ring molecule (called a base), (2) a five-carbon sugar molecule, and (3) at least one phosphate molecule (see Figure 10.10). When joined together in a chain, nucleotides form DNA—the code to make living things. Another particular nucleotide with three phosphates is ATP—an energy carrier that is made and consumed to drive chemical reactions in every living cell on Earth. These two sets of nucleotides alone are responsible for such a great many aspects of the way life on Earth *lives* that we must discuss them in more detail here.

10.3.3.1 Deoxyribonucleic Acid

Imagine if every electronic device stored information in the exact same format, on the exact same medium—a universal language encoded on a string—and that this form of information was used by your microwave, your laptop, your car, your calculator, your watch, your thermostat, your coffee maker, and every other such device in every country in the world. Now imagine that this universal language uses only four letters and that the string is only 2 nm thick. That is what DNA is to living things.

All living things are made of cells, and all cells contain DNA. Organisms store, use, and pass on their hereditary information using this well-known, double-stranded molecule. Because of this, a snippet of your DNA could be inserted into a bacterium and the information encoded within the DNA would be read, interpreted, and copied. The same would be true if the bacterium's DNA was inserted into one of the cells in your body.

To see the ways DNA can be used as a material is to look at its structure. DNA's well-known double helix (shaped like a spiral staircase) is made out of two nucleotide chains (see Figure 10.11). There are four bases in DNA: adenine, guanine, cytosine, and thymine. The nucleotide made with the cytosine base is cytosine monophosphate; at approximately 810 pm across, it is the smallest of the four bases. The largest of the four is guanine monophosphate, at about 860 pm. For short, we call the four bases A, G, C, and T. These bases form the "stairs" of the spiral staircase. The sequence of these bases is what gives an organism its genetic identity. A binds only to T, and C binds only to G, as shown in Figure 10.12. Human DNA contains about 3.4 billion base-pairs. Thus, storing the human genome on a computer requires 3.4 gigabytes (GB) of memory.

DNA is always a double-helix structure, no matter the order of the nucleotides. As with so many of the other cellular molecules we have discussed, the presence of water governs DNA's structure, with the hydrophobic base-pairs pushed toward the core, leaving the hydrophilic phosphates on the surface.

When cells divide and multiply, DNA replication occurs, in which fresh, complementary DNA strands are synthesized using the templated polymerization shown in part (c) of Figure 10.11. In this way, new cells take with them their own recipe, so that they too can divide and reproduce.

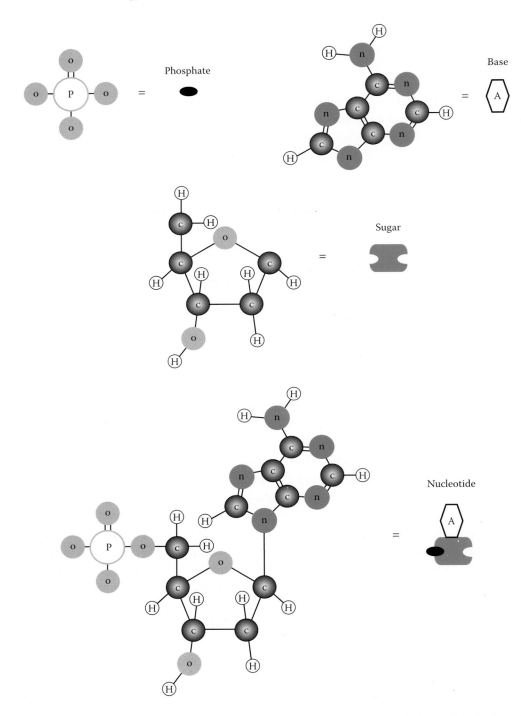

FIGURE 10.10 A nucleotide. All nucleotides have a nitrogen-containing ring molecule (usually called a base), a five-carbon sugar molecule, and at least one phosphate molecule. The base shown here is adenine; the sugar is deoxyribose. The bonding among the various atoms (oxygen, carbon, hydrogen, nitrogen, and phosphorus) is shown. Each component of the nucleotide, as well as the nucleotide itself, is simplified here using the representative icons for future reference.

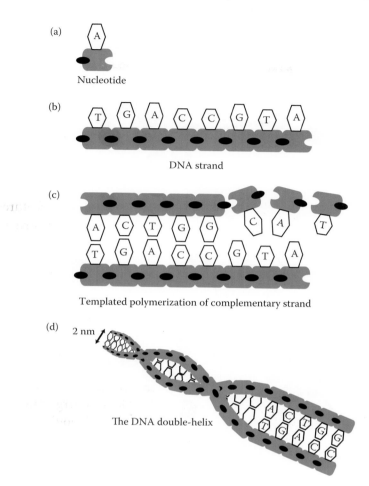

FIGURE 10.11 DNA. (a) The building blocks of DNA are nucleotides. Each nucleotide is made from three parts—a sugar, a phosphate, and a base. The base can be adenine, guanine, cytosine, or thymine (abbreviated A, G, C, or T). (b) The sugar–phosphate molecules link together with covalent bonds into a chain, forming DNA's backbone, with the bases poking out. You can see that each sugar–phosphate molecule in the chain is not symmetric; this endows the DNA strand with a polarity that governs the direction in which the code will be read—just as English is read from left to right. (c) The original, single strand serves as a template for the second. Because only T can pair with A, and only G with C, the nucleotide sequence of the second strand is said to be complementary to the first. This second strand also has an opposite polarity. The base-pairs are held together by weak hydrogen bonding (see Figure 10.12). (d) A DNA double helix.

DNA serves not only the purpose of archiving the genetic code, but it must also *use*, or express, this code. DNA guides the making of other molecules inside cells. Ribonucleic acid (or RNA) is very similar to DNA. During a process known as transcription, RNA makes a complementary copy of single-stranded DNA. This copy is expendable and used to guide the synthesis of proteins. DNA's incredibly long chain contains specific sequences of base-pairs, like segments of code. The segment that corresponds to one particular protein is known as a gene, and there are thousands of genes. It gives one a sense of kinship

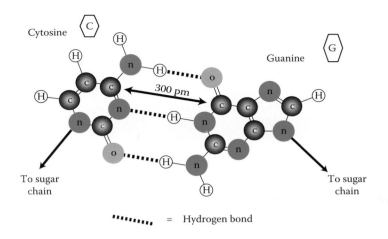

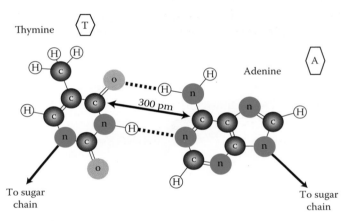

FIGURE 10.12 The structures and bonding configurations of DNA base-pairs. The four bases that protrude from the sugar–phosphate backbone of DNA (see Figure 10.11) are adenine (A), guanine (G), cytosine (C), and thymine (T). Their structures govern their pairings: A binds to T, and C binds to G. Each of these bases contains oxygen (O), carbon (C), hydrogen (H), and nitrogen (N). The bonds between the base-pairs are hydrogen bonds—a type of van der Waals bond, weaker than the covalent bonds that hold together DNA's sugar–phosphate backbone. Because of this, the bases can be pulled apart and the double helix can unwind without breaking the backbone and losing the sequence, to create two single strands of DNA. It is important that the sequence on this single strand be maintained for DNA replication and transcription.

with other forms of life to learn that a core set of over 200 genes is common to every living thing on Earth.

The applications of DNA for nanodevices are broad. DNA is conductive and can be used as a wire. It can be a catalyst. Or it can be used as part of a sensor to detect genetic diseases—a single strand of DNA "captures" complementary strands having a particular sequence that corresponds to the disease, for example. In self-assembly, DNA is an ideal functional material. Think of complementary strands of DNA as highly selective Velcro. By attaching strands of DNA to a particle, for example, and the complements of these strands to a particular location on a substrate, the DNA can be used to ensure that the particle will adhere to that location and no other. DNA can be used similarly to position and orient an object a certain way during construction of a nanometer-scale device. In addition, a DNA computer in which "calculation" is replaced by "molecular operation" based on a chemical reaction has been proposed. Imagine a spoonful of solution containing molecules of data to be calculated, a software molecule that dictates the program routine, and a hardware molecule to physically perform the calculations. A unique sequence of DNA can further be used as a tag for discrimination or personal identification—a type of DNA ink might someday be used to verify the true heirs in a will.

10.3.3.2 Adenosine Triphosphate

Everything cells do requires energy, and that energy is principally carried by the nucleotide called adenosine triphosphate (ATP). (Adenosine is made from a combination of adenine, which is one of the four bases in DNA, and ribose, a sugar.) The structure of ATP is given in Figure 10.13. This specialized molecule is the most convenient and efficient place to store and later deliver energy within living cells. Hundreds of cellular reactions require ATP. For example, it supplies energy to the many pumps transporting matter in and out of the cell and fuels the molecular motors that help muscles contract. It generates electricity as nerve cells transport messages. It is used to illuminate fireflies. Many crucial molecular interactions where two molecules are joined together are unfavorable from an energy conservation standpoint; without ATP, these reactions never occur. Life literally does not go on without ATP.

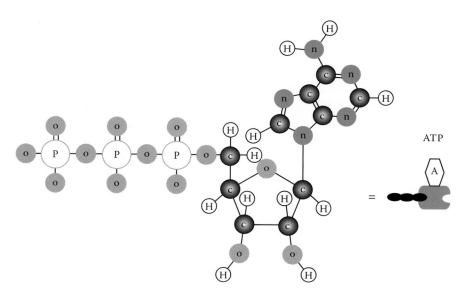

FIGURE 10.13 Adenosine triphosphate, ATP. Like all nucleotides, ATP has a base (adenine), a sugar (ribose), and at least one phosphate molecule (as its name suggests, it has three). For discussion purposes, we can simplify the chemical structure of nitrogen, carbon, oxygen, hydrogen, and phosphorus atoms to the ATP icon shown.

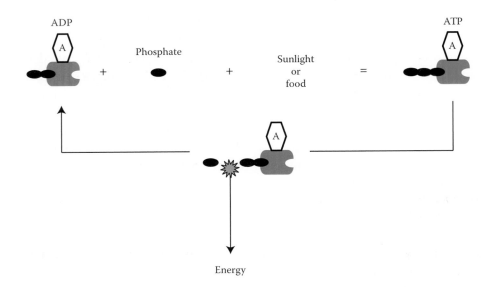

FIGURE 10.14 Like a battery, the ATP nucleotide can store and release energy to the cell. The reaction that forms ATP from ADP requires an extra phosphate molecule and energy. Animal cells get this energy from food they eat whereas plants get it from light (photons). Once ATP is formed, a rupture of the bond that holds its outermost phosphate molecule causes a release of energy that can be used by the cell. The by-product of this energy release is ADP, and the process starts again.

ATP stores its energy in the bonds that grip its outermost phosphate molecule. As we can see in Figure 10.14, the rupturing of this covalent bond releases energy that the cell can use (about 7 kcal per mole of ATP). The nucleotide left behind after the phosphate is gone is adenosine diphosphate (ADP), which has two phosphates instead of three. It is not much of a stretch to draw an analogy between ATP and rechargeable batteries, which store their potential energy for later use as heat and usable energy. Like a dead battery, ADP must be "recharged" with new energy into ATP. Such being the case, engineers would like to someday harness the rechargeable energy of this little nucleotide battery for use in nanoscale devices.

The creation of ATP from ADP and its subsequent conversion to energy happen often. For humans, an average daily intake of 2500 food calories (equivalent to 2500 kcal) leads to the turnover of about 400 pounds of ATP that day. What is even more amazing is that the body only has about one-tenth of a pound (50 g) of ATP in it at any one time. The same molecules are recycled over and over.

In 1945, almost the entire process of ATP synthesis was a scientific mystery. We know a great deal more today. The 1997 Nobel Prize in Chemistry was awarded to Paul Boyer, John Walker, and Jens Skou for discovering the mechanism by which ATP is synthesized. This process is carried out by a protein known as ATP synthase, which we will discuss in the next section.

10.3.4 Amino Acids Are Used to Make Proteins

All amino acids are built around a single carbon atom known as the α-carbon. Because carbon has four valence electrons, it can form four covalent bonds. The α-carbon forms

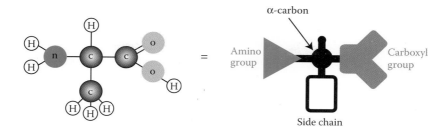

FIGURE 10.15   An amino acid. The 20 types of amino acid found in all life forms are made when a central carbon atom, called the α-carbon, forms bonds with a carboxyl group (–COOH), an amino group (H₂N), a hydrogen atom (H), and a side chain. The amino acid shown here is alanine, on which the side chain is CH_3.

one with a carboxylic acid group, one with an amino group, one with a side chain, and one with a hydrogen atom. This configuration is shown in Figure 10.15.

Amino acids have one monumental role: they link together as chains to make proteins. Because there are 20 different common side chains, there are 20 common amino acids used to make all the proteins in all the bacteria, plants, and animals on Earth. The chains are usually between 50 and 2000 amino acids in length. The largest amino acid is tryptophan, measuring about 670 pm across. The smallest is glycine, measuring about 420 pm.

Proteins are very large molecules, some with molecular masses as heavy as 3,000,000 grams per mole (g/mol)—compared with, say, 18 g/mol for water. They fold into beautifully intricate three-dimensional structures, depending on the order of amino acids in their chain. Like the other molecules we have been studying, water forces proteins into specific, compact shapes. The varying amino acid side chains along the length of the chain can be either hydrophobic or hydrophilic, so the chain will scrunch up and fold into a particular shape to keep the hydrophobic side chains at the core and the hydrophilic ones exposed.

Protein function closely follows form: reactive sites on the surfaces are shaped so as to bind with specific molecules and catalyze reactions in which covalent bonds are broken. Almost every chemical reaction that takes place inside living things is catalyzed by proteins. They also serve other functions, such as transporting nutrients, maintaining structures, creating movements, and detecting signals, making them of utmost importance and also of great interest for use in nanomechanical devices. For engineers seeking to build new, nanoscale methods for storing information, precise catalysis, and signal transduction at a molecular scale, proteins provide proof of concept. They already perform tasks reminiscent of machines.

Take, for example, a pair of remarkable motor proteins, myosin and kinesin, shown in action in Figure 10.16. Both of these proteins are linear motors that move along cellular filaments. They work like sailors hauling in a rope, heaving again and again, each time reaching out for a new grip. The difference is that myosin uses one "hand" and kinesin uses two.

(a)

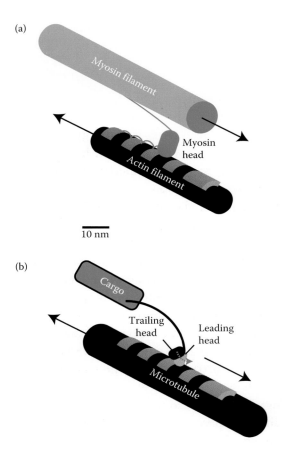

10 nm

(b)

FIGURE 10.16   Two molecular motors: the proteins myosin and kinesin. (a) Myosin's head attaches and detaches from binding sites along cellular filaments known as actin. Muscles work this way, with parallel filaments tugged by armies of myosin in opposing directions. (b) Kinesin has two heads and a tail attached to cellular cargo. It walks along the binding sites on a microtubule just like a pair of feet. Both motors use one ATP molecule per "step."

The myosin head measures 8.5 nm. It attaches and detaches, over and over again, from binding sites about 10 nm apart along cellular filaments called actin. As the head moves one binding site at a time with the help of energy it gets from a single ATP molecule, it attains speeds of between 0.2 and 60 μm/s. Myosin's tail connects to a thicker filament. This is how your muscles work: parallel filaments are pulled in opposite directions to contract your biceps or your heart.

Kinesin has two heads and a tail end that can be attached to large cellular cargo such as organelles (cell's organs) in order to transport them from one end of the cell to the other faster than would be possible by diffusion alone. The heads walk like feet, alternately attaching and detaching from binding sites about 8 nm apart along filaments called microtubules. Each step uses one ATP molecule. Kinesin can tow cargo at speeds of 0.02–2 μm/s.

Another astounding nanoscale protein machine found inside of cells is ATP synthase, which we will discuss in more detail in the next section.

10.3.4.1 ATP Synthase

As we have discussed, ATP is the universal currency of energy in living things. It is made with the help of a miraculous protein known as ATP synthase (or F_0F_1ATPase). This protein is truly a nanoscale machine, found in special energy-converting cellular organs called mitochondria. ATP synthase measures about 10 nm in diameter and weighs more than 500,000 g/mol. On the top of the protein is an electric motor (called F_0). The bulbous mass on the bottom is frequently described as a "lollipop head," and is made from a ring of six protein subunits (see Figure 10.17). The lollipop head is bound in place by a long arm, or stator, that extends down from the mitochondria membrane. The stator is in contact with the rotor of the electric motor, which is a ring of 10–14 protein subunits.

There are more protons outside the mitochondria than inside, and this gradient causes protons to flow across the membrane, through the narrow space separating the stator from the rotor. This charge migration makes the rotor ring spin. The attached axle spins with it. This motion can actually be observed. Scientists have "unbolted" these nanoscale motors from their moorings in the cell membrane, mounted them on treated surfaces, and put them to work. By attaching beads and filaments to the axle, it is possible to determine how fast it rotates using microscopes and ultra-high-speed cameras. Rotation rates of 480 rpm all the way up to 21,000 rpm have been measured.

During these rotations, proteins on the axle rub against the stationary ring of proteins in the lollipop head. The flow of protons has effectively been converted into mechanical energy in the electric motor. The lollipop head then behaves like a generator. Three of its six subunits have binding sites for ADP and a phosphate molecule. As we learned in the previous section, it is these two ingredients, plus energy, that are used to make ATP. The rubbing and reorienting of the proteins as the axle spins provides mechanical energy that can be converted into chemical bond energy, binding the crucial third phosphate to the ADP and creating ATP. Three or four protons must pass through ATP synthase to make one molecule of ATP, and it can pump out more than 100 ATP molecules per second. Reported measurements of the energy efficiency of this miraculous motor exceed 80%.

BACK-OF-THE-ENVELOPE 10.4

ATP synthase generates sufficient torque to produce three ATP molecules per revolution. This amount of work equates to about 20 $k_B T$ per ATP molecule, where k_B is the Boltzmann constant and T is the temperature in Kelvin. What is the estimated power output of one ATP synthase molecule at the temperature of the human body, 37°C?

$$\text{Power} = \frac{\text{Work}}{\text{Time}} = \frac{20\left(1.38 \times 10^{-23}\, m^2\, \dfrac{kg}{s^2 K}\right)(310K)(100\,\text{molecules})}{1\,s} = 8.56 \times 10^{-18}\, W$$

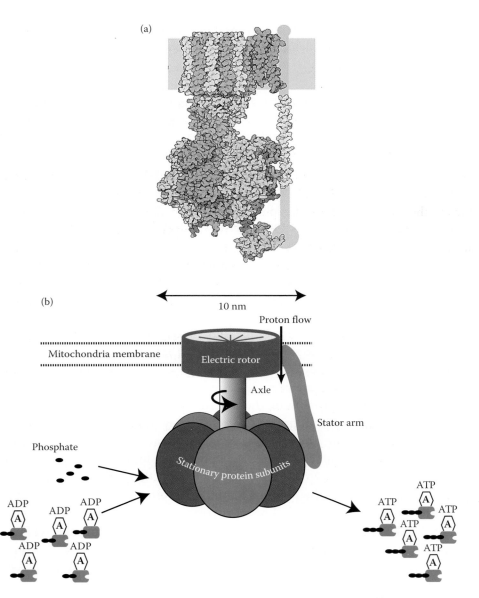

FIGURE 10.17   ATP synthase. (a) Here we see the three-dimensional structure of this indispensable protein, which builds the ATP to power living things. (b) A simplified drawing shows the primary components of ATP synthase. The lower portion contains a ring of six protein subunits, sometimes referred to as a "lollipop head." Fixed to this ring is a stator arm that reaches up into the mitochondrial membrane and makes contact with a rotor, formed by another ring of protein subunits. As protons flow through the narrow gap separating the stator and the rotor, the rotor turns and the axle turns with it. This causes proteins on the axle to rub against the stationary proteins in the lollipop head, half of which (indicated in blue) contain binding sites for ADP and phosphate. The rubbing introduces energy used to bind ADP and phosphate to make ATP. (The image in (a) is adapted from an image courtesy of the RCSB Protein Data Bank (http://www.pdb.org) Molecule of the Month by David S. Goodsell of The Scripps Research Institute.)

10.4 APPLICATIONS OF NANOBIOTECHNOLOGY

Now that we have looked at the machinery within a cell and have a better idea of how it works, how are we going to use any of it to create new technology? The machinery that makes up nature is so mature that it makes sense to use some of the existing tools to create new bio-tools. In this section, we will discuss a couple of examples of how we use existing biological machinery to create novel products, such as in the pharmaceutical and cosmetic industries.

10.4.1 Biomimetic Nanostructures

Water, as we have seen, is a predominant compound in all biological systems, and it is the hydrophobic effect that makes proteins and lipids fold into functional forms and also bind to one another. What if we use this property of water, along with the molecules in lipids and proteins, to create devices that "mimic" real biological functions? That is, we can use what we know about hydrophobicity and water to create biomimetic nanostructures through self-assembly in water. Such devices are known as *biomimetic nanostructures*.

One biomimetic nanostructure that has been developed is called the liposome, or polysome. Researchers have used micelle structures to create sacs capable of encapsulating certain biomolecules, or even drugs for drug delivery applications. (For more on micelles, see Chapter 4, Section 4.4.6.3 and Figure 4.23; the formation of a sac is shown in Figure 10.9.) Development of liposomes has been most prominent in the cosmetic and pharmaceutical industries. Liposomes are very much like fatty acids, but made of synthetic polymers. Recently, they have been used to carry anticancer drugs into the body and find the exact cells they need to affect—a much less expensive, more sterile, and more stable approach than current cancer medications. For cosmetic industries, nanostructures called novasomes are present in many of the topical creams we put on, and are used as emulsifiers to get the texture of the cream just right. In addition, researchers have tried to mimic the filamentous machinery of the cell, like cellulose, to create rod-like aggregates of material for strengthening materials such as adhesives.

10.4.2 Molecular Motors

Cells also have this amazing ability to move. Biomolecular motors are proteins that produce or consume mechanical energy. Motor proteins working together can exert forces as great as kiloNewtons—enabling behemoths such as whales and elephants to move. Motor proteins can also work individually, exerting picoNewton forces, to transport single molecules between different parts of a cell. And, as we are now struggling to make mechanical things "move" on a nanoscale, why not use what has been developed in the cell already? This field of study is concerned with trying to use the cell's motor proteins to create mechanical devices that can move: molecular motors.

Motors of many different shapes, sizes, speeds, efficiencies, functions, and designs already exist in nature. Motor proteins can be rotary motors, linear steppers, and screws. They can range in size from one to hundreds of nanometers. Right now, conventional "top-down" microelectromechanical systems (MEMS) processing cannot compete; however, there are some aspects of MEMS that can be used in combination with these biomolecular motors to create the most efficient device. In fact, the best of both worlds may result in some of the most useful applications of biomolecular motors.

10.5 SUMMARY

Nanoscale technology is already wildly successful in living things. Regardless of whether we find ways to harness this nanobiotechnology for our own uses, the molecular machinery of organisms is still technology. We just did not invent it.

The bioworld is different from the one in which engineers typically operate. Water is everywhere, and therefore hydrogen bonding plays a crucial role. Because so many organic molecules consist of both hydrophilic *and* hydrophobic parts, they rearrange themselves so as to take whatever shape is most energetically favorable. Hydrophobic parts are sequestered into the core of the molecule, leaving the hydrogen bonding to the hydrophilic parts on the outside of the molecule. Because the shapes of biological molecules often dictate their function within a cell, this simple love/fear relationship with water makes a huge impact.

Inertia and gravity are irrelevant compared to viscous forces at the cellular level. Brownian motion, which is the random motion of nanoscale objects due to thermal energy, keeps the many parts of a cell constantly bumping into one another and moving from one end of the cell to the other—enabling a "just-in-time-delivery" logistical solution for the many assembly lines running at all times inside a cell.

Although most cells contain thousands of different kinds of molecules, we can effectively organize this diverse set into four molecular families: sugars, fatty acids, amino acids, and nucleotides. Sugars, such as glucose, provide energy but can also be used to create structures like the cellulose in plant cell walls—the most abundant organic chemical on Earth. Fatty acids are used in cellular membranes, but can also store energy—as much as six times more per mass than sugars, although this energy "burns" more slowly. Amino acids are crucial because they make proteins—some of which, such as myosin and ATP synthase, are perfect nanoscale versions of linear and rotary motors. Finally, nucleotides are used to build some of life's most important molecules, including information-carrier DNA and energy-carrier ATP.

Finally, we touched on a couple of applications of nanobiotechnology, including biomimetic structures and biomolecular motors. Although these fields of study are both in their infancy, there is an open door of possibilities in using existing biological structures for technological applications.

HOMEWORK EXERCISES

10.1   What percentage of a cell's weight is water?

10.2   What dipole–dipole interaction gives water its unique properties and most of the "machines" inside a cell their shapes?

10.3   What types of molecules tend to be hydrophilic?

10.4   What atom is often found hidden in the core of cellular molecules, where interactions with water can be minimized?

10.5   Viscous friction causes a particle under the force of gravity to eventually fall with a constant velocity, v_{lim}, sometimes called terminal velocity, as long as the flow past the sphere is laminar (not turbulent).

a. Use the equation for terminal velocity in Chapter 2, Section 2.1.6 to determine the limiting velocity of a relatively spherical (1 μm diameter) *E. coli* bacterium falling through air. Assume the bacterium to have the same density as water. The viscosity and density of air are 1.8×10^{-5} N s/m^2 and 1.29 kg/m^3, respectively, and the density of water is $\rho_{water} = 1000$ kg/m^3.

b. How does this velocity compare to the velocity at which flagella propel the *E. coli* through water?

10.6 The random motion of molecules within a cell is driven by energy from heat.

a. At what temperature do most cells prefer to live?

b. The calorie is a unit of energy defined as the amount of energy needed to raise 1 g of water by 1°C. A bacterium (which is mostly water) weighs about 1 pg. Roughly how many grams of sugar does a cell have to use to warm back up to its preferred temperature if it is suddenly placed in a cooler environment that lowers the cell's internal temperature by 5°C?

c. If the sugar used for energy is glucose (molecular mass = 180.16 g/mol), how many molecules are used?

d. How much mass would be used as fuel to warm the cell if, instead of sugar, fatty acids served as the energy source?

10.7 Doubling the characteristic dimension of a cell means that a given molecule, when introduced inside the cell, will take _____ times longer to be found at any other location within the cell's volume.

10.8 Approximately how often do a pair of molecules, each 9 nm in diameter, each with a diffusion coefficient of 8^{-10} m^2/s in water, come in contact with one another inside a bacterium 5 μm in diameter?

10.9 True or false? Brownian motion is the sole means of transporting molecules inside of cells.

10.10 Name the four molecular families into which we can group the thousands of different kinds of molecules in a cell.

10.11 The human body's main energy source is _____.

10.12 What are the three products of glycolysis?

10.13 What chemical group found on sugars enables them to so readily form multiple linkages and large structures?

10.14 What polysaccharide tends to be larger: cellulose or glycogen? Which is more abundant?

10.15 True or false? Fatty acids are surfactants and form structures known as micelles.

10.16 What are the only two differences among the hundreds of fatty acids?

10.17 Which structure is more favorable from a conservation of energy standpoint: a planar phospholipid bilayer with its edges exposed to water or a sealed sphere with no exposed edges?

10.18 A phospholipid molecule is at least how much more likely to change places with another phospholipid in its own layer versus a switching to the opposing side of the bilayer?

10.19 What three molecules do DNA and ATP have in common?

10.20 Determine the complementary sequence to the following segment of single-stranded DNA: ATTCGGTAATTCTGC

10.21 Codons are like the words in DNA that govern which amino acids will be in a protein. How many three-letter words can be made using the four bases—adenine, cytosine, guanine, and thymine?

10.22 True or false? DNA is poorly suited to self-assembly processes.

10.23 There are 3.4 billion base-pairs lined up end to end in one human DNA molecule, which contains the entire genetic code for the makeup of an individual person. Each base-pair measures approximately 350 pm in length, and almost every cell in the body contains two complete DNA molecules.

   a. If one DNA molecule is stretched end to end, how far would it stretch?

   b. The human body is comprised of approximately 50 trillion cells. Assume all of these cells contain two complete DNA molecules. If the DNA in all of these cells were stretched out in one continuous strand, how many round trips could it make to the sun (93 million miles away)?

10.24 The useable energy in an ATP molecule is stored in the bond with its third phosphate.

   a. How many times during an average day is the third phosphate added and removed from a single ATP in your body?

   b. Approximately how many ATP molecules are there are in your body? (ATP = 507.18 g/mol.)

   c. How much energy (kcal) is created during an average day by the ATP in your body?

10.25 True or false? Covalent bonds are what bind the third phosphate to ADP to form ATP, as well as the backbone of DNA.

10.26 The width of a strand a DNA is _____ a single-wall carbon nanotube.

   a. About the same as

   b. Much smaller than

   c. Much larger than

10.27 The _____ is the segment of DNA sequence corresponding to a single protein.

RECOMMENDATIONS FOR FURTHER READING

1. B. Alberts, A. Johnson, J. Lewis, M. Raff, K. Roberts, and P. Walter. 2002. *Molecular Biology of the Cell*, 4th Edition. Garland Science.

2. David S. Goodsell. 2004. *Bionanotechnology: Lessons from Nature*. Wiley-Liss.

3. C. Niemeyer and C. Mirkin. 2004. *Nanobiotechnology: Concepts, Applications and Perspectives*. Wiley-VCH.

4. V. Vogel and B. Baird. 2003. *Nanobiotechnology: Report of the National Nanotechnology Initiative Workshop*. October 9–11, 2003, Arlington, VA.

Glossary

Absorption: The intake of molecules or particles of one substance (typically a liquid or gas) through minute pores or spaces in a second substance (typically a liquid or solid).

Acoustic phonon: A type of phonon present in all solid materials that carries sound waves through the crystal lattice.

Adenosine triphosphate (ATP): A nucleotide made of adenine, ribose, and three phosphate groups that serves as the principal carrier of chemical energy in living cells.

Adhesion: The tendency of dissimilar molecules to cling together due to attractive intermolecular forces (as opposed to cohesion, between similar molecules).

Adsorption: The binding of molecules or particles to a surface.

Advection: Transport of a substance by a fluid due to the fluid's bulk motion (flow) in a particular direction.

Amino acid: A molecule with an amino group and a carboxyl group. Amino acids are the building blocks of proteins.

Angstrom (Å): A unit of length equal to 0.1 nm, or 1×10^{-10} m.

Aspect ratio: The ratio of an object's long dimension to its short dimension (e.g., a wire's length divided by its diameter).

Assembler: A nanoscale machine capable of positioning atoms and molecules into useful configurations. Some consider certain biological systems to be assemblers. There is debate as to whether manmade assemblers are possible.

Atom: The basic unit of matter. The atom is the smallest component of an element having the chemical properties of that element.

Atomic force microscope (AFM): A type of scanning probe microscope (SPM) that detects atomic forces acting on a sharp probe tip as it interacts with the surface of a sample. The AFM can be used to generate topographic images, measure friction, manipulate surface atoms, and measure forces between molecules.

ATP synthase: The protein that makes adenosine triphosphate (ATP).

Backbone chain group (see also SAM): One of the three components of a self-assembled monolayer (SAM)—the other two being the tail group and the head group. The backbone chain group is the spacer between the head group (attached to the substrate) and tail group (the free surface of the SAM) and defines the thickness of the layer.

Band gap: The gap in allowed energy levels separating the top of the valence band and the bottom of the conduction band in semiconductors and insulators.

Binding energy: The strength of the chemical bond between a pair of atoms, determined by the energy needed to break it.

Biomimicry: The imitation of systems found in nature when designing new technology.

Boltzmann's constant, k_B: A physical constant relating thermal energy to temperature, $k_B = 1.38 \times 0^{-23}$ J/K.

Bottom-up engineering (fabrication): Creating something by starting with the smallest units of material (atoms) first, and building up to the final product—as opposed to top-down engineering.

Bravais lattices: The 14 types of crystal lattice. All crystalline materials assume one of these arrangements.

Brownian motion: The random motion of particles or molecules, driven by thermal energy.

Buckminsterfullerene (a.k.a. buckyball, C_{60}): A hollow, spherical fullerene made from 60 carbon atoms. It was named in honor of Buckminster Fuller, the architect who popularized the geodesic dome the molecule resembles.

Catalyst: A substance that reduces the energy required for a chemical reaction, thus increasing the reaction rate.

Cellulose: A chain made with thousands of glucose molecules that provides tensile strength in cell walls. It is the most abundant organic chemical on earth.

Characteristic dimension: A representative size (length) of something, chosen for comparison or simplifying purposes.

Chemical vapor deposition (CVD): A processing technique typically used for coating an object with a thin film. A common method involves flowing gas over a heated object. Chemical reactions on or near the hot surfaces result in the deposition of a thin film coating.

Chirality: Here, this refers to the three possible twist configurations of a nanotube—armchair, zigzag, and chiral. The chirality of a nanotube dictates some of its properties, including whether it is a semiconductor or a conductor.

Cohesion: The tendency of similar molecules to cling together due to attractive inter-molecular forces (as opposed to adhesion, between dissimilar molecules).

Colloid: A mixture of two phases of matter, typically solid particles dispersed in a liquid or gas.

Conduction band: The range of electron energies sufficient to free an electron from its individual atom and allow it to move freely throughout a material.

Conductor: A material with mobile electric charges (such as electrons or excitons). In a conductor, the conduction and valence bands overlap, so there is no band gap and electric current flows freely.

Continuum: A concept used to understand fluid behavior in which the fluid's properties are assumed to be the same throughout a given volume and the fluid to be infinitely divisible.

Convergent synthesis: Here, this refers to a technique used in creation of dendrimers in which preformed molecular wedges are coupled together and bonded to a central core. The opposite approach is divergent synthesis.

Correspondence principle: The principle that quantum mechanics and classical mechanics predict similar behavior in the case of large groups of atoms.

Coulomb blockade: The resistance to electron transport caused by electrostatic Coulomb forces in certain electronic structures, including quantum dots and single electron transistors.

Covalent bond: The connective link between atoms that are sharing electrons.

Crystal lattice: The geometric arrangement of the points in space at which the atoms, ions, or molecules in a crystal are located. When repeated over and over again, this arrangement forms a pattern, the smallest unit of which is called the unit cell.

Debye length, λ_D: The characteristic length of the diffuse layer of mobile ions in an electrolyte liquid flowing in a channel.

Dendrimer: A large artificial molecule made using branch structures of smaller molecules extending out from a common core.

Density of states function, $D_s(E)$: A mathematical function that gives the number of available electron energy states per unit volume, per unit energy, in a solid.

Deoxyribonucleic acid (DNA): A double-helix molecule made from nucleotides, used by living cells to store and pass on hereditary information.

Diffusion: The intermingling of molecules driven by thermal energy, causing the molecules to spread out from areas of higher concentration to areas of lower concentration.

Dipole–dipole force: The attractive van der Waals force between the positive end of one polar molecule and the negative end of another polar molecule. (This force is sometimes called the orientation force or Keesom force.)

Dipole-induced–dipole force: The attractive van der Waals force that results when a polar molecule induces a dipole in a nearby atom or nonpolar molecule. (This force is sometimes called the induction force or Debye force.)

Dispersion force: The attractive van der Waals force that results when a pair of atoms or molecules induces synchronous and opposite fluctuating poles in each other. (This force is sometimes called the London dispersion force.)

Divergent synthesis: Here, this refers to a technique used in creation of dendrimers in which molecular branches are added outward from a central core. The opposite approach is convergent synthesis.

Doping (semiconductor): The process of adding impurities into a semiconductor crystal in order to change its electrical properties.

Elastic scattering: A type of scattering in which the kinetic energy of the incident particle (such as a photon) is conserved, but the particle's direction is changed.

Electric double layer (EDL): The two layers of ions that form in a fluid near a solid surface—namely the fixed, Stern layer found against the surface and the mobile, diffuse layer found above it.

Electrokinetic flow: A type of fluid flow enabled by the electric double layer and useful for moving fluids at the nanoscale. The four major types of electrokinetic flow are: electrophoresis, electroosmosis, streaming potential, and sedimentation potential.

Electromagnetic radiation: Energy in the form of self-propagating waves. These waves can travel through both vacuum and matter, and are made of oscillating electric and magnetic fields perpendicular to each other and to the direction the wave oves.

Electromagnetic spectrum: The classification of electromagnetic radiation according to the frequencies (or wavelengths) of the electromagnetic waves.

Electromigration (fluids): A type of flux driven by charged ions moving through a flowing fluid.

Electron: A subatomic particle with negative electric charge. Electrons play defining roles in the structural and behavioral properties of matter.

Electron density, n_e: The number of free electrons per unit volume.

Electron-beam lithography (EBL): A fabrication technique that creates features on a surface using a focused beam of electrons to expose an electron-sensitive resist layer.

Electroosmosis: A type of electrokinetic flow in which fluid is driven by an applied electric field.

Electroosmotic pump: A pump consisting of a glass capillary packed with glass particles, creating a porous structure with numerous nanoscale flow channels. Electroosmotic flow is generated when an electric field is applied across the capillary. Or, a pressure gradient is formed when the flow is prohibited.

Electrophoresis: A type of electrokinetic flow in which charged bodies are moved through a fluid using an applied electric field.

Energy band: A defined range of energy that an electron can have in a solid. Between energy bands are band gaps corresponding to forbidden energy ranges.

Equilibrium separation: The distance at which the attractive and repulsive forces between a pair of interacting atoms balance for a net force of zero.

Exciton: The electron–hole pair created when an electron moves into the conduction band and becomes mobile, leaving behind an absence in the crystal lattice. The positively charged hole and the negatively charged electron attract one another.

Extinction coefficient, C_{ext}: A measure of light absorption defined as the fraction of photons lost per unit distance in the direction of propagation through a medium.

Fatty acids: Molecules with a carboxyl group head and a long hydrocarbon tail used by living cells for structure and also energy.

Fermi energy, E_F: The theoretical topmost occupied electron energy state in a material at absolute zero temperature.

Field effect transistor (FET): A switching device that uses an electric field to selectively enable the flow of electric current through a semiconducting channel.

Forced oscillation: The motion of an oscillating system acted upon by an external driving force.

Free oscillation: The motion of a oscillating system not acted upon by an external driving force.

Fullerenes: One of three crystalline forms of carbon. Fullerenes are hollow molecules in the shape of spheres, ellipses, or tubes such as the buckminsterfullerene (a.k.a. "buckyball" or C_{60}) and the nanotube.

Functionalization: The modification of a surface for a specific purpose. Commonly this entails attaching molecules of a specific chemical group to a surface to endow it with desired properties.

Gamma rays: The highest frequencies of electromagnetic radiation (typically greater than 10^{18} Hz).

Giant magnetoresistance (GMR): A phenomenon found in alternating layers of magnetic and nonmagnetic metals. Applying a magnetic field causes a localized change in electrical resistance that can be coded as a 1 or 0 and used to store information in computers.

Glucose: A sugar molecule that when broken down yields adenosine triphosphate (ATP). Glucose is crucial to the metabolism of living cells.

Glycogen: A molecule made from tens to hundreds of glucose rings. Glycogen is used for long-term storage of sugar energy in animals and humans.

Graphite: Covalently bonded carbon atoms arranged in flat sheets. Parallel sheets are held together by van der Waals forces.

Gray goo: An imagined substance created in a catastrophic scenario in which self-replicating assemblers convert everything they encounter into copies of themselves, consuming the earth and creating a thick, grayish liquid.

Head group (see also SAM): One of the three components of a self-assembled monolayer (SAM). The head group forms a chemical bond that binds the SAM to the substrate surface.

Heat carrier: One of the four nanoscale things that transport heat: atoms, electrons, photons, and phonons.

Hydrogen bonding: A type of dipole–dipole interaction in which a positively charged hydrogen atom forms a noncovalent bond with nearby, negatively charged atoms. Hydrogen bonding among H_2O molecules gives water its unique properties.

Hydrophilic: "Water loving." Used to describe molecules (or parts thereof) that are attracted to water, or materials that are soluble in it.

Hydrophobic: "Water fearing." Used to describe molecules (or parts thereof) that are repulsed by water, or materials that are insoluble in it.

Inelastic scattering: A type of scattering in which both direction and the kinetic energy of the incident particle (such as a photon) are changed.

Infrared: A portion of the electromagnetic spectrum with frequencies on the order of 10^{11}–10^{14} Hz.

Insulator: A material with minimal free electrons due to the wide band gap separating the valence and conduction bands. The material therefore resists the flow of electric current.

Interaction energy: The change brought about in the total energy level of atoms or molecules as they approach one another.

Interaction forces: The electrostatic forces that arise between atoms and molecules as they approach one another. These forces are often collectively called van der Waals forces.

Interband absorption: The phenomenon in which a semiconductor or insulator absorbs a photon, providing energy enough to boost an electron across the band gap into the conduction band.

Intraband absorption: The phenomenon in which a conductor absorbs a photon, providing energy that moves an electron slightly higher within its own band.

Ion: An atom or molecule with a positive or negative charge (due to a mismatch in the numbers of electrons and protons). Negative ions are called anions; positive ones are called cations.

Ionic bond: The connective link between oppositely charged atoms.

Kinetic energy: The energy of an object due to its motion.

Lab on a chip: A miniaturized version of laboratory equipment (such as for fluid analysis or disease screening) built on a microchip.

Laser: The acronym for Light Amplification by Stimulated Emission of Radiation (LASER). A laser uses this process to produce an intense beam of light. (See, e.g., quantum well laser.)

Lennard–Jones potential: A mathematical relationship used to describe the potential energy of a pair of atoms or molecules as a function of separation distance.

Liposome: A microscopic, fluid-filled sac whose walls are made using phospholipids like those in cell membranes. Liposomes can be used to hold and deliver medicine.

Long wave: The lowest frequency range of electromagnetic radiation (typically lower than 10^5 Hz).

Macromolecule: A large-scale molecule with a molecular mass greater than a few thousand daltons, such as a protein, nucleic acid, or polysaccharide. (*Note:* 1 Dalton equals about 1.7×10^{-24} g.)

Macroscale: Here, the dimensional range of about 1 mm and larger.

Mean free path, Λ: The average distance a heat carrier travels before losing its extra energy due to scattering.

Metallic bond: The connective link among metal atoms created by the "sea" of electrons moving throughout the lattice.

Micelle: Self-assembling structure formed from surfactant molecules.

Microcantilever: A microscale beam supported at one end and free at the other.

Microelectromechanical systems (MEMS): Microscale machines that can include mechanical elements, sensors, actuators, and electronics. MEMS are made using microfabrication techniques.

Microfluidics: The study of fluid behavior at the microscale (i.e., volumes thousands of times smaller than a common droplet), where fluids tend to be increasingly viscous.

Micrometer (micron), μm: A unit of length equal to 1×10^{-6} m.

Microscale: The dimensional range from about 1 μm to 1 mm.

Microwave: A portion of the electromagnetic spectrum with frequencies on the order of 10^8–10^{12} Hz.

Miniaturization: The trend toward ever-smaller products and devices.

Molecular electronics: The use of molecules as electronic devices in a circuit, as opposed to using more traditional electronic materials like silicon.

Molecular motor: Molecular-scale machines found in nature that convert energy into motion or work, such as the motor proteins myosin and kinesin.

Molecular recognition: A specific interaction between molecules via noncovalent bonding. This enables biological and chemical systems to distinguish molecules depending on how they "fit" together.

Molecule: Two or more atoms held together by covalent bonds.

Moore's Law: Intel co-founder Gordon Moore's prediction, made in 1965, that the number of transistors on a chip would double about every two years. This "law" has so far held true.

Nano: Metric prefix meaning one billionth, derived from the Greek word, *nanos*, for dwarf.

Nanobiotechnology: The study and application of nanotechnology already present in living things.

Nanoelectronics: The study and application of electronic behavior at the nanoscale in order to design, build, and use nanoscale circuits.

Nanomaterial: A material or structure with unique and useful features owing to its nanoscale size.

Nanomechanics: The study and application of motion and the forces that cause motion at the nanoscale in order to design, build and use nanoscale machines.

Nanometer, nm: A unit of length equal to 1×10^{-9} m.

Nanophotonics: The study and application of nanoscale interactions between photons and materials.

Nanoscale: The dimensional range from about 1 μm (1000 nm) down to 1 angstrom (0.1 nm).

Nanotechnology: Making use of the unique physical properties and interactions of nanoscale things in order to create novel structures, devices, and systems.

Nanotube: Hollow, cylindrical fullerenes with special mechanical, electrical, and material properties. Single-walled tubes are about 1.5 nm in diameter and typically a few hundred nanometers long.

Natural frequency, f_n: The frequency at which a simple harmonic system oscillates. (*Note:* Lower case n is used as the subscript in f_n to indicate the natural frequency. It is also sometimes used to represent the quantum number. These two n's are not the same, but it is common use for both natural frequency and quantum number to use this indication.)

Near-field light: Light found within about 200 nm (i.e., less than a single wavelength) of the object from which it was emitted or scattered.

Neutron: A subatomic particle with no net electric charge.

Newtonian fluid: An isotropic fluid whose shear stress is directly proportional to the rate of change in velocity between fluid layers. Water, air, oils, and solvents are Newtonian. Peanut butter and blood are not.

Nucleotide: A molecule made from a base, a five-carbon sugar molecule, and at least one phosphate molecule. Nucleotides are used by living cells to store information and carry energy.

Nucleus: In an atom, the positively charged central cluster of protons and neutrons.

Optical phonons: A type of phonon present in crystal lattices where the unit cell contains more than one type of atom. Optical phonons are excited by electromagnetic radiation.

Optical tweezers: One or more tightly focused laser beams used to grasp and move nano-scale objects.

Orbital: The regions of space in an atom where the electrons are most likely to be found, based on a probability function.

Particle (nanoparticle): A tiny piece of matter. Nanoparticles have diameters ranging from a few nanometers to a few hundred nanometers.

Particle-in-a-well model: A simplified mathematical model used in physics to describe the energy of a particle like a photon or electron.

Permittivity, ε: A measure of how strong an electric field a material can transmit, expressed in Farads per meter (F/m). Put another way, permittivity is the degree to which a material enables the propagation of photons through it.

Phonon: The quantum of vibrational energy.

Phospholipid: A molecule used to build membranes in living cells. Phospholipids are commonly composed of two fatty acids bonded to glycerol phosphate, which in turn is bonded to a small polar molecule such as choline.

Photoelectric effect: An energy transfer in which matter absorbs a photon and emits an electron.

Photolithography: A fabrication process in which a light-sensitive film, or photo resist, is selectively exposed to light in order to create a pattern in the film. Subsequent chemical treatments then engrave these patterns into the material below the remaining resist.

Photon: The quantum of electromagnetic radiation.

Photonic band gap: The range of photon energies forbidden within a particular material (while other energies are permitted). Photonic crystals are engineered to make use of this gap in order to route electromagnetic radiation, reflecting it or letting it pass freely where needed.

Photonic crystal: A material that can guide the motion of photons using the photonic band gap. This is analogous to the way the electronic band gap is used in semiconductor crystals to guide electrons.

Piezoelectric: A material that generates current when stress is applied, and vice versa.

Piezoresistor: A material whose conductivity changes when stress is applied.

Planck's constant, h: A physical constant equal to the ratio of a photon's energy to its frequency, $h = 6.626 \times 10^{-34}$ J/s.

Plasma: A gas made of ions. This is the fourth state of matter (the others being solid, liquid, and gas).

Plasma oscillation: The rapid oscillations of electron density in a conductor like plasma or metal. The frequency of these oscillations is the plasma frequency, f_p.

Plasmon: A coupling of a photon and a nearby electron that arises when these two are oscillating at similar frequencies and begin resonating together. The plasmon is considered the quantum of plasma oscillation.

Polariton: Typically, a coupling of a photon and a phonon that arises when these two are oscillating at similar frequencies and begin resonating together.

Polymer: A large molecule made out of smaller, covalently linked units (monomers).

Polysaccharide: A linear or branched polymer made from sugar molecules. Two key polysaccharides are glycogen and cellulose—both made from chains of glucose.

Potential energy: The stored energy of an object due to its position.

Potential well: A local minimum of potential energy.

Protein: A polymer made from amino acids linked in a specialized sequence. Proteins are the primary macromolecules in cells.

Proton: A subatomic particle with a positive electric charge.

Quantization: The process of constraining a continuous set of values to a discrete set.

Quantum (quanta): The smallest discrete unit. The plural form is *quanta*.

Quantum confinement: The effect achieved by reducing the volume of a solid so that the energy levels within it become discrete.

Quantum dot: An electrically isolated region, such as a particle or a portion of a bulk semiconductor, where electrons are constrained in all three dimensions, creating an artificial atom that exhibits quantum behavior.

Quantum harmonic oscillator: A quantum mechanical model of two atoms or molecules vibrating together, analogous to the classical harmonic oscillator of a mass connected to a spring.

Quantum mechanics: In physics, a theory describing the structure and behavior of atoms.

Quantum number: Typically, the positive integer used to specify the energy state of an atom, or of a larger system. It can also be used to specify other quantum states.

Quantum of thermal conductance, g_0: The smallest unit of heat that can be conducted through something, expressed in W/K.

Quantum well: An electrically isolated region, like a thin film, where electrons are constrained in one dimension. Such structures exhibit quantum behavior.

Quantum well laser: A quantum well structure used to confine excited electrons, which in turn stimulate the emission of photons, creating a laser beam.

Quantum wire: An electrically isolated region, like a nanotube or a nanoscale wire, where electrons are constrained in two dimensions. Such structures exhibit quantum behavior.

Radiation: The emission and propagation of waves or particles.

Radio wave: A portion of the electromagnetic spectrum with frequencies on the order of 10^5–10^9 Hz.

Rarefied gas: A gas in which the mean free path of the molecules is very large. One example is the Earth's outer atmosphere, where molecules are meters apart.

Repulsive forces: Forces that push molecules apart as their orbitals begin to overlap. Such forces include exchange repulsion, hard core repulsion, steric repulsion, and Born repulsion—all of which are types of van der Waals forces.

Resistance quantum: An invariant reference for resistance linked to Planck's constant. It is 25.813 kΩ.

Reynolds number: A nondimensional ratio of a fluid's momentum to its viscosity, often used to characterize the fluid flow as laminar (smooth) or turbulent, or to distinguish the type of flow, such as Stokes flow, common in nano- and microfluidic channels.

Scaling laws: Generalized equations used to estimate how physical characteristics vary depending on characteristic dimension.

Scanning probe microscopes (SPM): A family of microscopes that includes the AFM and STM. SPMs measure the properties and topography of a sample by scanning a sharp tip over its surface, and can also be used to manipulate surfaces with atomic resolution.

Scanning tunneling microscope (STM): A scanning probe microscope that operates by measuring the tunneling current between a sharp metal tip and a conducting surface in order to create a topographical image of the surface. It can also be used to manipulate surface atoms. The STM was the first scanning probe microscope.

Scattering: The rerouting of radiation or a particle caused by a collision with something in its way. One example is a photon forced by a reflective surface to deviate from its original path.

Schottky diodes: A diode formed at the junction between a metal and a semiconductor. Schottky diodes have a low forward voltage drop and a fast switching action, making them useful in computer processors.

Schrödinger equation: The fundamental equation for describing the behavior of quantum mechanical systems. It can be used to determine the allowed energy levels of atoms.

Self-assembled monolayer (SAM): A single layer of molecules that forms on its own under controlled conditions. SAMs are often used to functionalize surfaces.

Self-assembly: The spontaneous arrangement of atoms or molecules into structures when provided specific conditions.

Semiconductor: A material with more electrical conductivity than an insulator but less than a conductor. Its properties can be tuned to make it resist or enable the flow of electric current at desired locations and times, as in a transistor.

Single electron transistor (SET): A transistor made from a quantum dot that controls the current from source to drain one electron at a time.

Sol–gel: A process taking particles suspended in a liquid (sol) and aggregating them to form a networked structure (gel).

Spontaneous emission: The process by which an electron in an atom transitions to a ground energy state and emits a photon.

Standing wave: A wave confined to a given space, such as a vibrating piano string. The discrete, permissible wavelengths of such a wave are helpful in conceptualizing the quantized energy levels of atoms.

Steric hindrance: In fluid flow, the obstruction encountered by molecules flowing in channels that are similar in width to the molecules.

Stern layer: The fixed layer of ions within the electric double layer that forms in a fluid near a solid surface. Here, the potential drops from the wall potential to the zeta potential.

Stiction: A somewhat informal combo of the term "static friction." Stiction, or the threshold that must be overcome to unstick two objects in contact, is relatively large at the micro- and nanoscales due to the increased influence of van der Waals forces.

Stimulated emission: The process by which an incident photon perturbs an excited electron, causing the electron to transition to its ground state and emit a second photon having the same phase and trajectory as the first.

Stokes flow: A type of fluid flow, also called creeping flow, in which viscous forces dominate since the Reynolds number is much less than 1. Stokes flow is dominant for micro- and nanoscale systems.

Striction: The combination of adhesion and friction. This is different than stiction.

Sugars: Molecules with the general chemical formula $(CH_2O)_n$ used by living cells for energy and also structure.

Surface free energy, γ: A quantification of the work spent per unit area in forming a surface and holding it together.

Surfactant: A "surface active agent." Surfactant molecules reduce the surface tension between two liquids. Detergents use surfactants to enable oil to disperse in water, for example.

Tail group: One of the three components of a self-assembled monolayer (SAM). The tail group (or surface terminal group) is the functional, outermost part of the layer that physically interacts with the environment and determines the surface properties.

Thermal diffusivity, α: A property that indicates how fast heat propagates in a material. It can be used to determine the temperature distribution as a function of time in materials undergoing heating or cooling. The units are m^2/s.

Thermoelectric device: A device that develops a voltage bias when a temperature gradient exists from one side to the other. Or, a device that creates a temperature gradient when a voltage bias is applied.

Thin film: Nano- and microscale material layers typically used to modify an object's surface properties. Examples include scratch-resistant coatings for eyeglasses and low-friction coatings for bearings.

Top-down engineering (fabrication): The creation of something using the largest units of material first, removing material to create the final product—as opposed to bottom-up engineering.

Transistor: A semiconductor device with at least three terminals used to switch or amplify electronic signals. The current through one set of terminals controls the current through the other set. (One terminal of the control set is common to both the control and controlled set.)

Tunneling: The phenomena in which a particle, like an electron, encounters an energy barrier in an electronic structure and suddenly penetrates it.

Tunneling resistance, R_t: A measure of a junction's opposition to electron tunneling.

Ultraviolet: A portion of the electromagnetic spectrum with frequencies on the order of 10^{15}–10^{17} Hz.

Ultraviolet catastrophe: The problematic classical mechanics prediction of the late nineteenth century that an ideal blackbody—a hypothetical object that absorbs and emits all frequencies of radiation—would emit radiation of *infinitely* high intensity at high (ultraviolet) frequencies. This prediction was based on the assumption that thermal radiation is produced at all possible frequencies (a continuum). It proved false: radiation is in fact only produced at quantized frequencies, and higher frequencies are less likely because they require larger amounts of excitation energy at specific energy levels.

(Heisenberg) Uncertainty Principle: The principle, popularized by Werner Heinsenberg, which states that the more precisely one property in a quantum system is known, the less precisely a related property can be known—as with the position and momentum of an electron—because the act of measuring can affect what is being measured.

Unit cell: The smallest repeating element of a crystal lattice structure.

Valence band: The uppermost electron energy range that would contain electrons at a temperature of absolute zero. Each valence band electron is bound to an atom, as opposed to conduction electrons that move about freely.

Van der Waals bond: A weak link created by the attractive electrostatic forces that arise between atoms.

Van der Waals forces: Electrostatic forces—attractive and then repulsive—that arise between atoms and molecules as they approach one another.

Viscosity: The degree to which a fluid resists flow. In other words, viscosity is a measure of fluid "thickness."

Visible light: A portion of the electromagnetic spectrum with wave frequencies in the vicinity of 400–790 THz (or wavelengths of about 750–380 nm).

Wave function, ψ: A function used to describe the wavelike behavior of certain particles, such as electrons. The square of the wave function gives the probability of finding the particle at a given place and time.

Wave–particle duality: The way matter and light (electromagnetic radiation) can exhibit both wavelike and particle-like properties.

Work function, φ: The minimum amount of energy needed to remove an electron from a solid, usually a metal.

X-rays: A portion of the electromagnetic spectrum with frequencies on the order of 10^{16}–10^{20} Hz.

Zeolite: A high-surface-area material with very small pores (picometer to nanometer scale), often used in filtering and chemical catalysis processes.

Zeta potential, ζ: In a microfluidic channel, the potential difference between the shear plane near the channel wall (the location of the first moving liquid layer) and the bulk fluid.

Index